Turbulent Shear-Layer/ Shock-Wave Interactions

International Union of Theoretical
and Applied Mechanics

Turbulent Shear-Layer/ Shock-Wave Interactions

IUTAM Symposium, Palaiseau, France
September 9-12, 1985

Editor
J. Délery

Springer-Verlag
Berlin Heidelberg New York
London Paris Tokyo

Dr. J. Délery
Direction de L'Aérodynamique
ONERA
B.P. 72
29. Av. de la Division Leclerc
F-92322 Châtillon Cedex
France

ISBN-13: 978-3-642-82772-3 e-ISBN-13: 978-3-642-82770-9
DOI: 10.1007/978-3-642-82770-9

Library of Congress Cataloging-in-Publication Data
IUTAM Symposium (1985 : Palaiseau, France)
Turbulent shear-layer/shock-wave interactions.
At head of title: International Union of Theoretical and Applied Mechanics.
Sponsored also by the
Office national d'études et de recherches aérospatiales (ONERA).
1. Shock waves--Congresses.
2 Turbulent boundary layer--Congresses.
3. Shear flow--Congresses.
I. Délery, J. (Jean)
II. International Union of Theoretical and Applied Mechanics.
III. France. Office national d'études et de recherches aérospatiales.
IV. Title.
TL574.S4187 1985 629.132'32 86-17649

Softcover reprint of the hardcover 1st edition 1986

Offsetprinting: Color-Druck, G. Baucke, Berlin
Bookbinding: Luderitz & Bauer, Berlin
2161/3020-543210

Scientific Committee

S.M. Bogdonoff, USA
J. Délery, France (Chairman)
W. Fiszdon, Poland
P. Germain, France
R. Kawamura, Japan
E. Krause, FRG
B.E. Launder, UK
V.V. Sychev, USSR

Local Organizing Committee

J. Délery
B. Louis

Symposium Sponsor

International Union of Theoretical and Applied Mechanics (IUTAM)

Office National d'Etudes et de Recherches Aérospatiales (ONERA)

Preface

It was on a proposal of the late Professor Maurice Roy, member of the French Academy of Sciences, that in 1982, the General Assembly of the International Union of Theoretical and Applied Mechanics decided to sponsor a symposium on Turbulent Shear-Layer/Shock-Wave Interactions. This symposium might be arranged in Paris -or in its immediate vicinity-during the year 1985.

Upon request of Professor Robert Legendre, member of the French Academy of Sciences, the organization of the symposium might be provided by the Office National d'Etudes et de Recherches Aérospatiales (ONERA). The request was very favorably received by Monsieur l'Ingénieur Général André Auriol, then General Director of ONERA.

The subject of interactions between shock-waves and turbulent dissipative layers is of considerable importance for many practical devices and has a wide range of engineering applications. Such phenomena occur almost inevitably in any transonic or supersonic flow and the subject has given rise to an important research effort since the advent of high speed fluid mechanics, more than forty years ago.

However, with the coming of age of modern computers and the development of new sophisticated measurement techniques, considerable progress has been made in the field over the past fifteen years. The aim of the symposium was to provide an updated status of the research effort devoted to shear-layer/shock-wave interactions and to present the most significant results obtained recently.

The problems discussed can be classified into four main topics, namely :

- interactions in two-dimensional steady flows,
- interactions in three-dimensional steady flows,
- unsteady aspects of interaction phenomena,
- basic shock/turbulence interaction mechanism.

The program included 3 extended lectures each of a 60 minute duration and 29 papers each of a 30 minute duration. The symposium lasted four days from Monday morning to Thursday afternoon with a free afternoon on Tuesday dedicated to touristic activities and offering the occasion of new personal exchanges between participants.

This symposium brought together more than a hundred scientists coming from eight different countries with among them the best specialists in the field, including both theoreticians and experimentalists. The lectures provided a detailed and thorough presentation of the most recent progress in the physical understanding and in the modeling of strong turbulent interaction phenomena in compressible flows, the range of speed considered extending from transonic to hypersonic conditions. Each paper has given rise to stimulating and illuminating discussions allowing a deeper insight into the understanding of one of the most complex problems in fluid mechanics.

Obviously, the sucess of this hopefully fruitful symposium was due to the goodwill and active support of many persons and organizations.

We are grateful to the members of the Scientific Committee for their scientific assistance in the preparation of the symposium as well as for their precious help during the symposium itself.

Let us thank Session Chairmen, the Authors and all the participants for their active cooperation.

We would also like to express our gratitude to the General, Commander of Ecole Polytechnique, for his hospitality and the remarkable technical assistance provided by his school.

Generous and extremely efficient support was received from ONERA both before and during the symposium and I would like to express my best thanks to its President, Monsieur l'Ingénieur Général Jean Carpentier. Also, I would more especially like to acknowledge the enthusiastic help received from my colleagues of the Aerodynamics Department of ONERA.

Meudon, October 1985

J. Délery

Welcome to the Participants of the Symposium

R. LEGENDRE, Senior Scientific Advisor at ONERA, Member of the French Academy of Sciences.

ONERA, 29 Avenue de la Division Leclerc, 92320 CHATILLON, FRANCE

It is a great honour for me to open this session of an IUTAM Symposium dealing with the most intricate consequences of the accumulation of works of scientists of successive generations since the wonderful discovery of Ludwig Prandtl.

I very simply welcome you, well prepared to bring new progress in a domain where the use of computers cannot avoid a sound knowledge of empiricism. I am sure that your exchange of ideas will be fruitful.

It should have been the privilege of my best friend Maurice Roy to be in this place. He died last June after a severe motor car accident. France lost one of its greatest scientists well-known all over the world. We both belonged to the IUTAM General Assembly but it was him who, after reading an excellent paper of Professor Délery, proposed to the next meeting to hold this Symposium which will not be the last for this interesting topic.

Maurice Roy met Ludwig Prandtl already in 1922 and they became quickly good friends. Later in IUTAM, AGARD, ICAS, COSPAR, he became also the friend of almost all the great international scientists.

Your work will not only follow Ludwig Prandtl discovery and development, you will certainly comment on what is called Navier-Stokes equations.

Perhaps you don´t know that the great mathematician and physicist that was Stokes was about five years old when Navier presented his equations at the French Academy of Sciences. One year later Navier and Poisson presented a most clear explanation and later on, Saint-Venant brougth his contribution. So the equations were already classical when Stokes added his not negligible contribution.

This shows that exchanges of ideas are not only useful for scientists but also for History.

Je dis maintenant quelques mots en Français pour remercier l´Ecole Polytechnique d´avoir bien voulu abriter ce symposium.

Je remercie également l´Ingénieur Général Carpentier, Président Directeur Général de l´ONERA, d´avoir bien voulu aider le Professeur Délery.

Je suis sûr que l´Ecole Polytechnique et l´ONERA accroîtront ainsi leur renom.

Contributors

Andreopoulos, Y.	USA	Gas Dynamics Laboratory, Princeton University, Princeton, New Jersey 08512
Ashill, P.R.	USA	RAE, Dynamics Laboratory Bedford MK41 6AE
Aymer de la Chevalerie, D.	F	CEAT, 43 Rue de l'Aérodrome, 86000 Poitiers
Benay, R.	F	ONERA, Direction de l'Aérodynamique, B.P.72 92322 Châtillon Cedex
Bogdonoff, S.M.	USA	Department of Mechanical and Aerospace Engineering, Princeton University, Princeton, New Jersey 08544
Bohning, R.	FRG	Institut für Stromungslehre und Stromungsmaschinen, Universität (TH), Karlsruhe, Kaiserstrasse 12, D-7500 Karlsruhe
Brown, J.-L.	USA	NASA-Ames Research Center, Moffett Field, California 94035
Bushnell, D.M.	USA	NASA-Langley Research Center, Hampton, Virginia 23665
Cambier, L.	F	ONERA, Direction de l'Aérodynamique, B.P.72, 92322 Châtillon Cedex
Carrière, P.	F	ONERA, B.P.72, 92322 Châtillon Cedex
Carter, J.E.	USA	United Technologies Research Center, East Hartford, Connecticut 06108
Cheng, S.I.	USA	Department of Mechanical and Aerospace Engineering, Princeton University, Princeton, New Jersey 08544

Coakley, T.J.	USA	NASA-Ames Research Center, Moffett Field, California 94035
Collier, F	USA	Virginia Polytechnic Institute and State University, Blacksburg, Virginia 24061
Cusic, R.L.	USA	Grumman Aerospace Corporation, Bethpage, New York 11714
Debiève, J.-F.	F	IMST, 12 Avenue Général Leclerc, 13003 Marseille
Degrez, G.	B	IVK, 72 Chaussée de Waterloo, B-1640 Rhode-Saint-Genèse
Dolling, D.S.	USA	Department of Aerospace Engineering and Engineering Mechanics, The University of Texas at Austin, Austin Texas 78712-1085
Dor, J.-B.	F	CERT/DERAT, B.P.4025, 31055 Toulouse Cedex
Dussauge, J.-P.	F	IMST, 12 Avenue Général Leclerc, 13003 Marseille
Edwards, D.E.	USA	United Technologies Research Center, East Hartford, Connecticut 06108
Escande, B.	F	ONERA, Direction de l'Aérodynamique, B.P.72, 92322 Châtillon Cedex
Forester, C.K.	USA	Boeing Military Aircraft Co., P.O. Box 3707, Seattle, Washington 98124
Fulker, J.L.	UK	RAE, Dynamics Laboratory, Bedford MK 41 6AE
Ha Minh, H.	F	IMFT, Avenue du Professeur Camille Soula, 31400 Toulouse
Holden, M.S.	USA	Calspan-UB Research Center, P.O. Box 400, Buffalo, New York 14225

Hussaini, M.Y.	USA	ICASE, NASA-Langley Research Center, Hampton, Virginia 23665
Inger, G.R.	USA	Iowa State university, Ames Iowa 50011
Johnson, C.B.	USA	NASA-Langley Research Center, Hampton, Virginia 23665
Johnson, D.A.	USA	NASA-Ames Research Center, Moffett Field, California 94035
Kaufman II, L.G.	USA	Grumman Aerospace Corporation, Bethpage, New York 11714
King, L.S.	USA	NASA-Ames Research Center, Moffett Field, California 94035
Krogmann, P.	FRG	DFVLR-AVA, Bunsenstrasse 10, D-3400 Gottingen
Kussoy, M.I.	USA	NASA-Ames Research Center, Moffett Field, California 94035
Lacharme, J.-P.	F	IMST, 12 Avenue Général Leclerc, 13003 Marseille
Lee, P.	UK	RAE, Dynamics Laboratory, Bedford MK41 6AE
Liu, X.	UK	Cambridge University Engineering Dept., Cambridge University, Trumpington Street, Cambridge CB2 1PZ
Mabey, D.G.	U.K.	RAE, Dynamics Laboratory, Bedford MK41 6AE
Matoi, T.K.	USA	Boeing Commercial Airplane Co., P.O. Box 3707, Seattle, Washington 98178
Mc Lean, J.D.	USA	Boeing Commercial Airplane Co., P.O. Box 3707, Seattle, Washington 98178
Mc Murray, J.T.	USA	ETA Systems, Inc., St Paul, Minnesota 55112
Melnik, R.E	USA	Grumman Aerospace Corporation, Bethpage, New York 11714

Mignosi, A.	F	CERT/DERAT, B.P.4025, 31055 Toulouse Cedex
Muck, K.C.	USA	Gas Dynamics Laboratory, Princeton University, Princeton, New Jersey 08512
Pot, T.	F	ONERA, Direction de l'Aérodynamique, B.P.72, 92322 Châtillon Cedex
Purcell, C.J.	USA	ETA Systems, Inc., St Paul, Minnesota 55112
Rizzi, A.	S	FFA, The Aeronautical Research Institute of Sweden, S-161 11 Bromma
Saida, N.	JAP	Aoyama Gakuin University College of Science and Engineering, 6-16-1 Chitosedai, Setagayaku, Tokyo 157
Settles, G.S.	USA	303 Mechanical Engineering Building, Pennsylvania State University, University Park, Pennsylvania 16802
Seraudie, A.	F	CERT/DERAT, B.P.4025, 31055 Toulouse Cedex
Siclari, M.J.	USA	Grumman Aerospace Corporation, Bethpage, New York 11714
Smits, A.J.	USA	Department of Mechanical and Aerospace Engineering, Princeton University, Princeton, New Jersey 08544
Squire, L.C.	UK	Cambridge University Engineering Dept. Cambridge University, Trumpington Street, Cambridge CB2 1PZ
Strom, I.	USA	Boeing Military Aircraft Co., P.O. Box 3707, Seattle, Washington 98124
Tai, T.C.	USA	David Taylor Naval Ship Research and Development Center, Bethesda, Maryland 20084-5000

Thiede, P.	FRG	Transport Aircraft Division, MBB-TE 213, Postfach 10 78 45, D-2800 Bremen 1
Vandromme, D.	F	ONERA/IMFL, 5 Boulevard Painlevé, 59000 Lille
Whitfield, D.L.	USA	Department of Aerospace Engineering, Mississippi State University, Mississippi State, Mississippi 39762
Zierep, J.	FRG	Institut für Stromungslehre und Stromungsmashinen, Universität (TH) Karlsruhe, Kaiserstrasse 12, D-7500 Karlsruhe

Participants

Andreopoulos, Y.	USA	Gas Dynamics Laboratory, Princeton University, Princeton, New Jersey 08512
Ashill, P.R.	USA	RAE, Dynamics Laboratory Bedford MK41 6AE
Aymer de la Chevalerie, D.	F	CEAT, 43 Rue de l'Aérodrome 86000 Poitiers
Benay, R.	F	ONERA, Direction de l'Aérodynamique, B.P.72, 92322 Châtillon Cedex
Bogdonoff, S.M.	USA	Department of Mechanical and Aerospace Engineering, Princeton University, Princeton, New Jersey 08544
Bohning, R.	FRG	Institut für Stromungslehre und Stromungsmaschinen, Universität (TH), Karlsruhe, Kaiserstrasse 12 D-7500 Karlsruhe
Borel, M.	F	STPA, 4 avenue de la Porte d'Issy, 75996 Paris Armées
Borgetto, M.	F	ONERA, Direction de l'Aérodynamique, B.P.72, 92322 Châtillon Cedex
Brown, E.F.	USA	Mechanical Engineering Department, Virginia Polytechnic Institute and State Univer- sity, Blacksburg, Virginia 24061
Brown, J.L.	USA	NASA-Ames Research Center Moffett Field, California 94035

Cambier, L.	F	ONERA, Direction de l'Aérodynamique, B.P.72, 92322 Châtillon Cedex
Carrière, P.	F	ONERA, B.P.72, 92322 Châtillon Cedex
Caupennes, Ph.	F	AMD-BA, 78 Quai Carnot 92214 St Cloud Cedex
Chanetz, B.	F	ONERA, Direction de l'Aérodynamique, B.P.72, 92322 Châtillon Cedex
Chattot, J.-J.	F	Société MATRA, B.P. 1, 78146 Vélizy Villacoublay Cedex
Cheng, S.I.	USA	Department of Mechanical and Aerospace Engineering, Princeton University, Princeton, New Jersey 08544
Chevallier, J.-P.	F	ONERA, Direction de l'Aérodynamique, B.P.72, 92322 Châtillon Cedex
Costes, B.	F	ONERA, Direction de l'Aérodynamique, B.P.72, 92322 Châtillon Cedex
Courty, J.-C.	F	AMD-BA, 78 Quai Carnot, 92214 St Cloud Cedex
Cusin, B.	F	DRET, 26 Boulevard Victor, 75996 Paris Armées
D'Humières, C.	F	ONERA, Direction de l'Aérodynamique, B.P.72, 92322 Châtillon Cedex
Dang Khoa	F	ONERA, Direction de l'Aérodynamique, B.P.72, 92322 Châtillon Cedex
Debiève, J.-F.	F	IMST, 12 Avenue Général Leclerc, 13003 Marseille
Degrez, G.	B	IVK, 72 Chaussée de Waterloo, B-1640 Rhode-Saint-Genèse
Délery, J.	F	ONERA, Direction de l'Aérodynamique, B.P.72, 92322 Châtillon Cedex

Dériat, E.	F	ONERA, Direction de l'Aérodynamique, B.P.72, 92322 Châtillon Cedex
Dolling, D.S.	USA	Department of Aerospace Engineering and Engineering Mechanics, The University of Texas at Austin, Austin, Texas 78712-1085
Dor, J.-B.	F	CERT/DERAT, B.P.4025, 31055 Toulouse Cedex
Dubois, F.	F	Ecole Polytechnique, 91128 Palaiseau Cedex
Dupont, A.	F	ONERA, Direction de l'Aérodynamique, B.P.72, 92322 Châtillon Cedex
Durand, G.	F	DRET, 26 Boulevard Victor, 75996 Paris Armées
Dussauge, J.-P.	F	IMST, 12 Avenue Général Leclerc, 13003 Marseille
Dutouquet, L.	F	SNECMA, Centre de Villaroche, 77550 Moissy Cramayel
Edwards, D.E.	USA	United Technologies Research Center, East Hartford, Connecticut 06108
Elena, M.	F	IMST, 12 Avenue Général Leclerc, 13003 Marseille
Elsaesser, E.	F	Société MATRA, B.P.1 78146 Vélizy Villacoublay Cedex
Elsenaar, B.	NL	NLR, Anth. Fokkerweg 2, 1059 CM Amsterdam
Escande, B.	F	ONERA, Direction de l'Aérodynamique, B.P.72, 92322 Châtillon Cedex
Fernholz, H.H.	FRG	Technische Universität FB9, Hermann Fottinger Institut, Strasse des 17 Juni 135, D-1000 Berlin 12
Fourmaux, A.	F	ONERA, Direction de l'Aérodynamique, B.P.72, 92322 Châtillon Cedex

Freimuth, P.	FRG	DFVLR-AVA, Bunsenstrasse 10, D-3400 Gottingen
Fromentel, R.	F	AMD-BA, 78 Quai Carnot, 92214 St Cloud Cedex
Gauthier, S.	F	CEA, Centre d'Etudes de Limeil-Valenton, B.P.27, 94190 Villeneuve St Georges
Goutines, M.	F	SNECMA, Centre de Villaroche, 77550 Moissy Cramayel
Hass, J.-F.	F	CEA, Centre d'Etudes de Limeil-Valenton, B.P.27, 94190 Villeneuve St Georges
Ha Minh, H.	F	IMFT, Avenue du Professeur Camille Soula, 31400 Toulouse
Holden, M.S.	USA	Calspan-UB Research Center, P.O. Box 400, Buffalo, New York 14225
Horowitz, E.	F	Framentec, La Boursidière, R.N. 186, 92357 Le Plessis Robinson
Hoyt, A.R.	USA	EOARD, 223/231 Old Marylebone Road, London NW1 5TH
Hussain, S.	UK	Cranfield Institute of Technology, Cranfield, Bedfordshire MK43 0AL
Inger, G.R.	USA	Iowa State University, Ames, Iowa 50011
Johnson, D.A.	USA	NASA-Ames Research Center, Moffett Field, California 94035
Johnston, J.	UK	ARA, Menton Lane, Bedford MK41 7PF
Kaufman II, L.G.	USA	Grumman Aerospace Corporation, Bethpage, New York 11714
Krause, E.	FRG	Aerodynamisches Institut, RWTH Aachen, Wullnerstrasse zw. 5 u 7, D-5100 Aachen

Krogmann, P.	FRG	DFVLR-AVA, Bunsenstrasse 10 D-3400 Gottingen
Lacharme, J.-P.	F	IMST, 12 Avenue Général Leclerc, 13003 Marseille
Launder, B.E.	UK	Department of Mechanical Engineering, UMIST, P.O. Box 88, Manchester M60 1QD
Le Balleur, J.-C.	F	ONERA, Direction de l'Aérodynamique, B.P.72, 92322 Châtillon Cedex
Ledun, M.	F	Thomson-CSF, 52, Rue Guynemer, 92130 Issy les Moulineaux
Legendre, R.	F	ONERA, B.P.72, 92322 Châtillon Cedex
Lequette, L.	F	Société BERTIN, Z.I. Les Gâtines, B.P.3, 78273 Plaisir Cedex
Le Sant, Y.	F	ONERA, Direction de l'Aérodynamique, B.P.72, 92322 Châtillon Cedex
Leuchter, O.	F	ONERA, Direction de l'Aérodynamique, B.P.72, 92322 Châtillon Cedex
Liu, X.	UK	Cambridge University Engineering Dept., Cambridge University, Trumpington Street, Cambridge CB2 1PZ
Lock, R.C.	UK	74 Ormond Avenue, Hampton, Middx. TW122RX
Loisel, P.	F	ONERA, Direction de l'Aérodynamique, B.P.72, 92322 Châtillon Cedex
Louis, B.	F	ONERA, Direction de l'Aérodynamique, B.P.72, 92322 Châtillon Cedex
Mabey, D.G.	UK	RAE, Dynamics Laboratory, Bedford MK41 6AE

Mabey, D.G.	UK	RAE, Dynamics Laboratory, Bedford MK 41 6 AE
Melnik, R.E.	USA	Grumman Aerospace Corporation, Bethpage, New York 11714
Mignosi, A.	F	CERT/DERAT, B.P.4025, 31055 Toulouse Cedex
Mirande, J.	F	26 Quai du 4 Septembre, 92100 Boulogne
Monnerie, B.	F	ONERA, Direction de l'Aérodynamique, B.P.72, 92322 Châtillon Cedex
Mc Murray, J.T.	USA	ETA Systems, Inc., St Paul, Minnesota 55112
Nebbeling, C.	NL	Delft University of Technology, Kluyverweg 1, 2629 HS Delft
Neish, A.	UK	University College, 317 Front Lane, Upminster, Essex RM14 1LH
Nellner, P.-Ch.	FRG	Institut fur Stromungslehre und Stromungsmaschinen, Universitat Karlsruhe (TH), Herrenstrasse 28, D-7500 Karlsruhe 1
Neyland, V.Ya.	USSR	TSAGI, Ministry of Aviation Technology, Ul. Anskij 16, Moscow
Nicout, D.	F	ONERA, Direction de l'Aérodynamique, B.P.72, 92322 Châtillon Cedex
Niederdrenk, P.	FRG	DFVLR-AVA, Bunsenstrasse 10, D-3400 Gottingen
Pagan, D.	F	ONERA, Direction de l'Aérodynamique, B.P.72, 92322 Châtillon Cedex
Perrier, P.	F	AMD-BA, 78 Quai Carnot, 92214 St Cloud Cedex
Poisson-Quinton, Ph.	F	ONERA, B.P.72, 92322 Châtillon Cedex
Pot, T.	F	ONERA, Direction de l'Aérodynamique, B.P.72, 92322 Châtillon Cedex

Reijasse, P.	F	ONERA, Direction de l'Aérodynamique, B.P.72, 92322 Châtillon Cedex
Sagnier, P.	F	ONERA, Direction de l'Aérodynamique, B.P.72, 92322 Châtillon Cedex
Saida, N.	JAP	Aoyama Gakuin University, College of Science and Engineering, 6-16-1, Chitosedai, Setagayaku, Tokyo 157
Sajben, M.	USA	McDonnell Douglas Research Labs., P.O. Box 516, Saint Louis, Missouri 63166
Schmitt, V.	F	ONERA, Direction de l'Aérodynamique, B.P.72, 92322 Châtillon Cedex
Schneyer, G.P.	USA	S-CUBED, P.O. Box 1620 La Jolla, California 92038
Secuondov, A.	USSR	TSAGI, Ministry of Aviation Technology, Ul. Anskij 16, Moscow
Semezis, Y.	F	ONERA, Direction de l'Aérodynamique, B.P.72, 92322 Châtillon Cedex
Settles, G.S.	USA	303 Mechanical Engineering Building, Pennsylvania State University, University Park, Pennsylvania 16802
Smits, A.J.	USA	Department of Mechanical and Aerospace Engineering, Princeton University, Princeton, New Jersey 08544
Solignac, J.-L.	F	ONERA, Direction de l'Aérodynamique, B.P.72, 92322 Châtillon Cedex
Tai, T.C.	USA	David Taylor Naval Ship Research and Development Center, Bethesda, Maryland 20084-5000
Teupootahiti, E.	F	AMD-BA, 78 Quai Carnot, 92214 St Cloud Cedex
Thibert, J.-J.	F	ONERA, Direction de l'Aérodynamique, B.P.72, 92322 Châtillon Cedex

Thiede, P.	FRG	Transport Aircraft Division, MBB-TE 213, Postfach 10 78 45, D-2800 Bremen 1
Thomas, F.	F	ONERA/IMFL, 5 Boulevard Painlevé, 59000 Lille
Tjonneland, E.	USA	Boeing Military Aircraft Co., P.O. Box 3707, Seattle, Washington 98024
Vandromme, D.	F	ONERA/IMFL, 5 Boulevard Painlevé, 59000 Lille
Vasanta Ram, V.	FRG	Institut fur Thermo-und Fluiddynamic, Ruhr-Universitat Bochum, Postfach 10 21 48 D-4630 Bochum 1
Veuillot, J.-P.	F	ONERA, Direction de l'Aérodynamique, B.P.72, 92322 Châtillon Cedex
Vincent de Paul, M.	F	Alsthom-Atlantique, 55 Avenue Jean Jaurès, 93350 Le Bourget

Table of Contents

Session 3 Interactions in Two-Dimensional Steady Flows

Chairman : S.M. Bogdonoff

Session 4 Interactions in Three-Dimensional Steady Flows

Chairman : E. Krause

Session 5 Interactions in Three-Dimensional Steady Flows

Chairman : H. Fernholz

(*) Extended Lecture

Introduction

P. CARRIERE - Conseiller Scientifique à l'ONERA

ONERA, 29 Avenue de la Division Leclerc, 92320 CHATILLON, France

Aussi loin que je remonte dans mes souvenirs d'ingénieur, je retrouve des exemples d'applications dans lesquels le problème des interactions choc/couche cisaillée turbulente joue un rôle très important, souvent même déterminant.

L'ingénieur Général Gabeaud, professeur de Balistique en France avant 1940, avait coutume de souligner que le manque de précision dans le calcul de la traînée d'un simple obus d'artillerie provenait de la médiocre évaluation de la pression de culot et il ajoutait que c'était le scandale de l'Aérodynamique dont pourtant les progrès par ailleurs étaient déjà spectaculaires.

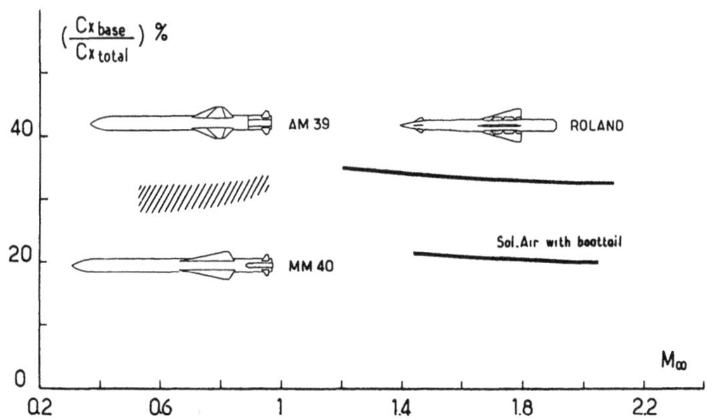

Fig. 1 - Base Drag of Tactical Missiles

Turbulent Shear Layer/Shock Wave Interactions
IUTAM Symposium Palaiseau 1985
Editor: J. Délery

Ce problème reste d'actualité, comme le montre la figure 1, pour les missiles modernes dont la traînée de culot - base drag - représente en croisière supersonique parfois jusqu'à 30 % de la traînée totale : son évaluation précise et, surtout, son optimisation imposent donc un effort très soutenu de la recherche.

Pour le vol transsonique des avions, je rappellerai que les travaux d'Ackeret, Feldmann et Roth furent les premiers, vers 1944, à fournir une analyse des phénomènes de choc transsonique sur un profil. La figure 2 est issue des travaux analogues du NPL qui, quelques années plus tard, contribuèrent à clarifier la nature des effets du choc sur le comportement aval de la couche limite.

Fig. 2 - Transonic Interaction on an Airfoil. Schlieren Photograph of the Flow with Severe Separation at the Foot of the Upper-Surface Shock (Pearcey, 1955)

Une cause fréquente d'apparition du pompage subcritique d'une prise d'air s'explique, comme l'a montré Ferri, par la géométrie des chocs et de la ligne de glissement : dans la figure 3 on montre l'éclatement de cette couche libre devant la carène sous l'effet d'un choc régressif, phénomène qui, modifiant le débit instantané capté, contribue à amorcer ou entretenir le pompage.

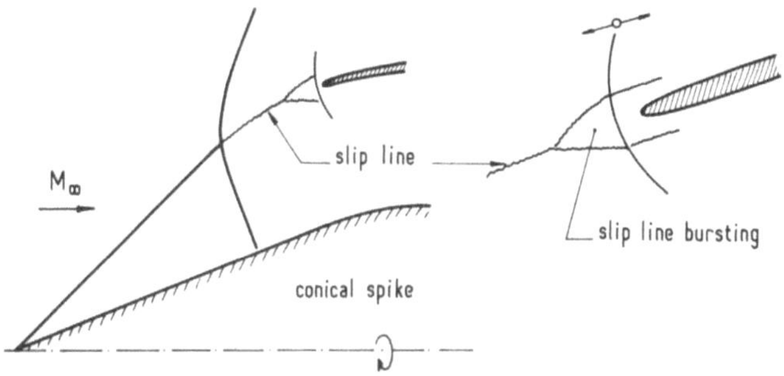

Fig. 3 - Subcritical Buzz of a Supersonic Air Intake

La poussée interne d'une tuyère propulsive (voir Fig. 4) en surdétente est fortement influencée par les décollements provoqués à la sortie par les chocs de recompression. Or l'on sait que dans un projet d'avion comme celui du Concorde une erreur de l'ordre de 1 % sur la poussée interne est considérée comme inacceptable par le constructeur. Une étude précise du décollement à la sortie est donc essentielle pour satisfaire à cette exigence.

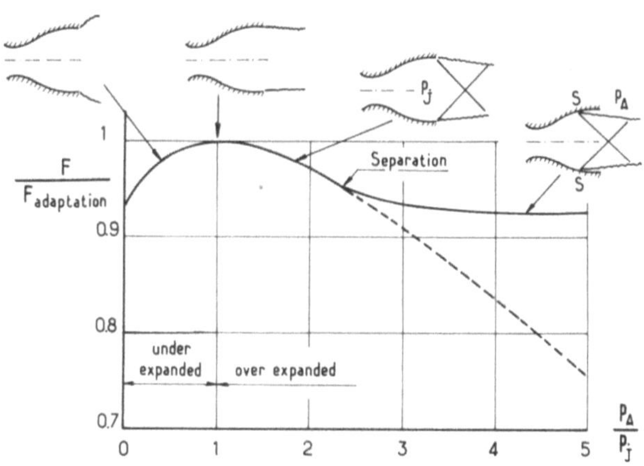

Fig. 4 - Nozzle Thrust as Function of Expansion Ratio

Je passerai maintenant en revue les divers problèmes fondamentaux soulevés par tous ces exemples et qui vont faire l'objet de votre symposium.

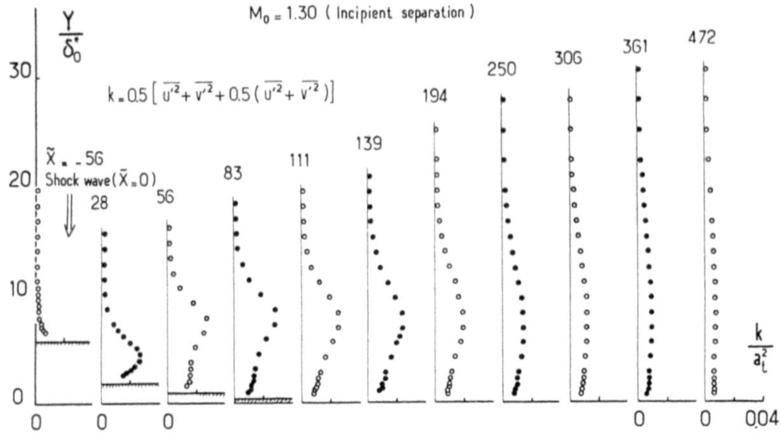

Fig. 5 - Turbulent Kinetic Energy Distributions in a Transonic Interacting Boundary-Layer

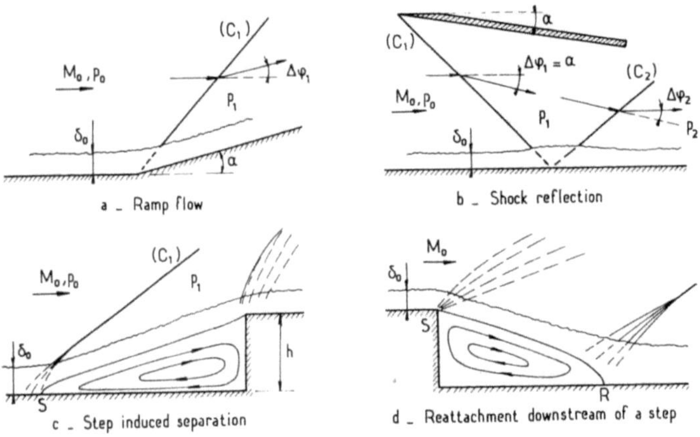

Fig. 6 - Basic Shock/Boundary-Layer Interactions in Supersonic Flow

En premier l'action d'un choc sur une couche limite turbulente de plaque plane : l'analyse de la figure 5 montre clairement la production d'énergie turbulente à la traversée du choc puis sa diffusion et sa dissipation vers l'aval. Si le problème de production était résolu, il semble que les méthodes récentes du type $(k, \mathcal{E})$ par exemple permettraient maintenant le calcul complet du phénomène. Les autres cas typiques qui impliquent un décollement suivi d'un recollement évoquent, sur la figure 6, les divers problèmes à résoudre sous leurs aspects <u>bidimensionnels</u> et <u>stationnaires</u> sans qu'il soit besoin de plus amples commentaires.

Chapman, Kuehn et Larson sont à l'origine d'une représentation phénoménologique très claire du décollement libre induit par une rampe dans un écoulement initial sans gradient. Nous l'avons, à l'ONERA, généralisée comme le montre la figure 7 en y introduisant explicitement les nombres de pression locaux ($\bar{\nu}$ avant perturbation et ν perturbé). On peut ainsi rendre compte des deux cas typiques a et b de la figure 6 mieux que par la formule linéarisée originale de Chapman et al.

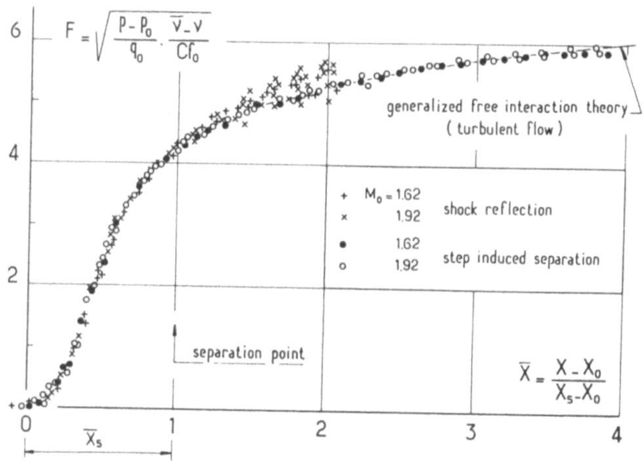

Fig. 7 - The Free Interaction Theory. Correlation of Wall Pressure Measurements. Uniform Incoming Flow

L'analyse de Korst relative à l'écoulement décollé derrière une marche (voir Fig. 8) a servi de base à de nombreux travaux - à l'ONERA notamment - et a été appliquée avec succès à de multiples problèmes techniques. Son point faible réside dans le critère de recollement qui, sous quelque forme qu'on le présente, implique une dépendance déterminée entre la pression de décollement (zône V) et la pression aval (zône IV, dite parfois de réhabilitation).

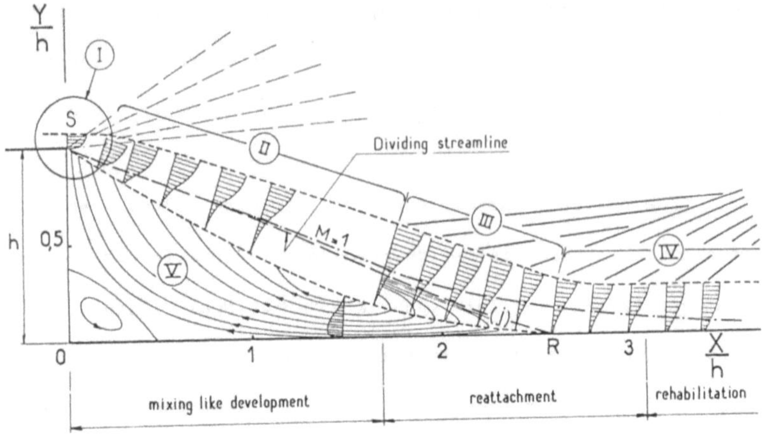

Fig. 8 - Supersonic Reattachment Downstream of a Rearward Facing Step. H. Korst's Modelization of the Flow Field

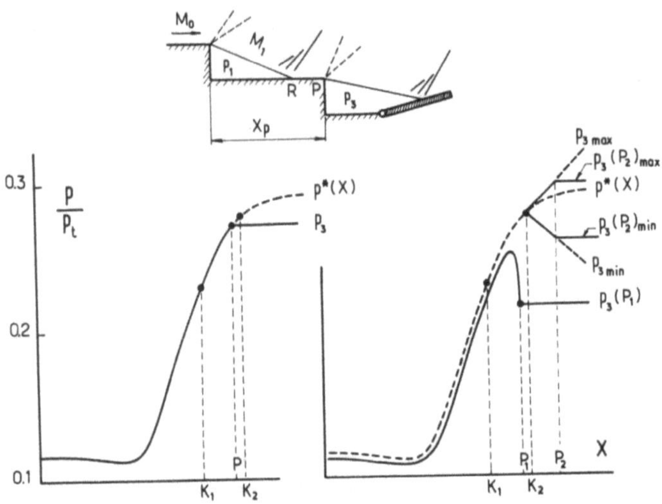

Fig. 9 - Critical Points at Reattachment

Nous avons en effet montré - précisant des travaux antérieurs de Bogdonoff - qu'en réalité il existe (voir Fig. 9) dans la zône de recollement un point critique K_1 en aval duquel peut être imposée une pression p_3 constante, inférieure à la pression dite de réhabilitation, sans aucune perturbation de la distribution amont pourvu que p_3 soit identique à la pression p*(x) existant au préalable à l'origine x de la plage isobare ; en aval du 2ème point critique K_2 on peut même imposer une pression p_3 très différente en plus ou en moins de la pression p*(x).

Ceci montre combien serait utile une analyse théorique beaucoup plus fine du problème de recollement dont le rôle est déterminant dans les phénomènes de décollement.

Le Professeur Ha Minh, à Toulouse, a déjà apporté dans ce sens une contribution intéressante à cette étude.

La figure 10 suivante évoque les très nombreuses études de décollement sur rampe, en hypersonique, problème pour lequel S.M. Bogdonoff et son laboratoire ont joué un rôle de pionniers. Les "overshoots" observés ici dans le cas extrême sont sans doute en relation avec l'existence dans la zône de recollement des points critiques K_1 et K_2, rappelée dans la figure 9.

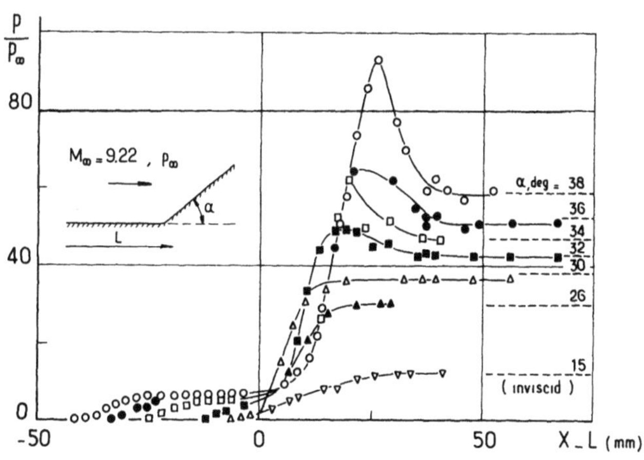

Fig. 10 - Downstream Overshoot at High Mach Number (Elfstrom, 1971)

Les problèmes que je viens de mentionner se situent dans l'optique "bidimensionnelle et stationnaire" : ils ont certes apporté un éclairage très utile. Mais, en fait, rares sont les cas où des aspects tridimensionnels et instationnaires ne viennent pas compliquer la situation. Le montage de la figure 11, suggéré notamment par Bogdonoff, devrait contribuer, grâce à sa relative simplicité, à compléter nos connaissances dans des cas nettement tridimensionnels (voir Fig. 12).

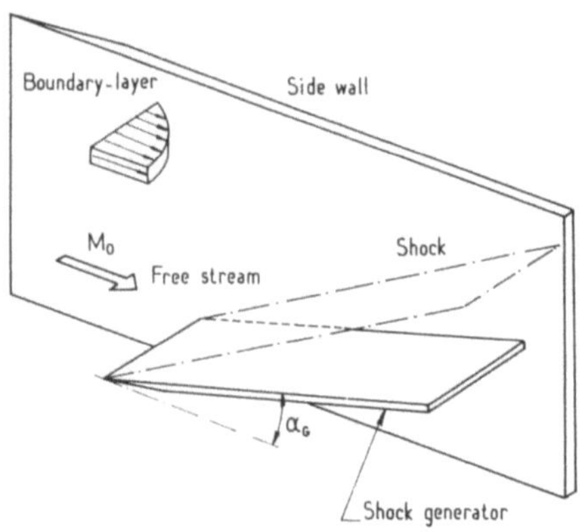

Fig. 11 - The Skewed Shock Wave (or Glancing Shock) Configuration

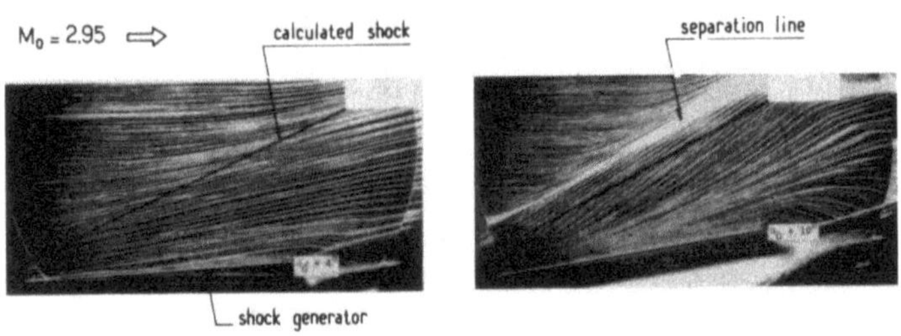

Fig. 12 - Skewed Shock Wave - Surface Flow Patterns on Side Wall (Oskam et al., 1976)

Pour ce qui concerne les aspects instationnaires, je me contenterai de prononcer le mot de buffeting pour rappeler un phénomène d'interaction particulièrement génant en Aéronautique et qui reste encore mal prévisible dans les conditions de vol, même après des essais en soufflerie où le nombre de Reynolds est encore insuffisant.

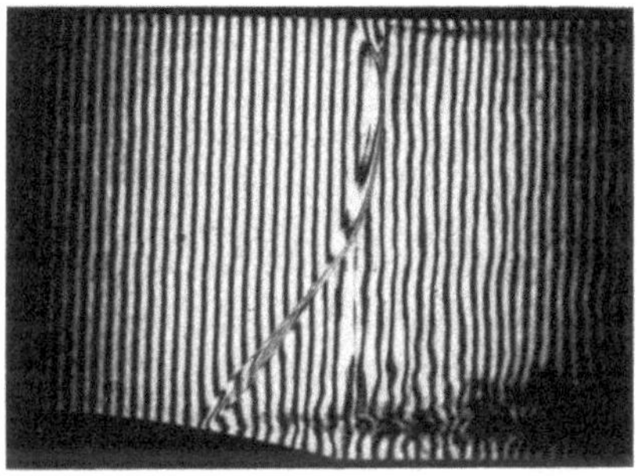

Fig. 13 - Short Exposure Time Interferogram of a Transonic Interaction with Separation.

L'interférogramme instantané de la figure 13 vient rappeler, pour terminer, l'existence de moyens d'analyse très fins - connus depuis longtemps en principe - mais devenus récemment exploitables commodément. L'expérience ici en cause rappelle les travaux de Pearcey et montre que le choc n'est stationnaire qu'en apparence : en réalité, ses oscillations induites, soit par des ondes acoustiques venant de l'aval, soit par les fluctuations de la couche limite incidente doivent être un facteur essentiel dans la production d'énergie turbulente au cours de l'interaction dont j'ai souligné plus haut l'importance essentielle. Ce problème est examiné à l'IMST, à Marseille (exposé de MM. Debiève et Lacharme) et à Princeton (exposé de MM. Dussauge, Muck et Andreopoulos).

Tout ces problèmes vont être maintenant largement débattus au cours de ce symposium, ce qui - j'en suis sûr - contribuera à de nouveaux progrès. C'est le succès que je vous souhaite.

Interactions in Two-Dimensional Steady Flows

Modelling of Compressible Turbulent Flows: Present Possibilities and Perspectives

H. HA MINH* & D.D. VANDROMME **

*Institut de Mécanique des Fluides, Laboratoire associé au C.N.R.S. LA 0005
Avenue du Professeur Camille Soula, 31400 Toulouse (France)

** Université des Sciences et Techniques de Lille, U.E.R. de Mathématiques,
Service de Mécanique,
59655 Villeneuve d'Asq Cédex (France)

Summary

The topics of this IUTAM Symposium is Turbulent Shear-Layer / Shock Wave interactions. It means that turbulence aspects play a determinant role in the physical analysis of compressible shear layer / shock-wave interactions. This contribution is devoted to describe briefly present possibilities of COMPRESSIBLE TURBULENT FLOW MODELLING and developments expected in the very near future.

1. INTRODUCTION

Since the pionnering theory of BOUSSINESQ [1] and the first application indicated by PRANDTL [2], turbulence modelling has been continuously improved in order to allow effective prediction of turbulence flows. Since these early contributions, various methodologies were worked out to account properly for "complex" effects on turbulence. Among others, these "complex" effects are due to strong curvatures, body forces, unsteadiness and variable density effects. In fact, in real flow situations, it is difficult to separate these effects and the prediction of an industrial interest flow configuration must always combine the simultaneous treatment of several effects. Depending on the application domain, some effects can be considered as dominant and allow to neglect others. As an example, geophysical flows in oceans are dominated by stratification and rotation while compressibility or solid boundary effects can be neglected. At the opposite, flows around an aircraft fuselage in real flight conditions are not influenced by gravitation or rotation. Because of the requirements imposed by a wide variety of industrial and aeronautical problems, one of the biggest challenge during the last few years in the field of turbulence modelling has been to include correctly the variable density effects. Unfortunately, so far, only a small fraction of modelling efforts has been devoted to these flows. Variable density can have different origins which can be, for instance : i) mixture of gases with different densities, ii) strong temperature gradients within an homogeneous fluid, iii) reactive flows, iv) strong distorsion by unsteady solid boundary (for instance, internal combustion engine piston), v) compressibility effects in high speed flows. Each origin gives rise to specific

Turbulent Shear Layer/Shock Wave Interactions
IUTAM Symposium Palaiseau 1985
Editor: J. Délery

aspects to be handled in any modelling effort. This contribution is devoted only to this last origin.

So far, most of the modelling attempts in compressible flows are based on extensions from incompressible turbulence modelling knowledge. This approach can be justified by past experiments : the first works concerning equilibrium boundary layer along adiabatic wall [3], [4], [5] or compressible wakes [6], led to MORKOVIN [5] to postulate that the dynamical field behaves like an incompressible one. From a different approach, LAUFER [7] came to an identical conclusion. Thus, for many years, compressible problems have been handled with incompressible models. But the limits of such extensions begin now to be better understood [8], [9], BRADSHAW [10] reviewed the various domains of validity for MORKOVIN's hypothesis and noticed that it applies to flows where the density fluctuations are moderate. Obviously, this excludes hypersonic boundary layer (Ma $>$ 5) and flows with strong pressure gradients such as shock-turbulent boundary layer interactions. BRADSHAW concluded also that this approximation is not well suited for the supersonic shear layers. One of the conclusions of the 1972 free shear-layer conference [11] confirmed that all existing turbulence models were not adapted to the prediction of these compressible flows.

Concerning the modelling, a limited number of significant results have been obtained so far. Most of the works consist in a more or less detailed analysis of existing incompressible models leading to an extension for the compressible flows, assuming that the existence of a variable density in the equations was sufficient to represent correctly the compressibility effects. Nevertheless, in presence of severe pressure gradients, these models do not produce satisfactory results [12], [13]. To date, two works have been reported on this topic. OH [14] proposed a one equation model including effects of the pressure-dilatation correlation and BONNET and LAUNDER [15] followed an analogous reasonning with a second-order closure model. Besides these two references, all attempts for compressible shear-layers modelling rely on incompressible models and thus, have a very limited domain of applicability.

When starting from incompressible flow models to extend into compressible turbulent flow studies, we meet two kinds of difficulties : the first one is due to the compressibility itself and the second one is caused by the turbulence phenomena.

From a mathematical point of view, compressible flows cannot be considered as straightforward extension of incompressible flows. This is a direct consequence of the new role of thermodynamic variables. The first difference comes naturally from the requirement for an energy equation. To allow temperature to be considered as an active variable rather than a passive scalar. Then the pressure field has a more complex role. Unlike incompressible flows where the pressure is only a force term and all disturbances propagate at infinite speed, pressure field " supports" also finite velocity propagation of disturbances. Finally the continuity equation is also governed by unsteady term and has therefore a fully hyperbolic character. Complexity due to compressible turbulence mainly concerns the fluctuating density, leading to more non-linear equations to be solved (note that variable mean density increases already the non-linear character of the mathematical model) and to more inextricable closure problem. This contribution is only devoted to the second point.

2. TURBULENCE ASPECTS AND THEIR MODELLING PROBLEMS

Turbulent flows have always been considered as fascinating on the one part from a fundamental point of view and on the other part for practical applications. As a consequence, theoretical and experimental means have been used by a large number of investigators attempting to solve the " turbulence mystery" that would lead to an understanding of its technologically important characteristics.

Until the fifties, turbulence theory depended primarily on analytical methods. Such studies required the statistical equations of motion to be transformed into ordinary differential equations. But, then the flows had to be well structured (turbulent energy equilibrium according to the Kolmogorov definition), self-similar or in a more general sense, in a thin shear layer form, generally far downstream from the turbulence generation location. The eddy viscosity concept was employed in most of this approach. It was implicitly understood that the turbulence was in a fully random state. Since the fifties, significant developments in the studies of fluid mechanics have been made both experimentally and computationally. Advances in flow field observational and measurement devices (multidimensional hot wires, multidimensional laser Doppler velocimeters, holography) and laboratory computers have resulted in a profusion of experimental data besides "pictures" of the overall flow fields, these data involve single and multipoint measurements that have been analyzed with both conventional and conditioned statistical methods. It has been found that the turbulence once thought to be entirely random, possesses distinct "coherent structures". Up to now, however, the importance of this structure in establishing the mean properties of the fluid flow has not been established unambiguously. As a consequence, most turbulence models for computational purposes have been confined to standard single-point statistical methods although length scale information has been derived from the multipoint measurements. A hierarchy of models has been developed ranging from simple mixing length theory to the rather complicated third order closures. The best of these models enjoyed considerable success during the seventies and the first half of the eighties. But it must be pointed out that so far i) incompressible flows with fully random characteristics were studied predominantly and ii) high order closure models have only been used within the boundary layer approximation due to their related numerical problems.

We are presently in the rather surprising following situation : on the one hand, sophisticated turbulence models are available but can be used mostly for thin shear flows where the flow complexity is not large enough to justify the use of such tools ; on the other hand, all the complex turbulent flows (for which sophisticated turbulence models would be needed) are usually calculated using very simple turbulence closures, e.g. mixing length theory which are basically unable to represent all the features of these flows.

It has been generally assumed that the Navier-Stokes equations are valid for turbulent flows because the scales of the smallest significant fluctuations of fluid particles identified with dissipation are much larger than the mean free path of fluid molecules. Provided that solutions to the equations in three dimensions and time exist and are unique (the latter has not been proved mathematically up to now), turbulent flow fields could be determined from solutions of these equations. One difficulty in doing that, however, is the correct and detailed prescription of the initial and time varying boundary conditions for all but the simplest of flow field. An additional difficulty for high Reynolds number flows is that the spectrum of the turbulent fluctuations spreads over several frequency decades and the required number of mesh points to resolve simultaneously all the characteristic scales is not even with the most powerful today computers.

For low or moderate Reynolds numbers and relatively simple flow fields, it is possible to use the three-dimensional time-dependent Navier-Stokes equations to resolve the largest structures and sub-grid scale modelling for the small dissipative scales of motions [16], [17]. Using the same approach, a compressible homogeneous turbulent shear flow has been numerically simulated [18] but this work was not followed by similar studies. For high Reynolds number flows or complex flow fields, however these approaches are not realistic at present and it is necessary to apply a statistical treatment to the Navier-Stokes equations before attempting any numerical solution. This approach, using a time or ensemble averaged set of equations gives rise to a new problem. Due to the non-linear convective terms, any statistical process that is employed leads to more unknown quantities than equations - this is the classic closure problem. These additional unknown quantities must be modelled, i.e. algebraïc relations or partial differential equations have to be established which allow to close the set of equations. This closure problem, already complex for incompressible flows, becomes much more complicated when density fluctuations are present. To overcome this difficulty, several decompositions of the turbulent variables (between a mean and a fluctuating part) have been suggested and used. Among them, the mass-weighted form, proposed by FAVRE [19], produces forms identical to the laminar equations except for the addition of Reynolds stresses or second moments of the turbulence quantities. An other advantage of the mass -weighted form is to keep the same form for exact Reynolds stress transport equation than the incompressible form, except the mean pressure gradient terms.

When buoyancy effects and fluctuations of diffusivity coefficients are neglected the open equations for the conservation of mass, momentum and total energy for compressible flow are :

$$\frac{\partial \bar{\rho}}{\partial t} + \frac{\partial \bar{\rho}\widetilde{U}_k}{\partial x_k} = 0 \tag{1}$$

$$\frac{\partial \bar{\rho}\widetilde{U}_i}{\partial t} + \frac{\partial}{\partial x_k}\left(\bar{\rho}\widetilde{U}_k\widetilde{U}_i + \bar{\rho}\widetilde{u_i u_k} - \bar{\sigma}_{ik}\right) = 0 \tag{2}$$

$$\begin{aligned} &\frac{\partial \bar{\rho}\widetilde{E}}{\partial t} + \frac{\partial}{\partial x_k}\Bigg[\left(\bar{\rho}\widetilde{E} + \bar{p}\right)\widetilde{U}_k + \bar{\rho}\widetilde{u_i u_k}\widetilde{U}_i + \bar{\rho}\frac{\widetilde{u_i u_i u_k}}{2} \\ &+ \gamma C_v \bar{\rho}\widetilde{\theta u_k} - \lambda\frac{\partial \widetilde{T}}{\partial x_k} - \mu\left(\widetilde{S_{ik}}\widetilde{U}_i + \overline{s_{ik}}\widetilde{U_i} + \overline{s_{ik}u_i} + \widetilde{S_{ik}}\overline{u_i}\right)\Bigg] = 0 \end{aligned} \tag{3}$$

$$p = (\gamma - 1)\,\rho e$$

$$\bar{\sigma}_{ik} = -\bar{p}\delta_{ik} + \overline{\mu S_{ik}}$$

$$S_{ik} = \left(\frac{\partial U_i}{\partial x_k} + \frac{\partial U_k}{\partial x_i}\right) - \frac{2}{3}\delta_{ik}\frac{\partial U_l}{\partial x_l}$$

$\widetilde{U}_k$ and u_k stand for mean and fluctuating velocities according to FAVRE's decomposition, $\overline{P}$ and p' represent mean and fluctuating pressures with conventional REYNOLDS sense.

There are two ways of expressing the Reynolds stress tensor :

i) the Reynolds stresses can be assumed as a local property of the mean flow and by analogy as a viscous stress written in the following form :

$$-\bar{\rho}\widetilde{u_i u_k} = \mu_t \widetilde{S_{ik}} - \frac{2}{3}\bar{\rho}\tilde{k}\delta_{ik} \tag{4}$$

Here, $\bar{\rho}\tilde{k} = \frac{1}{2}\bar{\rho}\widetilde{u_i u_i}$ is called the turbulent kinetic energy and μ_t is a "turbulent" viscosity coefficient. All the models using (4) belong to the class of "eddy viscosity models",

ii) the Reynolds stresses $-\bar{\rho}\widetilde{u_i u_k}$ are assumed to be field variables, then it is possible to obtain their values directly by solving suitable transport equations. Using the mass-weighted decomposition of FAVRE, they can be written as :

$$\begin{aligned}
&\frac{\partial}{\partial t}(\bar{\rho}\widetilde{u_i u_j}) + \frac{\partial}{\partial x_k}\left[\bar{\rho}\widetilde{U_k}\widetilde{u_i u_j} + \bar{\rho}\widetilde{u_i u_j u_k} + \delta_{ik}\overline{u_j p'} + \delta_{jk}\overline{u_i p'} - \left(\overline{\mu S_{ik} u_j} + \overline{\mu S_{jk} u_i}\right)\right] \\
&\quad = -\bar{\rho}\left[\widetilde{u_i u_k}\frac{\partial \widetilde{U_j}}{\partial x_k} + \widetilde{u_j u_k}\frac{\partial \widetilde{U_i}}{\partial x_k}\right] + \overline{p'\left(\frac{\partial u_i}{\partial x_j} + \frac{\partial u_j}{\partial x_i}\right)} \\
&\quad - \left[\overline{u_i}\frac{\partial \bar{p}}{\partial x_j} + \overline{u_j}\frac{\partial \bar{p}}{\partial x_i}\right] - \overline{\mu\left(S_{ik}\frac{\partial u_j}{\partial x_k} + S_{jk}\frac{\partial u_i}{\partial x_k}\right)}
\end{aligned} \tag{5}$$

It is worth pointing out that the mean pressure gradient terms in equation (5) are due to the non zero $\overline{u_i}$ in FAVRE's decomposition. In equation (5), new unknowns involving terms other than second order moments of the velocity fluctua tions appear and have to be modelled to close the system. Turbulence models that allow solution of equation (5) belong to the class of second order closure models.

3. THE CLOSURE PROBLEMS

3.1. Eddy viscosity concept models

The eddy viscosity concept is based on the assumption that the turbulent stresses depend linearly on the strain rate of the mean motion. This idea, originally due to BOUSSINESQ, is analogous to the formulation of viscous stresses in the kinetic theory of gases. To be dimensionally correct, the eddy viscosity μ_t in (4) is assumed proportional to the mean density to a turbulent characteristic velocity scale and to a turbulent length scale. For the molecular viscosity, the characteristic velocity and length scales are those of molecular motion and therefore do not depend on the macroscopic flow scales. Thus the molecular viscosity can be considered as an intrinsicproperty of the fluid. The situation is quite different for the eddy viscosity whose characteristic scales depend, in a direct manner, on the turbulent motion. As these scales are not distinct from the mean motion characteristics, a strong correlation between the mean and turbulent motion can be expected. Thus the eddy viscosity coefficient should be tightly linked to the features of the mean motion and depend on space and time in contrast with the molecular viscosity coefficient which depends solely on the fluid temperature.

The choice of the characteristic turbulent scales to define the eddy viscosity is not a simple problem for complex flows. Depending on the choice of these scales various types of eddy viscosity models can be distinguished :

* zero-equation models where the two scales are related algebraïcally to mean motion variables,
* one-equation models where the turbulent velocity scale is obtained from a transport equation of the turbulent kinetic energy whereas the characteristic length scale is evaluated algebraïcally,
* two-equation models where both characteristic scales are solutions of field equations.

3.1.1. Zero-equation models

These models use velocity and length scales given by the mean motion under the assumption that turbulence quantities are only related to the mean local properties. Implicit here is that the mean motion and the turbulent diffusivity have identical characteristic frequencies. This assumption however is not valid everywhere, particularly in fully turbulent zones, remote from solid walls where the turbulent diffusivity is more dependent on the integral quantities of the flow. For this reason, in compressible wall boundary-layer problems, the most widely used algebraic turbulence model which is due to CEBECI, SMITH and MOSINSKI [20] assumes different growth rates of the eddy viscosity coefficient for defined inner and outer regions. The inner part follows the standard mixing length theory where scales are dependent on the distance from a surface whereas the outer part is related to the displacement thickness of the boundary layer. As discussed by HUNG and Mac CORMACK [21] for compression corner flows and by BALDWIN and LOMAX [22] for transonic airfoil flows, large variations in the computed outer viscosity can occur as a result of the uncertainties in defining the boundary layer thickness in complex flows. In an attempt to overcome this difficulty, BALDWIN and LOMAX [22] proposed a new algebraic eddy viscosity model using, in the outer zone, a wake function based on CLAUSER's work and KLEBANOFF's experimental results. As this model does not require the determination of the boundary layer edge, it is claimed to be designed to separate flows as well as attached boundary layer. It has been observed that, in these models, the eddy viscosity strongly depends on the mean flow and therefore, on the geometric pattern of the flow. Since the experimental bases for these models are attached equilibrium boundary layers, computations of self similar flows have yielded rather satisfactory results. For use in the computation of complex separated flows however these models are used well outside the domain of their empirical base and are not expected to yield good quantitative results.

3.1.2. One-equation models

The one-equation or kinetic energy model uses, in addition to the mean flow equations, a partial differential equation for the turbulent kinetic energy, $\bar{\rho}\tilde{k} = \frac{1}{2}\bar{\rho}\widetilde{u_i u_i}$. In this class of models, the most extensively used are due to BRADSHAW [23] and RUBESIN [24]. The latter model has been widely investigated for shock/ boundary layer interaction flows by VIEGAS and HORSTMAN [25].

In the RUBESIN proposal, the turbulent energy equation is written as :

$$\frac{\partial}{\partial t}\left(\bar{\rho}\tilde{k}\right) + \frac{\partial}{\partial x_k}\left(\bar{\rho}\tilde{k}\tilde{U}_k - \mu_k\frac{\partial \tilde{k}}{\partial x_k}\right) = P_k + C_1\bar{\rho}\frac{\tilde{k}}{\gamma}M_a^2\frac{\partial \tilde{U}_k}{\partial x_k} - C_2\frac{\mu_t\tilde{k}}{L^2} \tag{6}$$

where M_a is the Mach number and P_k is the production term. The modelling constants and relationship between the eddy viscosity and k and L are given in [24]. Unfortunately, the length scale determination has to be tu ned to each flow configuration [25]. With the one-equation models, the turbulence character is only present through its absolute energy level while the length scale is determined algebraically. As for the algebraic models discussed previously for complex flows, including those with separation, the prescription of the algebraic length scale is again quite risky as there are very little data to guide choices of the dissipation length scale under these conditions.

3.1.3. Two-equation models

Two-equation models try to be more versatile than the one-equation models with the addition of partial differential equation for some distribution of the length scale variable. The equations admit advective-diffusive-productive and dissipative mechanisms and are not conservative like those for momentum, mass or total energy. This remark would be important for numerical point of view. Non conservative equations need more attention when initial conditions have to be prescribed and when realizability conditions have to be satisfied. Two-equation models are believed to provide more suitable velocity and length scales of the turbulence than those associated with the mean motion. Among others, the most widely used models are due to JONES and LAUNDER [25] and to WILCOX and RUBESIN [26]. The first of these intensively applied as well to incompressible flows as to compressible ones, uses the turbulent kinetic energy and its dissipation rate as characteristic turbulent scales. The second model more applied to compressible flows uses the specific dissipation rate as a second field variable.

As an illustration, the modelled equations for the first model are written for compressible flows as :

$$\frac{\partial}{\partial t}\left(\bar{\rho}\widetilde{k}\right)+\frac{\partial}{\partial x_k}\left(\bar{\rho}\widetilde{k}\widetilde{U}_k-\mu_k\frac{\partial\widetilde{k}}{\partial x_k}\right)=P_k-\rho\epsilon-2\mu\left(\frac{\partial\sqrt{\widetilde{k}}}{\partial x_n}\right)^2-\overline{u}_i\frac{\partial\bar{p}}{\partial x_i}+\overline{p'\frac{\partial u_i}{\partial x_i}} \quad (7)$$

$$\frac{\partial}{\partial t}\left(\bar{\rho}\widetilde{\epsilon}\right)+\frac{\partial}{\partial x_k}\left(\bar{\rho}\widetilde{\epsilon}\widetilde{U}_k-\mu_\epsilon\frac{\partial\widetilde{\epsilon}}{\partial x_k}\right)=C_{\epsilon1}\frac{\widetilde{\epsilon}}{\widetilde{k}}P_k-C_{\epsilon2}f_2\frac{\widetilde{\epsilon}}{\widetilde{k}}\rho\epsilon^*+2\mu\frac{\mu_t}{\rho}\left(\frac{\partial^2\widetilde{U}}{\partial x_n^2}\right)^2$$
$$-C_{\epsilon3}\frac{\widetilde{\epsilon}}{\widetilde{k}}\overline{u}_i\frac{\partial\bar{p}}{\partial x_i}+C_{\epsilon4}\frac{\widetilde{\epsilon}}{\widetilde{k}}\overline{p'\frac{\partial u_i}{\partial x_i}} \quad (8)$$

Details of the modelling constants and relationship of the eddy viscosity and $\widetilde{k}$ and $\widetilde{\varepsilon}$ are found in [25].

3.2. Second order moment closures

The second order moment closures (also sometimes called Reynolds stress models) are currently the most general models from the point of view of physical theory. These models require the solution of numerous additional field equations, for example, equations for each of the individual Reynolds stresses $\bar{\rho}\widetilde{u_i u_j}$ and a scale equation such as the dissipation rate of turbulent kinetic energy similar to equation (8). These models are therefore much more complicated than eddy viscosity formulations. One of the important physical properties contained in these models is a stress relaxation property which is only partially treated in the advanced eddy viscosity models through the $\widetilde{S}_{ik}$ in equation (4). This could be an important feature for flow fields where large anisotropies develop as a result of shock waves or sudden expansions. The Reynolds stress transport equations are modelled by extending the incompressible flow model given by the most important stone corner paper of LAUNDER, REECE and RODI [26] completed by HANJALIC LAUNDER [27] to compressible flow conditions using FAVRE formalism reintroducing the non-zero divergence terms that were eliminated in the original model and also RUBESIN [24] for the non-zero mass weighted fluctuating velocities. Furthermore, some of the modelling decisions of HANJALIC and LAUNDER, relating the values of the Reynolds stresses to the turbulent kinetic energy in their two-equation model can be avoided when all stresses are computed individually.

The modelled Reynolds stress equation is written as [28] :

$$\frac{\partial}{\partial t}\left(\bar{\rho}\widetilde{u_iu_j}\right)+\frac{\partial}{\partial x_k}\left(\bar{\rho}\widetilde{u_iu_j}\widetilde{U_k}-\mu\frac{\partial}{\partial x_k}\widetilde{u_iu_j}-C_s'\frac{\widetilde{k}}{\widetilde{\epsilon}}\widetilde{u_ku_l}\frac{\partial}{\partial x_l}\widetilde{u_iu_j}\right)$$
$$=P_{ij}-\frac{C_2+8}{11}\left(P_{ij}-\frac{2}{3}\delta_{ij}P\right)-\frac{8C_2-2}{11}\left(D_{ij}-\frac{2}{3}\delta_{ij}D\right)$$
$$-\frac{30C_2-2}{55}\bar{\rho}\widetilde{k}\widetilde{S_{ij}}-C_1\bar{\rho}f_1\frac{\widetilde{\epsilon}}{\widetilde{k}}\left(\widetilde{u_iu_j}-\frac{2}{3}\delta_{ij}\widetilde{k}\right)$$

$$+\left(C_3\bar{\rho}\frac{\widetilde{\epsilon}}{\widetilde{k}}\left(\widetilde{u_iu_j}-\frac{2}{3}\delta_{ij}\widetilde{k}\right)+C_4\left(P_{ij}-D_{ij}\right)+C_5\bar{\rho}\widetilde{k}\widetilde{S_{ij}}\right)\frac{\widetilde{k}^{\frac{3}{2}}}{\widetilde{\epsilon}x_m}$$
$$-\bar{\rho}\frac{\widetilde{\epsilon}}{\widetilde{k}}\left(\widetilde{u_iu_j}f_s+(1-f_s)\frac{2}{3}\delta_{ij}\widetilde{k}\right)-\overline{u_i}\frac{\partial\bar{p}}{\partial x_j}-\overline{u_j}\frac{\partial\bar{p}}{\partial x_i} \tag{9}$$

where

$$P=D=-\bar{\rho}\widetilde{u_iu_j}\frac{\partial\widetilde{U_i}}{\partial x_j}\quad ;\quad \overline{u_i}=\frac{\widetilde{U_k}\widetilde{u_iu_k}}{(n-1)C_p\widetilde{T}}$$

$$P_{ij}=-\bar{\rho}\left(\widetilde{u_iu_k}\frac{\partial\widetilde{U_j}}{\partial x_k}+\widetilde{u_ju_k}\frac{\partial\widetilde{U_i}}{\partial x_k}\right)\quad ;\quad D_{ij}=-\bar{\rho}\left(\widetilde{u_iu_k}\frac{\partial\widetilde{U_k}}{\partial x_j}+\widetilde{u_ju_k}\frac{\partial\widetilde{U_k}}{\partial x_i}\right)$$

The total energy equation is modelled as :

$$\frac{\partial\bar{\rho}\widetilde{E}}{\partial t}+\frac{\partial}{\partial x_k}\left[\left(\bar{\rho}\widetilde{E}+\bar{p}+\frac{2}{3}\mu\frac{\partial\widetilde{U_l}}{\partial x_l}\right)\widetilde{U_k}+\left(\bar{\rho}\widetilde{u_iu_k}-\mu\frac{\partial\widetilde{U_i}}{\partial x_k}\right)\widetilde{U_i}\right.$$
$$-\mu\frac{\partial}{\partial x_k}\left(\frac{\widetilde{U_i}\widetilde{U_i}}{2}+\frac{\widetilde{u_iu_i}}{2}\right)-0.5C_s'\frac{\widetilde{k}}{\widetilde{\epsilon}}\widetilde{u_ku_l}\frac{\partial}{\partial x_l}\widetilde{u_iu_i}$$
$$\left.-\left(C_{se}C_v\gamma\bar{\rho}\frac{\widetilde{k}}{\widetilde{\epsilon}}\widetilde{u_ku_l}+\delta_{lk}\lambda\right)\frac{\partial\widetilde{T}}{\partial x_l}\right]=0 \tag{10}$$

In equation (10), the turbulent heat fluxes have been modelled as a generalized gradient diffusion term. The turbulence variables are scaled with the dissipation rate of turbulent kinetic energy which is the solution of the modelled equation :

$$\frac{\partial}{\partial t}(\bar{\rho}\widetilde{\epsilon})+\frac{\partial}{\partial x_k}\left(\bar{\rho}\widetilde{\epsilon}\widetilde{U_k}-(\bar{\mu}\delta_{kl}+C_\epsilon\bar{\rho}\frac{\widetilde{k}}{\widetilde{\epsilon}}\widetilde{u_ku_l})\frac{\partial\epsilon}{\partial x_l}\right)$$
$$=-C_{\epsilon1}\bar{\rho}\frac{\widetilde{\epsilon}}{\widetilde{k}}\widetilde{u_iu_k}\frac{\partial\widetilde{U_i}}{\partial x_k}-C_{\epsilon2}\bar{\rho}\frac{\widetilde{\epsilon}}{\widetilde{k}}f_\epsilon\left(\widetilde{\epsilon}-2\nu\left(\frac{\partial\sqrt{\widetilde{k}}}{\partial x_m}\right)^2\right)$$
$$-C_{\epsilon3}\bar{\mu}\frac{\widetilde{k}}{\widetilde{\epsilon}}\widetilde{u_ju_k}\left(\frac{\partial^2\widetilde{U_i}}{\partial x_j\partial x_l}\right)\left(\frac{\partial^2\widetilde{U_i}}{\partial x_k\partial x_l}\right)-C_{\epsilon4}\frac{\widetilde{\epsilon}}{\widetilde{k}}\overline{u_i}\frac{\partial\bar{p}}{\partial x_i} \tag{11}$$

The modelling coefficients used in these equations are given in [28]. The sparseness of existing data, especially for the individual Reynolds stresses, $\bar{\rho}\widetilde{u_iu_j}$ and the heat fluxes, makes it difficult at present to assess the validity of all the assumptions employed in the closures.

3. THE TURBULENCE MODELLING CHOICE

In our opinion and with our rather modest experiences in computation of com-

pressible turbulent flows, there is no universal turbulence model to be recommended for all compressible cases. Each turbulent flow has dominant specific characters and the turbulence model chosen must be able to describe correctly all of them. When choosing an adequate turbulence model for complex turbulent flow computations, it is necessary to compromise between performance and cost. Accordingly in most of complex flow calculations, eddy viscosity concept models are generally simpler to use than second order closure models using Reynolds stress transport equations. Certainly, it is shown that in most turbulent shear flows, eddy viscosity concept models (for example, the very popular JONES-LAUNDER model) are able to provide results as good as those given by the second order closure model. Moreover, in some circumstances, for example, for equilibrium boundary layers, mixing length theories seem to give more satisfactory results than the second order closure models. In fact, numerical considerations enforce the previous choice. Obviously, much computing time may be saved by the use of zero-equation or two-equation models but the main reason of this success is much more numerically fundamental. When using an eddy viscosity concept model, the local Peclet number (convective effects / diffusive effects) is actually small or moderate. Therefore, numerical schemes performed for low or moderate Reynolds number flows (generally in incompressible cases) are well adapted for the calculation of turbulent flows using eddy viscosity model although the nominal Reynolds numbers of these flows are generally very high. When second order closure models are used, this convenient arrangement is no longer available : the main momentum fluxes (the Reynolds stresses) appear as pressure terms and not as gradient fluxes like they do for laminar flows. Accordingly, the Reynolds equations are more hyperbolic (like Euler equations) than elliptic (like the classical Navier-Stokes viscous equations). This shows the necessity of using numerical methods well adapted to very high Reynolds number flows like those used for Euler equations. This remark does not mean that high Reynolds number flows are "inviscid" and can be governed by Euler model. In fact, although numerically Reynolds stress are handled as a pressure term, physically, the importance of turbulent shear stress conserves real diffusive nature of flow and precludes the use of inviscid approach even for very high Reynolds number flows.
We must not forget that the choice of turbulence models to be used must be decided by physical considerations and not only by numerical or economical aspects. Turbulence models are not only intended for the computation of global properties. They have also to be considered as "numerical experiments" developed for contributing and to the analysis of complex turbulent flows. For this purpose, it is essential that the turbulence models used for turbulent flow calculations

must be able to reveal the most significant physical aspects and frequently the simplicity of the model used is spoilt by the scarcity of available informations.

4. CONCLUDING REMARKS

Although higher sophisticated models exist presently (large eddy simulation, direct simultion of turbulence...), it is reasonable to conclude that statistical turbulence models are still the more adapted way to predict numerically turbulent complex incompressible or compressible flows in realistic or industrial geometries.
For equilibrium flows and in thin shear layer approximation (without separation), the zero-equation models seem to be sufficient.
In a study where the turbulence field variables are not in equilibrium with the mean flow, only the transport equation turbulence models are able to provide detailed information regarding the turbulence. The two-equation model gives an adequate estimate of the mean flow features, particularly the length and thickness of the separated zone but the turbulence level is still underpredicted in the interaction region. In fact, the model cannot produce simultaneously a correct level of turbulence with an adequate level of the crosswise turbulent fluxes ($\widetilde{v^2}$).
The use of the Reynolds stress equation model relaxes the previous constraint and improves the estimation of the anisotropy. The integration of the Reynolds stress equation to a surface shows eventually a mechanism of turbulent anisotropy inversion which, in high Reynolds number flows today, has not been accessible to either experimental investigations or large eddy simulation studies.
Although the work above has shown advantages of employing the more complex models in many cases, it must be stated that these advantages are not free of cost. The introduction of these models into the Navier-Stokes codes while not entirely straightforward are certainly achievable with some perseverance. In addition, the cost of achieving converged solution with the increasingly complex models, namely algebraic, two-equation and second order closure models are roughly in the order of 1 : 2 : 5. Thus, the increased cost of the more advanced models is currently quite reasonable and advances in the algorithm currently under consideration may well reduce the cost of second order closure calculation to less than those of algebraïc models to-day.

References

1. Boussinesq, J. : Théorie de l'écoulement tourbillonnant et tumultueux des liquides dans les lits rectilignes à grande section. Tome I-II, Gautier-Villars, Paris (1807).

2. Prandtl, L. : Uber die ausgebieldete Turbulenz. Z. Angew. Math. Mech., vol. 5, (1925) 136-139.

3. Van Driest, E.R. : Turbulent boundary layer in compressible fluids. J. Aero. Sci., vol. 18, n° 3 (1951).

4. Kovasznay, L.S.G. : Turbulence in supersonic flow. J. Aero. Sci., vol. 20 (1953).

5. Morkovin, M. : Effects of compressibility on turbulent flows. Mécanique de la Turbulence, C.N.R.S. A. Favre (ed.) 1962.

6. Demitriades, A. : Turbulence measurements in supersonic two-dimensional wake. Phys. of Fluids, vol. 13, n° 7 (1970).

7. Laufer, J. : Thoughts on compressible turbulent boundary layers. Memorandum R.M. 5946-PR, Rand. Corp. or NASA-SP-216 (1969).

8. Kistler, A.L. : Fluctuations measurements in a supersonic turbulent boundary layer. Phys. of Fluids, vol. 2, n° 290 (1959) 69-72.

9. Kistler, A.L. and Chen, W.S. : The fluctuating pressure field in a supersonic turbulent boundary layer. J.F.M., vol. 16, n° 41 (1963) 72.

10. Bradshaw, P. : Compressible turbulent shear layers. Ann. Rev. of Fluid Mech., (1977) 33-64.

11. Morkovin, M.V. et al. : Report of conference evaluation committee. Free Turbulent Shear Flows Conf. Proceed., Hampton, Virginia, NASA-SP-321, (1972).

12. Wilcox, D.C. and Alber, I.E. : A turbulence model for high speed flows. Proceed. of the 1972 Heat Transfer and Fluid Mechanics Institute, Stanford Univ. Press (1972) 231-252.

13. Bradshaw, P. : The effect of mean compression or dilatation on the turbulence structure of supersonic boundary layers. J.F.M., vol. 63-3 (1974).

14. Oh, Y.H. : Analysis of two-dimensional free turbulent mixing. A.I.A.A. Paper 74-594 (1974).

15. Bonnet, J.P. and Launder, B.E. : Case 85M : 1980-1981, AFOSR-HTTM Stanford Conference on Complex Turbulent Flows. Kline et al (eds.), Stanford Univ. (1981).

16. Schumann, U. ; Grotzbach, G. and Kleiser, L. : Direct numerical simulation of turbulence in prediction methods for turbulent flows. Hemisphere Publ., New York (1980).

17. Moin, P. and Kim, J. : Numerical investigation of turbulent channel flow. NASA TM81309 (1981).

18. Feiereisen, W.J. ; Reynolds, W.C. and Ferziger, J. : Numerical simulation of a compressible homogeneous turbulent shear flow. Tech. Rep. n° TF-13, Thermosciences division, Stanford Univ. (1981).

19. Favre, A. : Equations des gaz turbulents compressibles. J. de Méc., vol. 4, n° 3, 361-390, vol. 4, n° 4, 391-421 (1965).

20. Cebeci, T.O. ; Smith, A.M. and Mosinski, S.G. : Calculations of compressible adiabatic turbulent boundary layer. A.I.A.A. Journal, vol. 8 (1970) 1974-1982.

21. Hung, C.M. and Mac Cormack, R.W. : Simulation of supersonic and hypersonic turbulent compression corner flows. A.I.A.A. Journal, vol. 15 (1977) 410-416.

22. Baldwin, B. and Lomax, H. : Thin layer approximation and algebraïc model for separated turbulent flows. A.I.A.A. Paper 78-257 (1978).

23. Bradshaw, P. ; Ferris, D.H. and Atwell, N.P. : Calculation of boundary layer development using the turbulent energy equation. J.F.M., n° 28 (1967) 593.

24. Rubesin, M.W. : A one-equation model of turbulence for use with the compressible Navier-Stokes equations. NASA TM X-73-128 (1976).

25. Viegas, J.R. and Horstman, C.C. : Comparison of multiequation on turbulence models for several shock boundary layer interaction flows. A.I.A.A. Journal, vol. 17, n° 18 (1979) 811-820.

26. Launder, B.E. ; Reece, G.R. and Rodi, W. : Progress in the development of a Reynolds stress turbulence closure. J.F.M., vol. 60/3 (1975) 537-566.

27. Hanjalic, K. and Launder, B.E. : Contribution towards a Reynolds stress closure for low Reynolds number turbulence. J.F.M., vol. 74/4 (1976) 593-610.

28. Vandromme, D. ; Ha Minh, H. ; Vieags, J.R. ; Rubesin, M.W. and Kollmann, W. : Second order closure for the calculation of compressible wall bounded flows with an implicit Navier-Stokes solver, 4th T.S.F., Karlsruhe (1983).

Transonic Separated Flow Prediction Based on a Mathematically Simple, Nonequilibrium Turbulence Closure Model

D. A. JOHNSON AND L. S. KING

NASA Ames Research Center
Moffett Field, CA 94035

Summary

A mathematically simple, turbulence closure model designed to treat transonic airfoil flows even with massive separation is described. Numerical solutions of the Reynolds-averaged, Navier-Stokes equations obtained with this closure model are shown to agree well with experiments over a broad range of test conditions.

Introduction

The greatest limiting factor in the accurate numerical prediction of airfoil flows has been the lack of an adequate turbulence closure model. For airfoil flows with sufficiently mild pressure gradients and no separation, simple algebraic turbulence models have been shown to yield good results. As speeds in the transonic regime and angles of attack are increased, however, adverse pressure gradients become stronger and separated flow regions make their appearance. Under these harsher conditions, the algebraic turbulence models do not do well. Such cases also present a severe test for more sophisticated turbulence models. In Ref. 1 for example, relatively poor predictions of surface pressure were obtained for a series of transonic, inviscid-viscous interaction flows (all with some separation) with both the Cebeci-Smith [2], algebraic model and the Jones-Launder [3], $k - \varepsilon$ model.

Recently (Ref. 4), a mathematically simple closure model was proposed that showed promise in treating the turbulent boundary layer development even under conditions where the inviscid-viscous interaction results in massive separation. Subsequent to this work, further evaluations of this closure model (Refs. 5 and 6) have been made using time-marching, Navier-Stokes methods.

This paper first describes the turbulence closure model of Ref. 4, and presents some of the results obtained to date with this closure model.

Turbulent Shear Layer/Shock Wave Interactions
IUTAM Symposium Palaiseau 1985
Editor: J. Délery

Closure Model

The intent of the study (described in Ref. 4) was to develop a turbulence closure model for a particular class of flows: wall-bounded shear flows subjected to large and rapidly changing streamwise pressure gradients. The impetus was the prediction of turbulent boundary layer development on airfoils under conditions of strong inviscid-viscous interaction. The model was developed under the assumption that the major difficulty encountered in the prediction of wall-bounded shear flows subjected to adverse pressure gradients sufficiently strong to cause separation has been due to the lack of (or an improper accounting of) "history effects" owing to convection.

The closure model can best be described as a hybrid eddy-viscosity/ Reynolds-shear-stress model. To account for the strong "history effects" present in flows with large and rapidly changing streamwise pressure gradients, a simplified Reynolds-shear-stress equation (an ordinary differential equation for the maximum Reynolds shear stress) is used to determine eddy-viscosity changes in the streamwise direction. Eddy-viscosity models tend to predict too rapid a change in the Reynolds shear stress when the mean flow field is rapidly distorted. The present closure model was designed to overcome this weakness.

An algebraic eddy viscosity distribution

$$\nu_t = \nu_{to}(1 - \exp - \nu_{ti}/\nu_{to}) \tag{1}$$

$$\nu_{ti} = D^2\ 0.4y(\overline{-u'v'_m})^{1/2} \tag{2}$$

$$\nu_{to} = \sigma(x) \bullet 0.0168\ u_e \delta_i^* / [1 + 5.5(y/\delta)^6] \tag{3}$$

is used to describe the variation of the Reynolds shear stress normal to the shear layer under both attached and separated flow conditions. "History effects" are taken into account through the parameter $\sigma(x)$ in Eq. (3). The streamwise distribution of this parameter is established in the solution procedure (details are given in Ref. 5) to give a streamwise distribution of $\overline{-u'v'}$ which satisfies the following ordinary differential equation:

$$\bar{u}_m \frac{d\tau_m}{dx} = \frac{a_1}{L_m} \tau_m \left[\left(\tau_{eq}\Big|_m \right)^{1/2} - \tau_m^{1/2} \right] - a_1 \mathcal{D}_m \tag{4}$$

In this equation, τ is used to represent $-\overline{u'v'}$, which, for convenience, will be referred to as the Reynolds shear stress. The subscript m and vertical slash m denote that the quantity is evaluated at y_m, the y location where $-\overline{u'v'}$ is a maximum. This equation which describes the development of the Reynolds shear stress along the path of maximum kinetic energy is obtained from the turbulence kinetic energy equation. In Eq. (4), $a_1 = -\overline{u'v'}_m/k_m$ (where k is the turbulence kinetic energy) and is assumed to be a constant. L_m is the dissipation length scale, which is modeled algebraically. $\mathcal{D}_m$ is the turbulent diffusion and τ_{eq} is the Reynolds shear stress based on the eddy viscosity given by Eqs. (1)-(3) with $\sigma(x)$ equal to unity. $\mathcal{D}_m$ is assumed to always be positive and to depend on the nonequilibrium state of the flow (i.e., the degree $\sigma(x)$ deviates from unity). The modeled form of this term and its formulation can be found in Ref. 5. When the convective term in Eq. (4), $\bar{u}_m d\tau_m/dx$ becomes small, $\sigma(x)$ approaches unity and τ_m approaches $\tau_{eq}\big|_m$.

Results

Results are now presented which demonstrate the present closure model's ability to describe wall-bounded shear flows over a range of conditions. All of the test cases to be presented have some separated flow; most have massive separation zones.

The first flow to be considered is the low-speed diffuser flow of Ref. 7. In Fig. 1, mean velocity profiles predicted with the present closure model and the Cebeci-Smith model are compared with the experimental measurements. These predictions were obtained with an inverse-boundary-layer method. Figure 1 illustrates the basic inadequacy of the (equilibrium) Cebeci-Smith model in predicting separated flows, and the improvements achievable with the present simple nonequilibrium model. The different characteristic shapes of the mean velocity profiles predicted by these two closure models are primarily a result of the eddy viscosity expressions used for the inner part of the boundary layer. The present closure model also did much better than the Cebeci-Smith model in predicting the surface pressure distribution (Ref. 4).

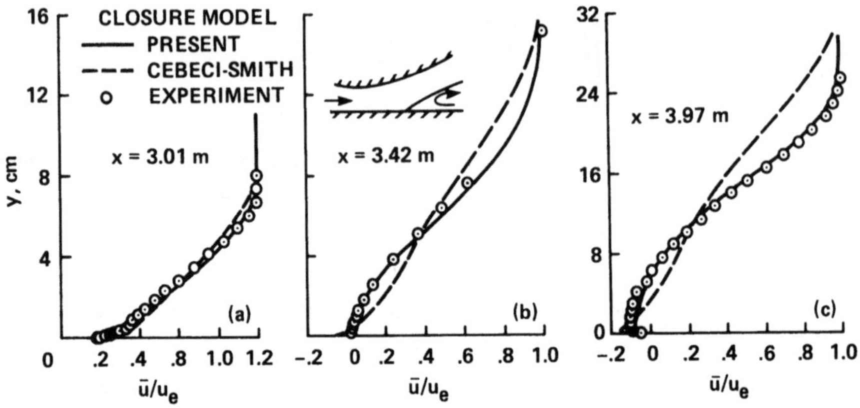

Fig.1. Mean-velocity comparisons for the low-speed diffuser of Simpson et al. [7] a) Upstream of separation; b) at separation; c) downstream of separation.

The remaining examples are all for transonic conditions. The first of these is the flow developed around an axisymmetric bump (Ref. 8). In Fig. 2, experimental surface pressure distributions for three freestream Mach numbers (M_∞ = 0.6, 0.875, and 0.925) are compared with computed distributions obtained with three different closure models. The experimental separation and reattachment locations are also shown in this figure. The computations were obtained with a computer code based on MacCormack's [9] explicit-implicit algorithm. The numerical details can be found in Refs. 1 and 5. As evident from Fig. 2, the pressure distributions obtained with the present model are in much better agreement with experiment than those obtained with either the Cebeci-Smith or Jones-Launder model.

The predicted mean velocity and Reynolds shear stress profiles for M_∞ = 0.875 based on the present model are compared to experiment in Figs. 3 and 4. At the stations x/c = 0.69 to 0.94, the calculated results are for stations slightly upstream of the experimental stations. This shift was applied to approximately account for the predicted shock location being 0.02 chord forward of the experiment. The Reynolds shear stresses are compared in shear-layer coordinates. The importance of comparing in shear-layer coordinates when the flow is highly nonisotropic and the closure model is capable of describing the Reynolds shear stresses accurately but not the Reynolds normal stresses is discussed in Ref. 5.

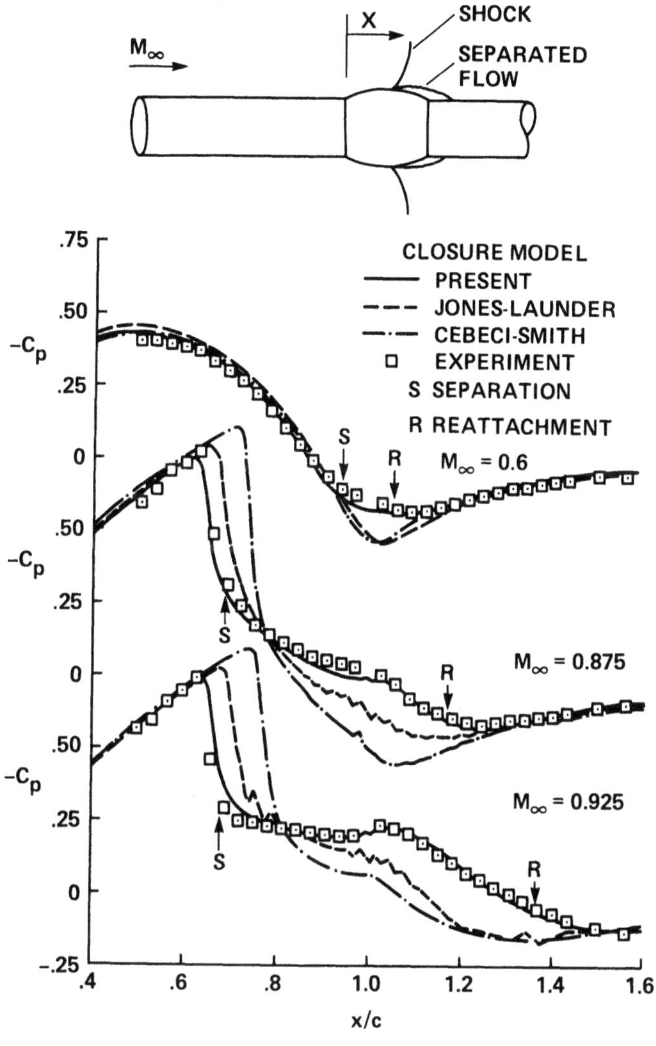

Fig.2. Surface pressure comparisons for the axisymmetric-bump flow of Bachalo and Johnson [8].

The mean velocity and Reynolds shear stress profiles predicted by the Cebeci-Smith and Jones-Launder models (not shown here) do not agree nearly as well with the experimental results (Ref. 5). Both of these models predicted more rapid rises in the Reynolds shear stresses at the shock and, as a result, underpredicted the momentum losses incurred by the boundary layer.

There is some disagreement between the present calculations and the experiment as evident from Figs. 3 and 4; (1) the predicted Reynolds shear stresses are larger than the experimental stresses in the vicinity of the shock (x/c = 0.69 and 0.75), and (2) the predicted mean velocity recovery is less than the experiment downstream of reattachment. The former disagreement may be due to possible measurement errors in $(\overline{u'^2} - \overline{v'^2})$. At x/c = 0.69 and 0.75, the untransformed shear stresses were as large or slightly larger than the computed values. But, since the measured values

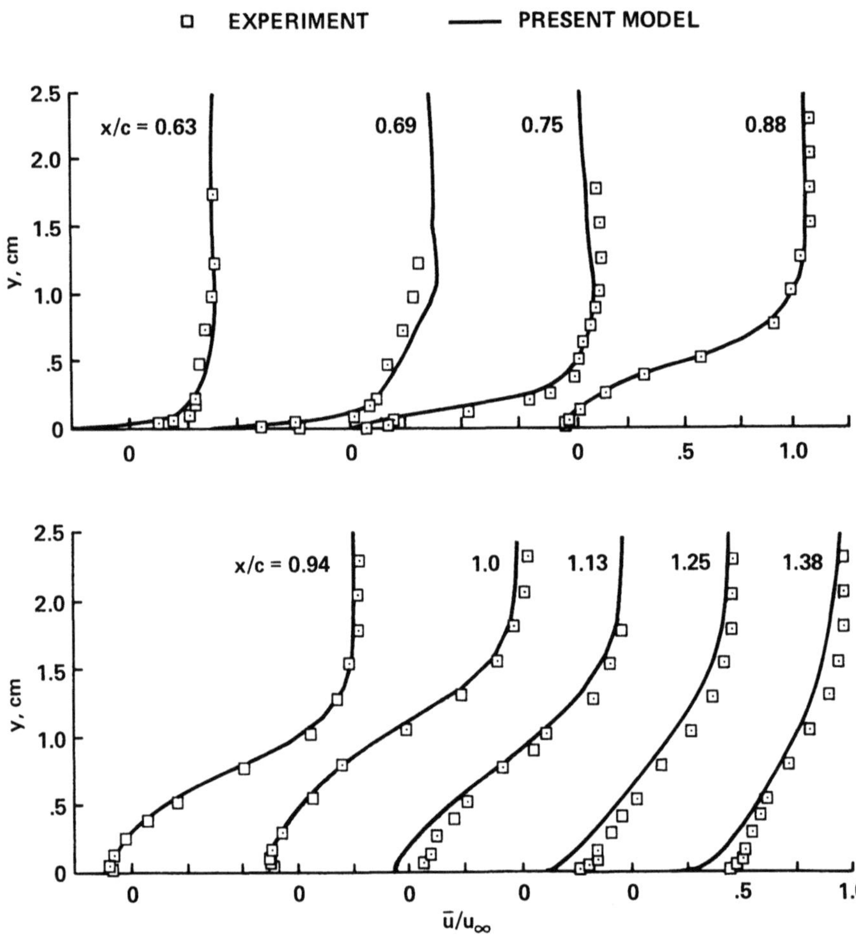

Fig.3. Mean-velocity comparisons for the axisymmetric-bump flow of Bachalo and Johnson [8]; $M_\infty = 0.875$.

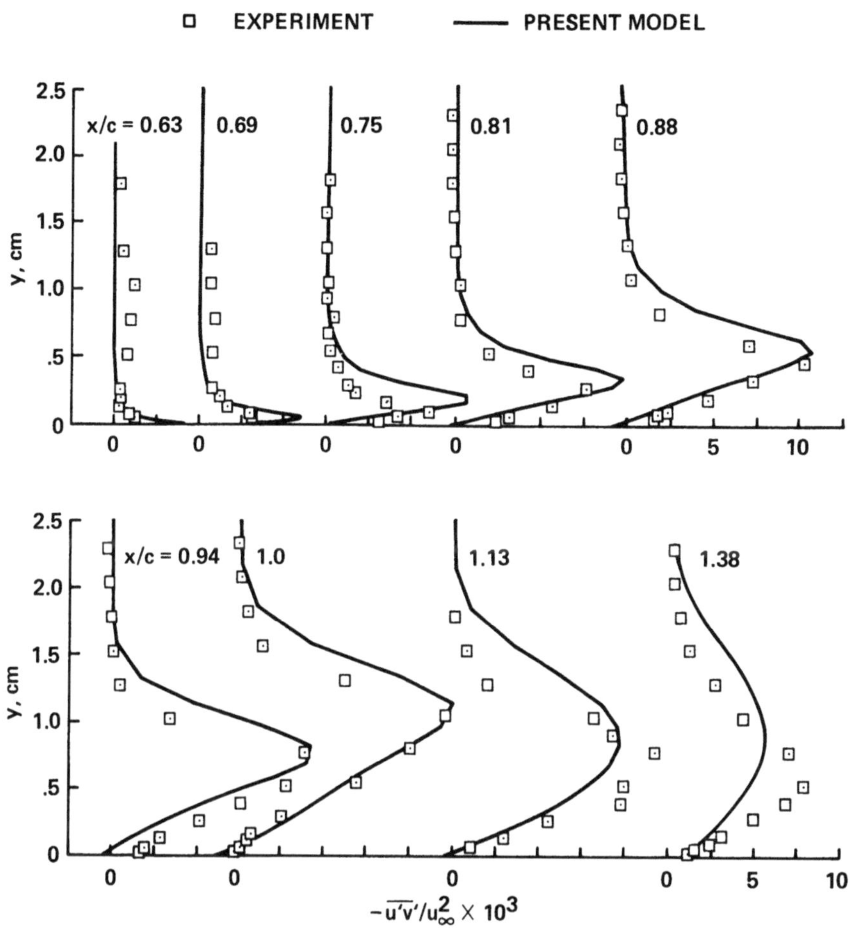

Fig.4. Reynolds shear-stress comparisons for the axisymmetric-bump flow of Bachalo and Johnson [8]; M_∞ = 0.875.

of $(\overline{u'^2} - \overline{v'^2})$ were so large in this region, a coordinate rotation of only a few degrees resulted in the much smaller shear stresses shown in Fig. 4. The latter disagreement downstream of reattachment appears to be a result of the actual inner length scales being larger than was modeled. The experimental eddy viscosity profiles suggest that the length scales in this region were closer to 0.8y than the 0.4y assumed in Eq. (2). (These discrepancies between the computations and the experiment are discussed in more detail in Ref. 5.)

The next two examples are transonic airfoil flows. Solutions for these flows were obtained using a Navier-Stokes airfoil code based on the algorithm of Beam and Warming [10]. The two examples are a supercritical airfoil (DSMA 671) section at transonic cruise conditions, and a NACA 64A010 airfoil section under shock-induced stall conditions. The experimental results for these two airfoils are given in Refs. 11 and 12, respectively. The calculations were performed at the set angles of attack with tunnel wall effects taken into account by using static pressure measurements at one chord above and below the airfoils as boundary conditions. The numerical details are given in Ref. 6.

In Fig. 5, predicted surface pressure distributions obtained with the Baldwin-Lomax [13] model and the present model for the supercritical airfoil case are compared with the experiment. The shock wave was weak and had little effect on the boundary layer, but the inviscid-viscous interaction near the airfoil trailing edge was quite strong. For example, the boundary-layer shape factor H grew to a value of 4 on the airfoil upper

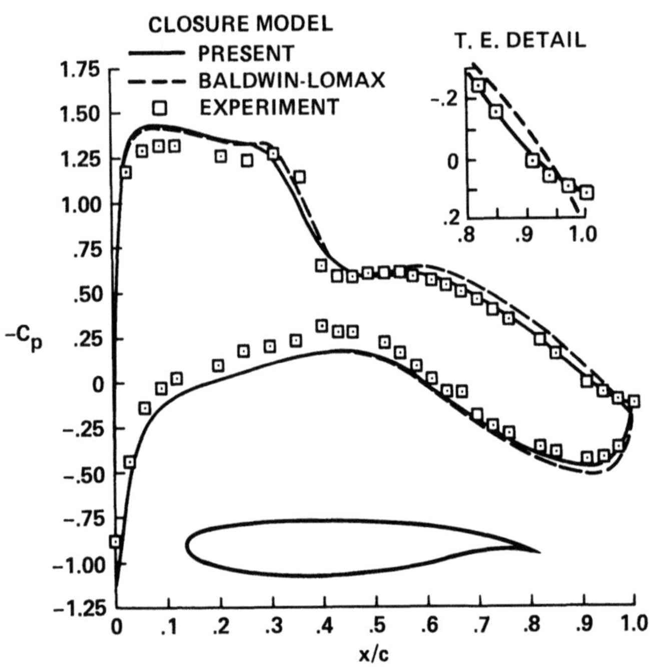

Fig.5. Surface pressure comparisons at $M_\infty = 0.72$ and $\alpha = 4.3°$ for the DSMA 671 supercritical airfoil section (data from Johnson and Spaid [11]).

surface (separation occurred at x/c = 0.98). The present model was better able to predict the observed rapid thickening of the boundary layer near the trailing edge than the Baldwin-Lomax model. As a consequence, the present model as seen in Fig. 5 gave better predictions of surface pressure on the rearward part of the airfoil. Notice that the present model was able to predict the pressure plateau just upstream of the trailing edge caused by boundary layer separation.

The last example is the NACA 64A010 at shock-induced stall conditions. In Fig. 6, predicted surface pressures are compared with the experiment. For this test case, a Cebeci-Smith model solution is shown since a steady solution could not be obtained with the Baldwin-Lomax model. As evident

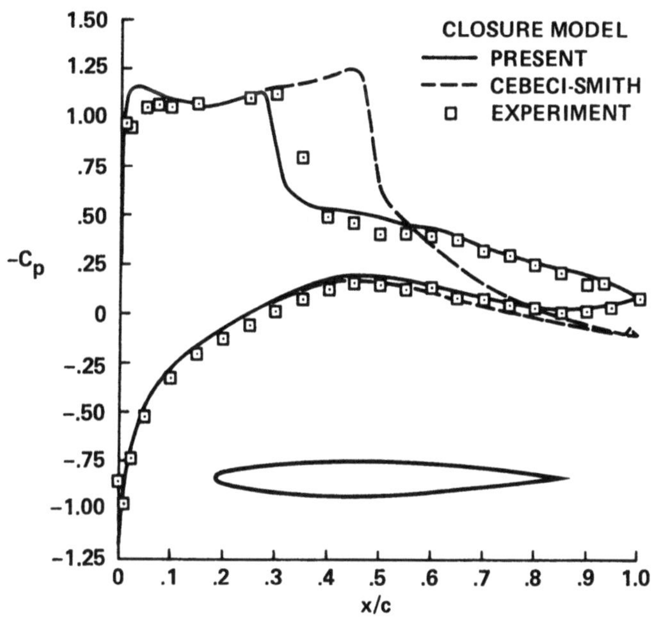

Fig.6. Surface pressure comparisons at $M_\infty = 0.8$ and $\alpha = 6.2°$ for the NACA 64A010 airfoil section (data from Johnson and Bachalo [12]).

from Fig. 6, the surface pressure distribution obtained with the present model is in substantially better agreement with the experiment than that obtained with the Cebeci-Smith model. In Fig. 7, predicted mean velocity and Reynolds shear stress profiles are compared with the experimental data in shear-layer coordinates. The predicted results based on the present model compare quite favorably with the experiment, especially considering the difficulty of this test case and the experimental uncertainties.

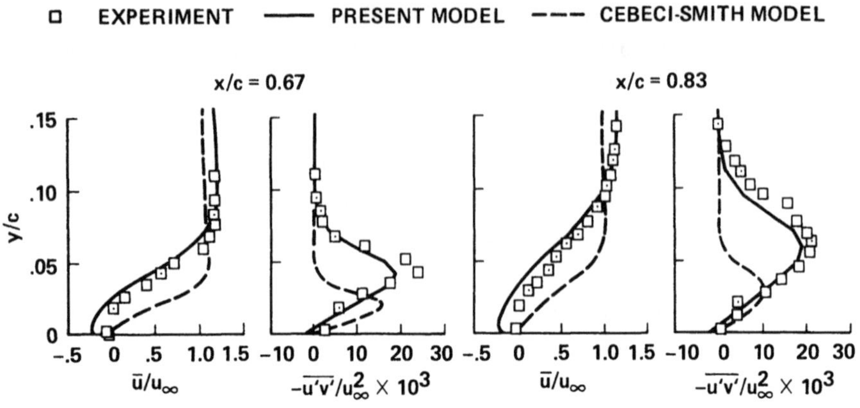

Fig.7. Mean-velocity and Reynolds shear stress comparisons at $M_\infty = 0.8$ and $\alpha = 6.2°$ for the NACA 64A010 airfoil section (data from Johnson and Bachalo [12]).

Concluding Remarks

As illustrated by the foregoing examples, a simple closure model has been developed which is capable of describing wall-bounded shear flows developing on airfoils at transonic conditions even when massive separation is present. The closure model requires little more computational effort than equilibrium algebraic closure models, and does not introduce any numerical stability problems.

References

1. Horstman, C.C.; Johnson, D.A.: Prediction of Transonic Separated Flow. AIAA Journal, vol. 22, no. 7 (July 1984) 1001-1003.

2. Cebeci, T.; Smith, A.M.O.: Analysis for Turbulent Boundary Layers. New York, Academic Press, 1974.

3. Jones, W.P.; Launder, B.E.: The Prediction of Laminarization with a Two-Equation Model of Turbulence, International Journal of Heat Transfer, vol. 15 (Feb. 1972) 301-314.

4. Johnson, D.A.; King, L.S.: A New Turbulence Closure Model for Boundary Layer Flows with Strong Adverse Pressure Gradients and Separation, AIAA Paper 84-0175, Reno, Nev. (1984).

5. Johnson, D.A.: Predictions of Transonic Separated Flow with an Eddy-Viscosity/Reynolds-Shear-Stress Closure Model, AIAA Paper 85-1683, Cincinnati, Ohio (1985).

6. King, L.S.; Johnson, D.A.: Separated Transonic Airfoil Calculations with a Nonequilibrium Turbulence Model, NASA TM-86830, 1985.

7. Simpson, R.L.; Chew, Y.T.; Shivaprasad, B.G.: The Structure of a Separating Turbulent Boundary Layer. Part I. Mean Flow and Reynolds Stresses, Journal of Fluid Mechanics, vol. 113 (1981) 23-51.

8. Bachalo, W.D.; Johnson, D.A.: An Investigation of Transonic Turbulent Boundary Layer Separation on an Axisymmetric Flow Model, AIAA Paper 79-1479, Williamsburg, Va. (1979).

9. MacCormack, R.W.: A Numerical Method for Solving the Equations of Compressible Viscous Flow. AIAA Journal, vol. 20 (Sept. 1982) 1275-1281.

10. Beam, R.; Warming, R.F.: An Implicit Factored Scheme for the Compressible Navier-Stokes Equations, Journal of Computational Physics, vol. 15 (1974) 299-319.

11. Johnson, D.A.; Spaid, F.W.: Supercritical Airfoil Boundary-Layer and Near-Wake Measurements, Journal of Aircraft, vol. 20 (Apr. 1983) 298-305.

12. Johnson, D.A.; Bachalo, W.D.: Transonic Flow Past a Symmetrical Airfoil--Inviscid and Turbulent Flow Properties, AIAA Journal, vol. 18 (Jan. 1980) 16-24.

13. Baldwin, B.S.; Lomax, H.: Thin Layer Approximation and Algebraic Model for Separated Turbulent Flows, AIAA Paper 78-257, Huntsville, Ala. (1978).

Turbulence Modeling in Transonic Interactions

B. ESCANDE and L. CAMBIER

ONERA - 29, Avenue de la Division Leclerc - 92320 CHATILLON (France)

Summary

The paper presents comparisons of turbulence models implemented in a 2-D Navier-Stokes solver, for the prediction of transonic shock wave turbulent boundary-layer interactions. A mixing length model and a (k,ε) model are considered, and the results are compared to experimental data obtained with LDV.

1. Introduction

The prediction of shock wave turbulent boundary-layer interactions in transonic flows is of great practical interest for many industrial applications. The performance of modern computers now allows a more current use of Navier-Stokes solvers in conjunction with grids sufficiently fine to accurately represent viscous effects. At the present time, a special effort is made to develop numerical codes able to predict the behaviour of a turbulent boundary-layer interacting with a shock wave. Until now such prediction was most often unsatisfactory mainly because of the imperfections of the turbulence models generally used.

The aim of the present work is to compare turbulence models employed to simulate shock wave turbulent boundary-layer interactions of various strength, accompanied with the development of more or less extended separated region. These computations were performed by solving the time-averaged full Navier-Stokes equations and by considering two different turbulence models : an algebraic mixing-length model and a two-transport equation model . Two flow configurations are considered, which are both two-dimensional transonic channel flows. These flows have been studied experimentally by Laser Doppler Velocimetry [1]. The very detailed set of experimental data thus obtained allows a thorough discussion of the validity of the numerical results.

Turbulent Shear Layer/Shock Wave Interactions
IUTAM Symposium Palaiseau 1985
Editor: J. Délery

2. Turbulence modeling

The present work is based upon the solving of the time-averaged 2-D Navier-Stokes equations written in conservative form, and upon the following classical hypotheses : perfect gas, constant specific heat ratio, constant Prandtl number. The molecular viscosity is computed from Sutherland's law.

Because of the non-linearity of the Navier-Stokes equations, the Reynolds averaging shows off correlations between turbulent fluctuations. These unknown quantities, which are the turbulent stresses for the momentum equations, and the turbulent heat fluxes for the energy equation, have to be evaluated by mathematical representations.

The two turbulence models tested in the present work are both based upon the eddy-viscosity concept : the turbulent stresses and heat fluxes are expressed in a similar manner to the viscous quantities, corresponding to a newtonian fluid, with Stokes hypothesis.

Hence, one assumes that the Reynolds stress tensor $\bar{\bar{\sigma}}_R$ and the turbulent heat flux vector q_t are expressed as (with classical notations) :

$$\bar{\bar{\sigma}}_R = -\frac{2}{3}(\rho k + \mu_t \operatorname{div} \bar{V})\, 1 + \mu_t (\operatorname{grad} \bar{V} + \operatorname{grad} \bar{V}^{\,t}) \tag{1}$$

$$\bar{q}_t = -\gamma \frac{\mu_t}{Pr_t} \operatorname{grad} e,$$

and the turbulent Prandtl number Pr_t is here, like in most aerodynamical flows, supposed constant and equal to 0.9. Then, the closure of the Navier-Stokes equations consists in modeling two quantities : the eddy-viscosity μ_t and the turbulent kinetic energy k.

2.1. Mixing-length model

The first model used in our computations is an algebraic model of mixing-length type, relating the eddy-viscosity to the mean velocity field. This model, proposed by Michel et al. [2] for equilibrium turbulent boundary-layers, is given by :

$$\mu_t = \rho\, l^2\, F^2(\xi)\, |\bar{\Omega}| \tag{2}$$

The turbulent kinetic energy k in Eq. 1 is here neglected, and the initial model has been modified by replacing the normal derivative $\partial u/\partial y$ of the streamwise velocity component by the vorticity modulus $|\bar{\Omega}|$.

The mixing length l is given by the following relationship :

$$\frac{l}{\delta} = 0.085 \tanh\left(\frac{K}{0.085}\frac{y}{\delta}\right), \text{ where } K = 0.41.$$

Thus, this turbulence closure requires the evaluation of the boundary-layer thickness δ.

The Van Driest damping function F is here given by the following expression :

$$F(\xi) = 1. - \exp\left(-\frac{\sqrt{\xi}}{26\ K}\right), \text{ where } \xi = \rho\ l^2 \frac{\mu_t + \mu}{\mu^2} |\bar{\Omega}|.$$

This formula shows that the relation giving the eddy-viscosity leads to an implicit equation for μ_t which is solved by a fixed-point method in the present case.

2.2. Two-equation model

This turbulence model is directly obtained from the classical (k,ε) model [3] by rewriting the transport equations for a compressible fluid with non-fluctuating density. Thus, the modeled equations for the turbulent kinetic energy and for its dissipation rate are :

$$\frac{\partial}{\partial t} \rho k + \text{div}\,(\rho \bar{V} k) = \bar{\bar{\sigma}}_R : \text{grad}\ \bar{V} - \rho\varepsilon + \text{div}\left(\frac{\mu_t}{\alpha_k}\text{grad}\ k\right) \tag{3}$$

transport along the mean flow | production | dissipation | diffusion

$$\frac{\partial}{\partial t} \rho \varepsilon + \text{div}\,(\rho \bar{V} \varepsilon) = C_1 \frac{\varepsilon}{k} \bar{\bar{\sigma}}_R : \text{grad}\ \bar{V} - C_2\, \rho \frac{\varepsilon^2}{k} + \text{div}\left(\frac{\mu_t}{\alpha_\varepsilon}\text{grad}\ \varepsilon\right) \tag{4}$$

Whereas each term of the modeled transport equation for the turbulent kinetic energy is either exact or based upon reasonable physical hypotheses, the modeling of the transport equation for ε results in great part from a functional similitude assumption between the two turbulent transport equations.

The eddy-viscosity is then computed by :

$$\mu_t = C_\mu\, \rho\, \frac{k^2}{\varepsilon} \tag{5}$$

The model constants have the following classical values :

$C_\mu = 0.09$, $\alpha_k = 1$, $\alpha_\varepsilon = 1.3$, $C_1 = 1.57$, $C_2 = 2.$

Equations (3) and (4) are valid only in fully turbulent regions, that is to say for high values of the turbulent Reynolds number $R_t = \rho k^2/\mu\varepsilon$. Consequently, in the near wall region, the (k,ε) model is replaced by the mixing-length model, as long as the turbulent Reynolds number is less than 100.

In the outer part of the boundary-layer, k and ε tend simultaneously to zero. Thus, to avoid numerical difficulties, they are bounded to small positive values k_{min} and ε_{min}, corresponding to a very small value of the eddy-viscosity ($\mu_t/\mu \simeq 10^{-3}$).

3. Numerical method

In this section, we present the main features of the numerical method which has been described in detail elsewhere (see for instance [4]).

3.1. Multi-domain approach

The numerical applications (§ 4) consist in computing two shock wave boundary-layer interactions in a 2-D symmetrical channel and the computational domain (see Fig. 2) is bounded below by the wall, and above by the symmetry axis. The computational domain is divided into two sub-domains D_p and D_v. In the first one, near the wall, viscous effects are important and the time-averaged Navier-Stokes equations completed by a turbulence model are solved. The remaining part of the flow (sub-domain D_v) where the inviscid approximation is justified, is computed from the Euler equations. The interface between the two sub-domains is located in a region where the flow is supposed to be inviscid, and the matching of the two sub-domains is achieved by a technique based on the use of selected compatibility relations associated with the hyperbolic system of the Euler equations.

Both the Navier-Stokes equations in D_v and the Euler equations in D_p are solved by a time-marching method with an explicit predictor-corrector type scheme. The equations are discretized using MacCormack's finite difference scheme directly in the physical plane on an arbitrary curvilinear mesh. The turbulence transport equations (case of the (k,ε) model) are very similar to the Navier-Stokes equations and therefore they are discretized by the same numerical scheme.

3.2. Boundary conditions

On the wall, we impose a no-slip condition and a zero heat flux. Wall pressure is derived from the normal momentum equation at the wall.

On the symmetry axis (boundary of D_p), we impose a slip condition which is equivalent in perfect fluid to the symmetry condition.

On the upstream boundary of D_p and D_v, the stagnation pressure, the stagnation temperature and the direction of the velocity are prescribed. The flow properties are derived from these three boundary conditions supplemented by the compatibility relation corresponding to a propagation of information from the computational domain to the outside.

On the downstream boundary of D_p and D_v, the static pressure is imposed and the other flow properties are deduced from the three compatibility relations to be used. The shock location essentially depends on the static pressure level.

In the case of the (k,ε) model, the values of k and ε on the upstream boundary are set equal to the values k_{min} and ε_{min} (see § 2.2) and they are extrapolated on the downstream boundary. On this boundary, the imposed quantity is then a "pressure" p^* defined as the mean normal stress, that is to say : $p^* = p + \frac{2}{3}\rho k$. In the regions where k is negligible, p^* is equal to the static pressure. On the interface between D_P and D_V, either the fluid flows from D_P into D_V and k and ε are set equal to k_{min} and ε_{min}, or the fluid flows from D_V into D_P and k and ε are then obtained from the numerical scheme.

3.3 Zonal grid refinement technique

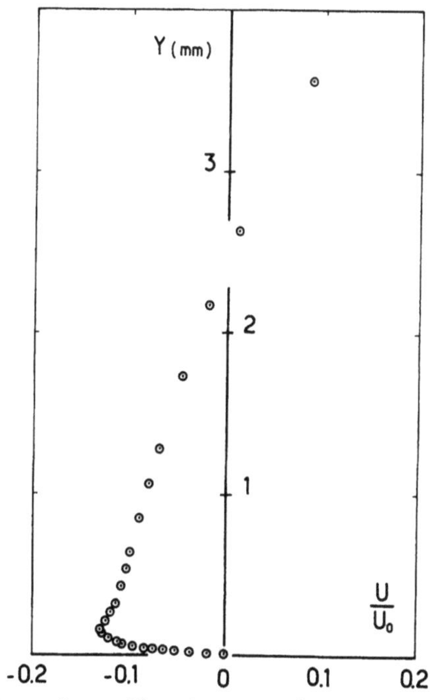

Fig. 1 - Example of mesh refinement in the near wall region.

Near the wall, the very important mesh refinement required to represent the rapid variations of the flow properties in the viscous sublayer is obtained by a zonal grid refinement technique consisting of successive halvings of the mesh size from one zone to the following one. This technique of dichotomy allows to obtain a rapid transition from a relatively coarse mesh in the outer flow to a very fine mesh near the wall with a smoothly varying mesh size in each zone. Coupling of adjacent zones is ensured by overlapping. The time-sweeping of the zones is carried out in a way which allows to increase largely the performance of the method with regard to the standard implementation of an explicit scheme. To illustrate the mesh refinement achievable by the present dichotomy technique, Figure 1 shows a plotting of the mesh points on a computed streamwise velocity profile in the region close to the wall. The profile located in a separated flow region results from a calculation described in § 4 (flow ⓑ, (k,ε) model). Figure 1 shows the very fine mesh obtained near the wall in the separated region by a subgrid with 8 zones. The mesh size at the wall is $y/\delta = 5.10^{-4}$. Numerical experiments have proved that a greater mesh refinement does not change the solution, and therefore the solution can be considered as accurate enough to allow a valid comparison between turbulence models.

4. Numerical applications

4.1. Description of the computations

The two computed flows correspond to shock turbulent boundary-layer interactions studied experimentally [1]. Detailed measurements of the mean and turbulent quantities were performed by a two-component Laser Doppler Velocimeter.

The Reynolds number Re calculated with the height of the symmetric channel ($H = 10^{-1}$ m) and with stagnation conditions is equal to $2.078.10^{6}$. The first configuration (flow ⓐ) corresponds to a shock boundary-layer interaction with incipient separation at the shock foot and the Mach number M_o at the beginning of the interaction is then equal to 1.3. In the second configuration (flow ⓑ), a well developed separated region forms, M_o being now equal to 1.45.

For each flow, a first computation has been carried out with the algebraic turbulence model. The initial conditions correspond to an inviscid one-dimensional flow with a shock located at its experimental location. On the downstream boundary, the pressure is imposed equal to a value ($p_2 = 0.655\ p_{i_o}$ for flow ⓐ and $p_2 = 0.667\ p_{i_o}$ for flow ⓑ) adjusted to give the experimental shock location at convergence. This value is found to be different from the experimental value ($p_{2_e} = 0.636\ p_{i_o}$ for flow ⓐ and $p_{2_e} = 0.648\ p_{i_o}$ for flow ⓑ), but one important part of the difference $p_2 - p_{2_e}$ is due to side effects resulting from the presence of lateral boundary-layers in the experiment.

Then, a second computation has been carried out for each flow with the (k,ε) turbulence model. The initial state for this computation corresponds to the solution obtained with the mixing length model, the values of the two quantities k and ε being computed from the eddy-viscosity field by the following relations :

$$k = \frac{-\overline{u'v'}}{0.3} \qquad \varepsilon = -\overline{u'v'}\left(\frac{\partial u}{\partial y} + \frac{\partial v}{\partial x}\right) \tag{6}$$

with $-\overline{u'v'} = \frac{\mu_t}{\rho}\left(\frac{\partial u}{\partial y} + \frac{\partial v}{\partial x}\right)$, x, y being here cartesian coordinates.

The downstream pressure is imposed equal to the same value as in the first computation.

The computations with the two models are performed on the same grids. The sub-domain meshes used for the computations of flow ⓐ are represented in Fig. 2. They are made up of (141 x 16) points in D_p and (141 x 48) points in D_v (total number of mesh points : 9024). The ratio R of the sizes in the y-direction of the cell near the wall and of the cell near the symmetry axis is equal to 0.003.

The meshes used for the computations of flow ⓑ present the same ratio R, and they are made up of (181 x 13) points in D_p, and (181 x 53) points in D_v (total number of mesh points : 11986). The larger number of points in the x-direction is due to the fact that the downstream boundary is located at x = 350 mm, instead of x = 250 mm for flow ⓐ.

4.2. Presentation of results for flow ⓐ

Figures 3 and 4 which represent the iso-Mach lines for flow ⓐ computed with the two turbulence models show that there is a perfect continuity across the interface between D_p and D_v. A comparison between the two figures points out that the extent of the interaction region is located more upstream with the (k,ε) model.

Due to the discrepancies between the two computed origins x_o of the interaction (defined as the abscissa corresponding to the wall pressure minimum value), a seemingly more meaningful quantitative comparison will be done by considering a relative abscissa $x-x_o$.

The wall pressure distributions thus plotted in Fig. 5 show that the compression predicted by the mixing length model is too steep, the (k,ε) model giving a better agreement with experiment. However the displacement thickness distributions $\delta^*(x-x_o)$ (Fig. 6) given by the (k,ε) model is in good agreement with experiment only in the first part of the interaction, the return to equilibrium being ill predicted. For the mixing length model, the observed good agreement of $\delta^*(x-x_o)$ with the experimental data is due in fact to a balance of the imperfections of the computed velocity profiles, as can be seen in Fig. 7. This figure also confirms that the (k,ε) model gives results nearly coincident with experiment in the first part of the interaction (see the profile at $x-x_o$ = 11 mm) whereas the downstream relaxation process is not correctly predicted (see the other profiles). The turbulent shear stress profiles plotted in Fig. 8 show that both models give a too rapid increase of the maximum shear stress, and do not correctly represent its slow decrease in the interaction downstream part.

4.3. Presentation of results for flow ⓑ

The main conclusions derived from the comparison of the two turbulence models, obtained for flow ⓐ, are fully corroborated in the case of flow ⓑ, which corresponds to a stronger interaction leading to a noticeable separation. The iso-Mach lines plotted in Figs. 9 and 10 show that the interaction region is far more extended with the (k,ε) model than with the mixing length model. The (k,ε) model gives a much larger separated bubble associated with a well defined lambda shock pattern in the inviscid outer flow. This shock location is clearly more upstream than with the algebraic model.

The distributions (Fig. 11) of wall pressure as function of the relative abscissa $x-x_o$ demonstrate the improvement obtained with the (k,ε) model. In particular, the characteristic plateau of the experimental distribution is well represented by the (k,ε) calculation whereas it is inexistant in the mixing length calculation. For flow ⓑ the displacement thickness distributions (Fig. 12) lead to the same comments as for flow ⓐ : i) the apparent good prediction obtained with the mixing length model is due to the balancing of the imperfections in the velocity profiles (Fig. 13) ; ii) the good prediction given by the (k,ε) model is observed only in the first part of the interaction domain. Further downstream this model leads to an unrealistic increase in δ^* corresponding to an overprediction of the separated region (clearly shown also by the velocity profiles in Fig. 13). The turbulent shear stress profiles (Fig. 14) show that the predicted increase of the maximum shear stress at the interaction beginning is far too rapid with both models. On the other hand, in the downstream part of the interaction, the mixing length model leads to a too rapid decrease of this maximum, whereas the (k,ε) model predicts a too slow relaxation process.

5. Conclusion

The numerical solution of the time-averaged Navier-Stokes equations applied to two shock boundary-layer interactions occurring in a two-dimensional transonic channel allows an instructive comparison between the mixing length model and the (k,ε) model. In order to eliminate discrepancies due to the numerical method, the computations have been carried out in very fine meshes allowing to obtain mesh-independent solutions.

It was confirmed that the mixing length model is inadequate to represent large separated regions. For that type of flow, the (k,ε) model improves greatly the wall pressure distribution prediction, and gives a fair representation of the first part of the separation phenomenon. However, the downstream evolution obtained with the (k,ε) model shows large discrepancies with experiment : in particular this model predicts a too slow relaxation process towards a new equilibrium state. Such discrepancies are largely imputable to an unsatisfactory description of the turbulent stresses, the eddy-viscosity concept itself being questionable in such strong interactions.

A more satisfactory solution of this problem will probably necessitate the use of transport equations for the full Reynolds tensor or of a simplified form of these equations, as in the ASM model [5].

References

1. Délery, J. : Investigation of strong shock turbulent boundary-layer interaction in 2-D transonic flows with emphasis on turbulence phenomena. AIAA Paper 81-1245 (1981). See also AIAA Journal, Vol. 21, n° 2 (Feb. 1983).

2. Michel, R. ; Quémard, C. and Durant, R. : Application d'un schéma de longueur de mélange à l'étude des couches limites turbulentes d'équilibre. ONERA-NT 154 (1969).

3. Jones, W.P. and Launder, B.E. : The prediction of laminarization with a two-equation model of turbulence. Int. J. Heat and Mass Transfer, Vol. 15, pp. 301-314 (1972).

4. Cambier, L. ; Ghazzi, W. ; Veuillot, J.P. and Viviand, H. : A multi-domain approach for the computation of viscous transonic flows by unsteady type methods. Recent Advances in Numerical Methods in Fluids, Vol. 3 ; Habashi, W.G., Ed. ; Swansea : Pineridge Press (1984).

5. Launder, B.E. : A generalized algebraic stress transport hypothesis. AIAA Journal, Vol. 20, N° 3 (1982).

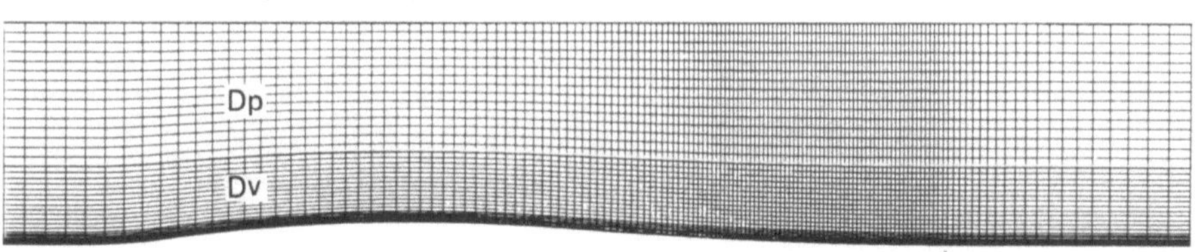

Fig. 2 - Flow ⓐ - Sub-domain division and meshes.

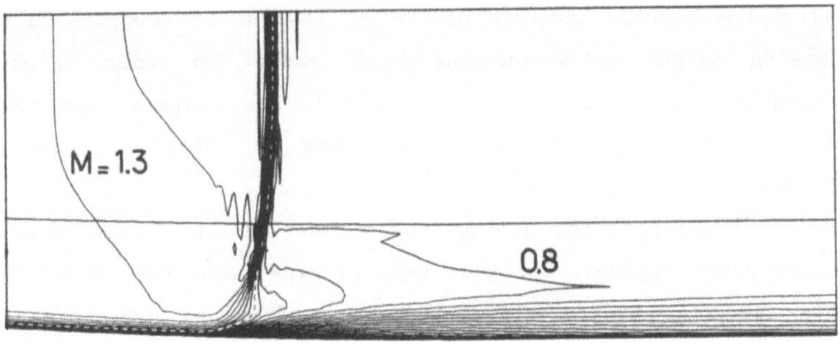

Fig. 3 - Flow ⓐ - Iso-Mach lines (mixing-length model) $\Delta M = 0.5$.

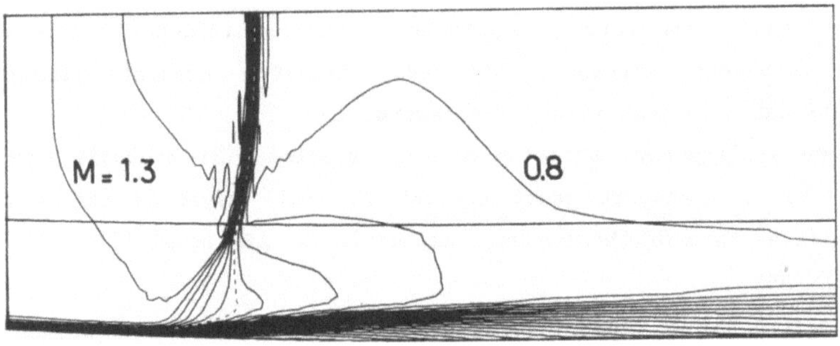

Fig. 4 - Flow ⓐ - Iso-Mach lines (k-ε model) $\Delta M = 0.5$.

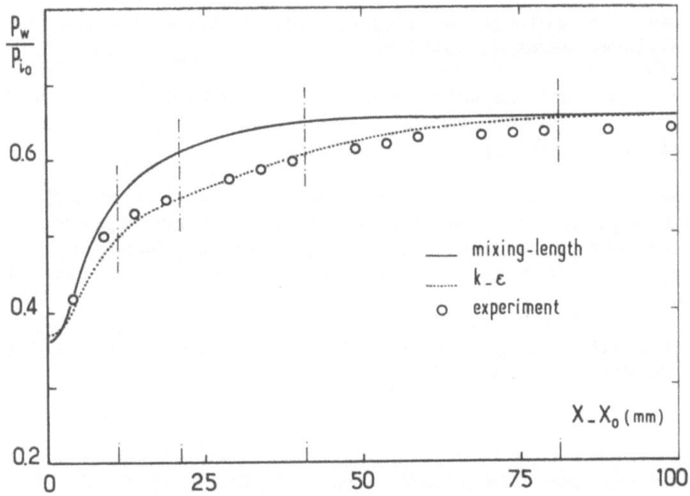

Fig. 5 - Flow ⓐ - Wall pressure distributions.

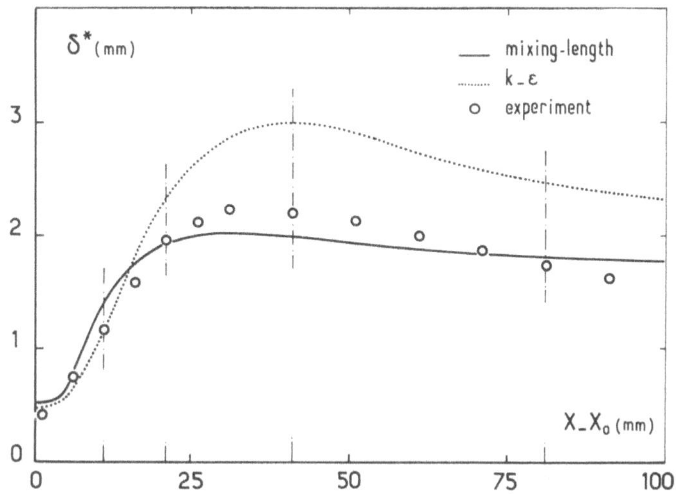

Fig. 6 - Flow ⓐ - Displacement thickness distributions.

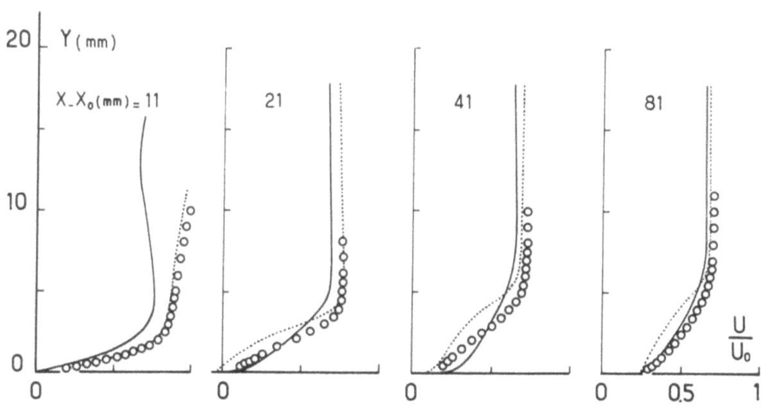

Fig. 7 - Flow ⓐ - Velocity profiles (—— mixing-length ; --- k-ε ; ∘ experiment).

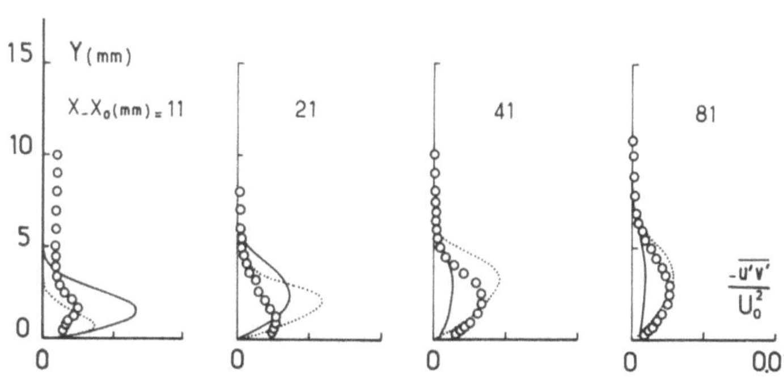

Fig. 8 - Flow ⓐ - Turbulent shear stress profiles (—— mixing-length ;--- k-ε ; ∘ experiment).

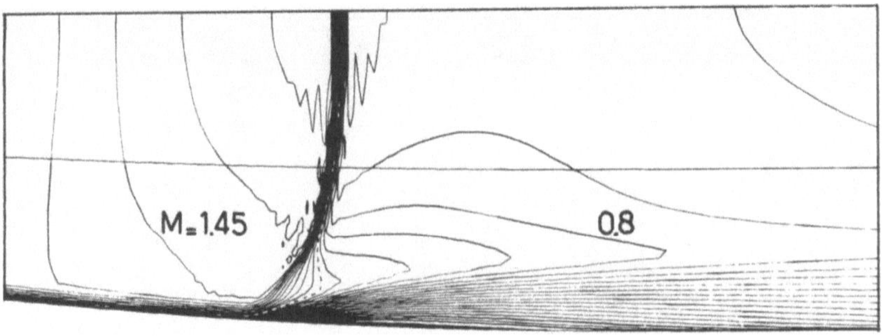

Fig. 9 - Flow ⓑ - Iso-Mach lines (mixing-length model).

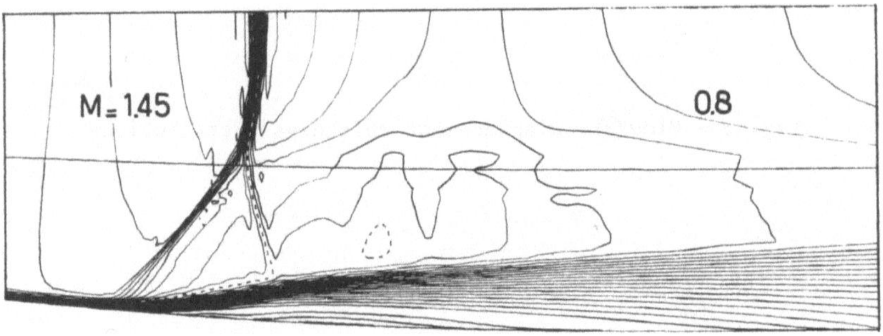

Fig. 10 - Flow ⓑ - Iso-Mach lines (k-ε model).

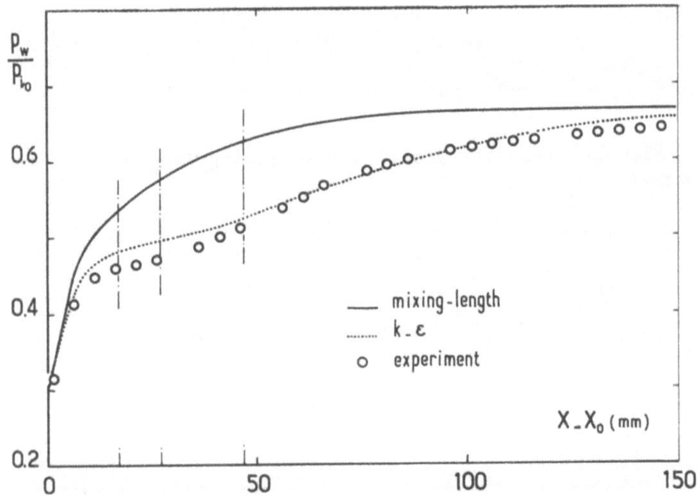

Fig. 11 - Flow ⓑ - Wall pressure distributions.

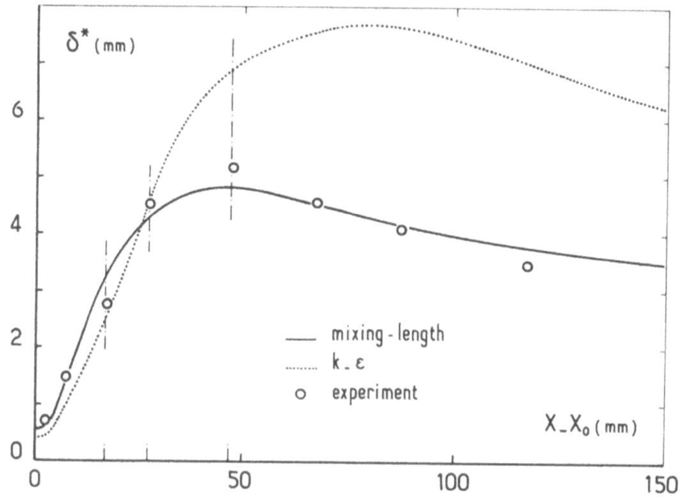

Fig. 12 - Flow ⓑ - Displacement thickness distributions.

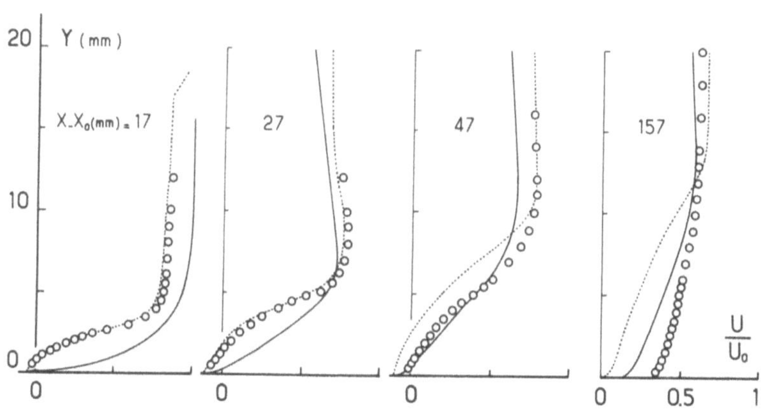

Fig. 13 - Flow ⓑ - Velocity profiles (—— mixing-length ; --- k-ε ; ∘ experiment).

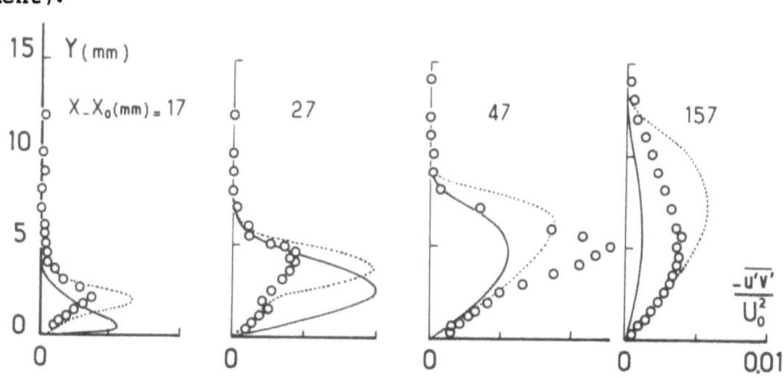

Fig. 14 - Flow ⓑ - Turbulent shear stress profiles (—— mixing-length ; --- k-ε ; ∘ experiment).

Implicit Navier-Stokes Calculations of Transonic Shock/Turbulent Boundary-Layer Interactions

G. DEGREZ

Von Karman Institute for Fluid Dynamics, Belgium.

D. VANDROMME

Institut de Mecanique des Fluides de Lille, France.

Summary

Three different shock boundary layer interaction flows are calculated. Two different computers codes are used on the same grid systems with identical initial conditions. The numerical schemes, which are both implicit, are due respectively to Beam and Warming and to MacCormack. The turbulence models are the algebraic models of Cebeci, Smith and Mosinski, Baldwin and Lomax and the k-ε two equation models. The performances of the numerical schemes are compared and a significant improvment is brought in with the two equation model.

I. Introduction

In the transonic regime, viscous effects can largely affect both internal and external flows as those currently encountered on modern supercritical airfoils or in turbomachinery designs. Of particular importance is the viscous interaction produced by the impigment of a shock wave on a turbulent boundary layer, which may, in some instances, lead to flow separation.

The advent of implicit time-dependent techniques to solve the Navier-Stokes equations (e.g. ref. /1/-/2/) made possible to predict such complicated flowfields. Unfortunately, poor results have been observed for separated flow, mostly due to deficiencies in the turbulence models. However, the respective influences of the turbulence models and computational techniques themselves as causes of discrepancies have been little studied.

To provide a firm basis of comparison for numerical predictions, Delery (ref. /3/) has performed an extensive experimental study of the shock wave turbulent boundary layer interactions in transonic nozzles. Both mean and turbulence values were measured using laser doppler anemometry.

In this study, these three experimental configurations were

Turbulent Shear Layer/Shock Wave Interactions
IUTAM Symposium Palaiseau 1985
Editor: J. Délery

simulated numerically. Three computations of each configuration were performed using the Beam-Warming /1/ and the MacCormack /2/ methods and different turbulence models, but using the same grids and initial conditions to allow for a direct comparison of the methods and turbulence models.

In section II, the governing equations and turbulence models are recalled and the mesh system and boundary conditions are presented. In section IV, the flowfield is qualitatively described and the results are discussed.

II. Governing equations and turbulence models

The flow field is described by the set of partial differential equations known as the time-dependent compressible Navier-Stokes equations. The hypothesis used in conjunction with these equations, include the perfect gas assumption, constant specific heats and Prandtl number, the Sutherland law for molecular viscosity and Stokes assumption (no bulk viscosity). Three models were choosen for turbulence closure: the mixing length (algebraic) models of Cebeci-Smith-Mosinski /4/ and Baldwin-Lomax /5/ and a 2-equation model derived from the k-ε model /6/.

II.1. The Navier-Stokes equations.

The two-dimensional Reynolds averaged Navier-Stokes equations can be written in conservation form as (cartesian coordinates):

$$\frac{\partial U}{\partial t} + \frac{\partial F}{\partial x} + \frac{\partial G}{\partial y} = 0 \tag{2-1}$$

with

$$U = \begin{pmatrix} \rho \\ \rho u \\ \rho v \\ \rho E \end{pmatrix} \quad F = \begin{pmatrix} \rho u \\ \rho u^2 + \sigma_x \\ \rho u v + \tau_{xy} \\ (\rho E + \sigma_x) u + \tau_{xy} v - K\frac{\partial T}{\partial x} \end{pmatrix} \quad G = \begin{pmatrix} \rho v \\ \rho u v + \tau_{yx} \\ \rho v^2 + \sigma_y \\ (\rho E + \sigma_y) v + \tau_{yx} u - K\frac{\partial T}{\partial y} \end{pmatrix} \tag{2-2}$$

For the numerical analysis, it is customary to divide the flux vectors F and G into convective and viscous parts, e.g.: $F = F_c + F_v$. The stress terms are written as:

$$\sigma_x = p - 2\mu \frac{\partial u}{\partial x} + \frac{2}{3}\mu\left(\frac{\partial u}{\partial x} + \frac{\partial v}{\partial y}\right)$$

$$\sigma_y = p - 2\mu \frac{\partial v}{\partial y} + \frac{2}{3}\mu\left(\frac{\partial u}{\partial x} + \frac{\partial v}{\partial y}\right)$$

$$\tau_{xy} = \tau_{yx} = -\mu\left(\frac{\partial u}{\partial y} + \frac{\partial v}{\partial x}\right) \tag{2-3}$$

where μ is the viscosity coefficient, E is the specific total energy

$$E = c_v\, T + \frac{u^2 + v^2}{2} \tag{2-4}$$

and K is the heat conductivity coefficient.

To account for arbitrary geometries and clustered grid systems near solid boundaries, two procedures are avalaible: either use coordinate mapping and transform the equations in curvilinear coordinates, or use the finite volume technique. The former procedure was used in the Beam-Warming code whereas the latter was choosen with the MacCormack method.

II.2. Turbulence models

Due to the high Reynolds number value of the flows under study in this work, the Navier-Stokes equations need to be supplemented with a turbulence model. To account for the turbulent fluctuations, the basic flow variables are splitted into a fluctuating component and a mean value, the latest is provided by time averaging. Then, when time averaging is applied to the motion equations, the non linear convective terms produces extra unknowns, the turbulent fluxes. As these fluxes have dimensions of stresses in the momentum equation, it is convenient to evaluate them with a constitutive relationship similar to the Newton's law of viscosity for viscous stresses. This relationship uses a turbulent viscosity coefficient, which can be scaled on the mean motion or on suitable turbulence parameters, such as the turbulent kinetic energy or a given characteristic scale. Both manners of scaling are used here.

II.2.1. Mixing length turbulence models

The first way to evaluate the turbulent viscosity coefficient is to relate it only to the mean motion and the flow geometry.

Relying on experimental observation for a flat plate boundary layer, Cebeci, Smith and Mosinski /4/ (refered in the following as CSM) proposed a complete algebraic model. Then Baldwin and Lomax /5/ (refered as BL) modified it to account for separated flows.

a) The CSM model

Two different zones are distinguished in the boundary layer. In the inner zone, the classical Prandtl mixing length hypothesis is used, including the Van Driest near wall treatment:

$$\mu_{t_{in}} = \rho\, l^2 \left| \frac{\partial U}{\partial y} + \frac{\partial V}{\partial x} \right| \tag{2-5}$$

with

$$l = \chi y \left(1 - \exp\left(- \frac{y\sqrt{\tau_w/\rho}}{\nu A}\right)\right) \tag{2-6}$$

where χ is the Von Karman constant, τ_w is the wall shear stress value and A is a constant equal to 26.

In the outer zone, the Clauser eddy viscosity formulation is used, including an intermittency factor evaluated from experimental results for a flat plate boundary layer. Then, this component of the turbulent viscosity can be written as:

$$\mu_{t_{out}} = 0.0168\, \bar{\rho}\, U_{max}\, \delta_i^* \left(1 + 5.5 \left(\frac{y}{\delta}\right)^6\right)^{-1} \tag{2-7}$$

where U_{max} is the maximum velocity achieved in the boundary layer and δ_i^* is the kinematic displacement thickness. The matching location of the two zones occurs where the two formulas (2-5) and (2-7) produce equal values for the inner and outer eddy viscosities.

b) The BL model.

An alternative formulation for the eddy viscosity is an adaptation of the previous model for separated flows. The two layer concept is conserved. The Prandtl-Van Driest formulation for the inner region is written now as:

$$\mu_{t_{in}} = \bar{\rho}\, l^2\, |\omega| \tag{2-8}$$

where l is given by (2-6) and $|\omega|$ is the magnitude of the vorticity, i.e.

$$|\omega| = \sqrt{\left(\frac{\partial U}{\partial y}-\frac{\partial V}{\partial x}\right)^2+\left(\frac{\partial V}{\partial z}-\frac{\partial W}{\partial y}\right)^2+\left(\frac{\partial W}{\partial x}-\frac{\partial U}{\partial z}\right)^2} \tag{2-9}$$

The outer viscosity formulation is based on an arbitrary function F_{wake} which is extensively discussed in ref /5/. This avoids the use of the displacement thickness as length scale, when the boundary layer edge cannot be easily located, as in separated flow configurations.

Both models, wich are basically simple algebraic mixing length models assume that, within the boundary layer, the turbulence is everywhere in or close to equilibrium. This is obviously not verified for the interaction flows studied in this paper. Therefore, it is necessary to account specifically for the turbulence field, by solving transport equations for some characteristic turbulence quantities. Doing so, memory effects (especially the turbulence relaxation) are hoped to be better described.

II.2.2. Transport equation turbulence model

This two equation turbulence model is an extension for compresible flows of the standard k- model, which has been developped by the Imperial College group, during the early seventies. Adaptation for compressible flows allows variable density and pressure. Furthermore, turbulent kinetic energy is accounted for in the conservation of total energy /6/.

The modelled equations for the turbulence variables k and are:

$$\frac{\partial}{\partial t}(\bar{\rho} k)+\frac{\partial}{\partial x_\alpha}(\bar{\rho} U_\alpha k)= P_k - \bar{\rho}\varepsilon + \frac{\partial}{\partial x_\alpha}\left(\mu_k \frac{\partial k}{\partial x_\alpha}\right) - \overline{u'_\beta}\frac{\partial \bar{P}}{\partial x_\beta} + \overline{p'\frac{\partial u''_\alpha}{\partial x_\alpha}} - 2\mu\left(\frac{\partial \sqrt{k}}{\partial x_n}\right)^2 \tag{2-10}$$

$$\frac{\partial}{\partial t}(\bar{\rho}\varepsilon)+\frac{\partial}{\partial x_\alpha}(\bar{\rho}\varepsilon U_\alpha) = +C_{\varepsilon 1}\frac{\varepsilon}{k} P_k - C_{\varepsilon 2} f_2 \frac{\varepsilon}{k}\rho\varepsilon^* + \frac{\partial}{\partial x_\alpha}\left(\mu_\varepsilon \frac{\partial \varepsilon}{\partial x_\alpha}\right) - 2\mu\frac{\mu_t}{\rho}\left(\frac{\partial^2 U}{\partial x_n}\right)^2 - C_{\varepsilon 3}\frac{\varepsilon}{k}\overline{u''_\alpha}\frac{\partial \bar{P}}{\partial x_\alpha} + C_{\varepsilon 4}\frac{\varepsilon}{k}\overline{p'\frac{\partial u'_\alpha}{\partial x_\alpha}} \tag{2-11}$$

in which P_k represents the production of turbulent kinetic energy under the action of mean shear. A complete discussion of the modelling of the density and pressure correlation terms is given in ref. /6/. The turbulent viscosity is determined as:

$$\mu_t = c_\mu f_\mu \bar{\rho} \frac{k^2}{\varepsilon} \qquad (2-12)$$

in which C_μ (=0.09) is a constant and f_μ is a turbulent Reynolds number dependent function /6/. Unlike the algebraic models, the turbulent kinetic energy appears in the constitutive relationship for the turbulent stresses as:

$$- \bar{\rho}\, \widetilde{u''_\alpha u''_\beta} = \mu_t \left(\frac{\partial U_i}{\partial x_j} + \frac{\partial U_j}{\partial x_i} - \frac{2}{3} \delta_{ij} \frac{\partial U_\ell}{\partial x_\ell} \right) - \frac{2}{3} \delta_{ij} \bar{\rho} k \qquad (2-13)$$

This verifies that the Reynolds stress tensor trace is different from zero.

II.3. <u>Numerical procedures</u>

Two numerical procedures were used. The first is the implicit approximate factorization scheme of Beam and Warming /2/. The second is the implicit predictor-corrector finite volume method of MacCormack /2/. The Beam-Warming code was run with the BL turbulence model (method 1) whereas the MacCormack code was run with the CSM model (method 2) and the two-equation model (method 3).

Calculations are identified by a two-digit number, the first digit indicating the flow case and the second digit indicating the numerical method. All calculations were performed using 80*40 mesh system with points clustered near the bottom wall and near the shock wave as experimentally determined. The mesh system for case 2 is shown in fig. 1. All three methods use the same mesh system.

Initial flow quantities are supposed constant, equal to their inlet value. No account is made of the nozzle geometry. Only the downstream pressure is imposed on the last grid column.

Characteristic theory for 2-D inviscid flow equations indicates that the number of boundary conditions to be imposed from outside depends whether or not, the characteristic lines cross the boundary. For the inlet boundary, both numerical method use three fixed parameters from outside (for example: total pressure, total temperature and transverse velocity) and allow the propagation of inside information to travel across the boundary by a characteristic relationship (method 2-3) or a discretised form of the continuity equation (method 1).The downstream pressure is fixed at the outlet boundary, and standard extrapolation formulas are used for the other flow variables.

III. Test cases

III.1. Flow description

The experimental arrangement consists of a variable geometry nozzle followed by a constant area section and an adjustable second throat. The flow is accelerated through the nozzle to supersonic conditions. The second throat area is adjusted so as to produce a standing normal shock wave of desired intensity in the diverging section of the nozzle. Depending on the shock intensity and the nozzle geometry, the flow exhibits incipient, moderate or extended separation.

III.1.1. The incipient separation case (case 1)

The shock wave impiges normally on the boundary layer. The upstream Mach number is 1.3. The interaction results mostly in a spreading of the pressure rise on the tunnel wall and in a very small separation.

III.1.2. The moderately separated case (case 2)

The upstream Mach number is 1.45 for this case. The larger shock intensity causes the boundary layer to separate, forming a rather large separated bubble. The presence of this bubble is felt in the outer flow by the outer shock wave curving upstream in the wall vicinity.

III.1.3. The extensively separated case (case 3)

In this case, the nozzle geometry is such that curvature effects produce a more extensive separated region associated with a lambda shock structure of the outer inviscid flow. This is shown in the interferogram (figure 2) taken from /3/. Separation induces the weak oblique shock wave C1. The supersonic flow downstream of C1 is terminated by a strong normal shock C2. The two shocks coalesce to form C3, the strong outer shock wave. From the triple point, emanates a contact discontinuity separating different entropy zones.

III.2. Results and discussion

III.2.1. Case 1

Fig. 3 presents the wall isentropic Mach number distributions for computations 1-1, 1-2 and 1-3, together with the experimental data. The agreement is seen to be very good in the attached flow region, upstream of the shock. Downstream, one first observes that, to fix the shock wave at the correct experimental position, the outlet pressure had to be adjusted to a value somewhat higher than the experimental value for case 1-1, but lower for cases 1-2 and 1-3. The difference is small in relative value (3%), but seems to be systematic for method 1, as will be shown for case 2 and 3. It is conjectured that this is due to the treatment of the downstream boundary conditions and deserves further study. Apart from this downstream pressure discrepancy, the major differences between computed results and experiments, appear in the interaction region as expected. This is particularly pointed out by the much improved prediction obtained by the two-equation model with respect to the mixing length model. This improvement appears even more clearly on fig. 4, where mean velocity profiles are plotted. Upstream of the interaction, all methods agree with the experimental results, but in the interaction region, the two-equation model follows the experimental data much more closely than the CSM model. Also, the BL model, more specifically designed for separated flows, gives results comparable to the $k-\varepsilon$ model for this mildly separated case.

III.2.2. Case 2

The different behaviours of the models are amplified for this moderately separated case, as appear in fig. 5. The CSM model produces a very sharp pressure rise at the shock location and does not capture the downstream spreading of the interaction due to turbulence relaxation phenomena. In contrast, the BL model reproduces this phenomenon through a plateau-like region associated with the separated zone. No stable solution was obtained with the k-ε model for that case. In fig. 6, the velocity profiles for this case are displayed. The BL method follows the experimental results very well up to X=0.21 m, but predicts a separated zone extended too far downstream, because of underestimated mixing in the reattachment region. The CSM model, which assumes turbulence local equilibrium produces better results in this region since the flow has returned to nearly equilibrium state.

III.2.3. Case 3

Case 3 is the most severe test case as the inviscid structure of the flow is altered by the viscous interaction. As discussed earlier (fig. 2), the CSM model exhibits its limitations in fig. 7, since the pressure rise is unique whereas the BL and k-ε models reproduce the first pressure rise associated with the separation shock wave, followed by a plateau region and a second pressure rise. however, since the plateau level corresponds to subsonic flow conditions, the second pressure rise cannot be a shock wave, unlike the experimental observations. The velocity profiles for this case are shown in fig. 8. Again the CSM method profiles do not exhibit the flow recirculation. The BL profiles follow correctly the separation process, but do not relax fast enough to capture the reattachment process, whereas the k-ε method shows its superiority in this respect.

IV. <u>Conclusions</u>

A numerical study of three transonic shock wave turbulent boundary layer interaction was performed. It was aimed to evaluate the respective advantages of various numerical methods and turbulence models. As far as convergence speed is concerned, no major difference was noted between the Beam-Warming and MacCormack methods. The converged solutions were obtained after around 1000 time steps, with a maximum CFL number around 1200. The turbulence models were shown to improve the results when passing from the CSM to the BL and then to the k-ε models These results are preliminary as further study should be made of the issue of the downstream pressure and of the turbulence field.

Part of this work has been supported by the Direction des Recherches, Etudes et Techniques of the french ministry of defense.

References

1. Beam, R.M.; Warming, R.F.: An implicit factored scheme for the compressible Navier-Stokes equations, AIAA J. 16, (1978), 393-402.

2. MacCormack, R.W.: Numerical solution of the equations of compressible viscous flow, Transonic, Shock and Multidimensional Flows, (Advances in scientific computing), Academic Press, 1982.

3. Delery, J.: Investigation of strong shock turbulent boundary layer interaction in 2-D transonic flows with emphasis on turbulence phenomena, AIAA p 81-1245.

4. Cebeci, T.; Smith, A.M.O.; Mosinski, S.G.: Calculations of compressible adiabatic turbulent boundary layers, AIAA J. 8, (1970), 1974-1982.

5. Baldwin, B.S.; Lomax, H.: Thin layer approximation and algebraic model for separated turbulent flows, AIAA p 78-257

6. Vandromme, D.: Contribution à la modélisation et la prédiction d'écoulements turbulents à masse volumique variable, Thèse de Doctorat, Université de Lille, 1983.

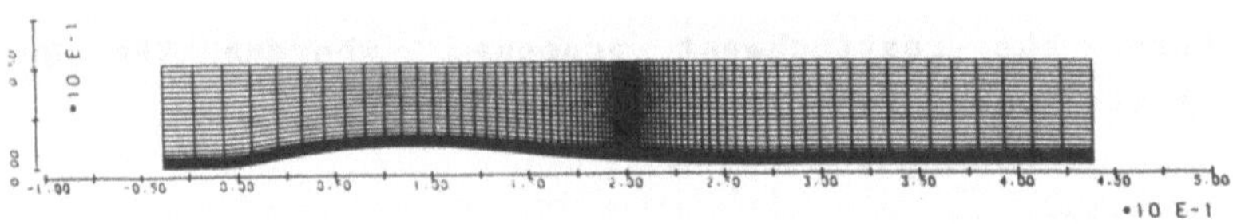

Fig. 1: Mesh system

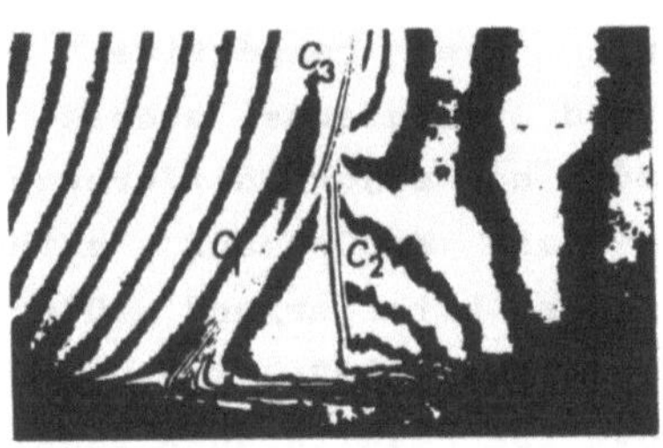

Fig. 2: Flow interferogram

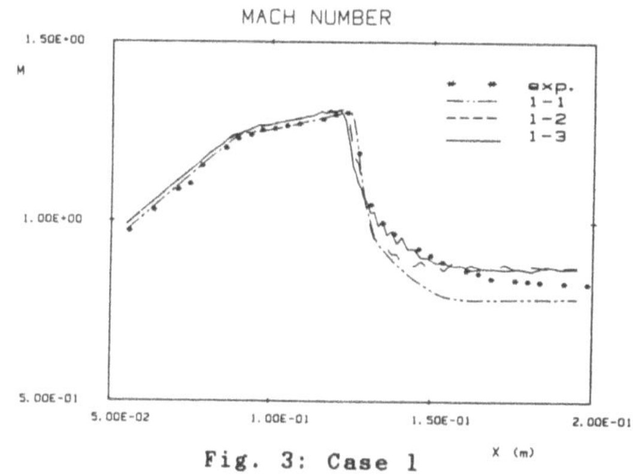

Fig. 3: Case 1

VELOCITY PROFILES

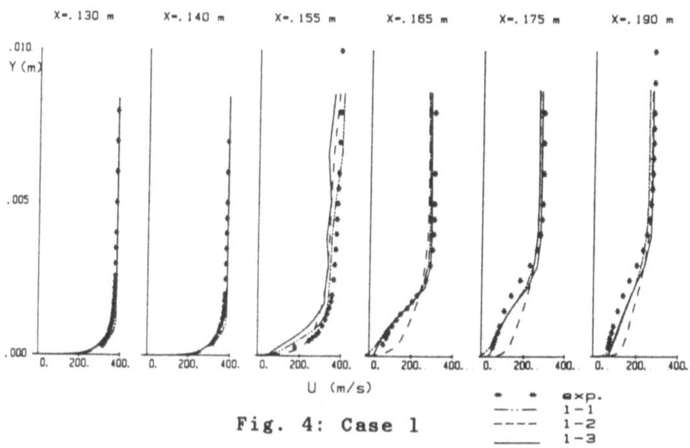

Fig. 4: Case 1

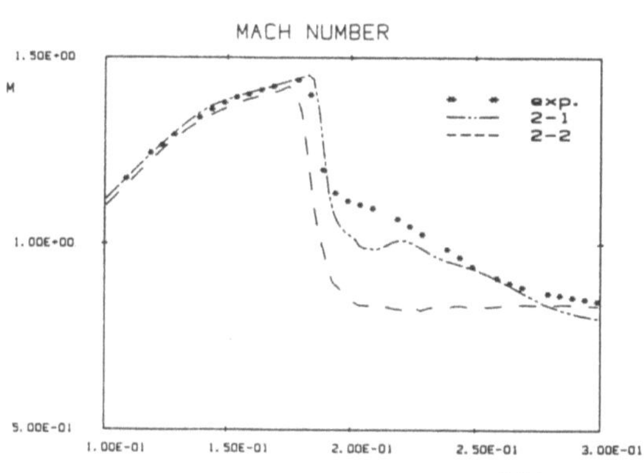

Fig. 5: Case 2

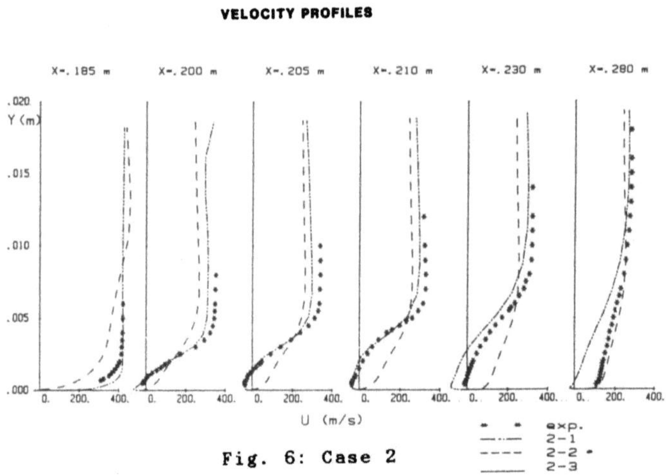

Fig. 6: Case 2

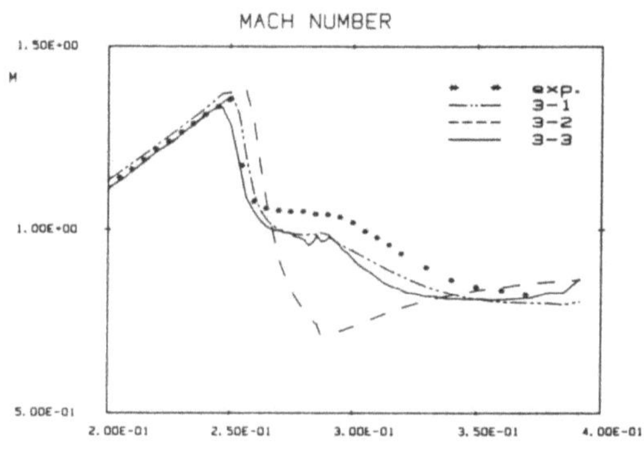

Fig. 7: Case 3

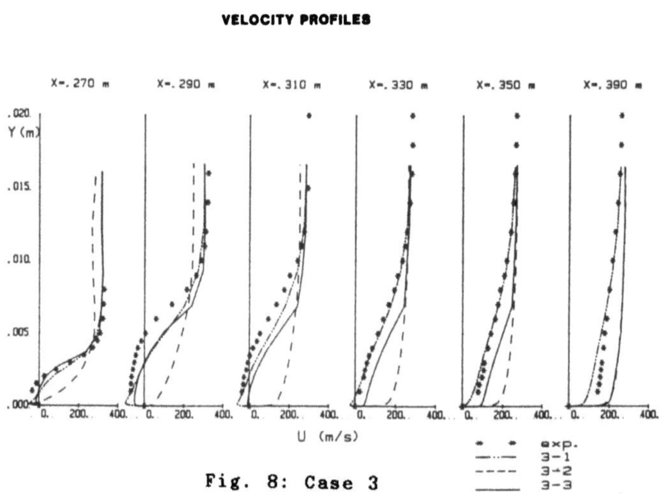

Fig. 8: Case 3

Interactions in Two-Dimensional Steady Flows

Strong versus Weak Interaction System for Transonic Flow Computation

TSZE C. TAI*

David Taylor Naval Ship Research and Development Center
Bethesda, Maryland 20084-5000 USA

Summary

The direct coupling of the viscous and inviscid flows (Strong Interaction) and the boundary layer correction (Weak Interaction) procedures are described, and a comparison of the results of the two procedures is made. Based on the boundary layer integral method, the Strong Interaction formulation that renders a continuous velocity decrease throughout the shock wave region seems to properly simulate the decelerating flow of the interaction phenomenon. The Weak Interaction procedure, on the other hand, always converges to an abrupt pressure rise at the shock, which discounts the smearing effect due to viscosity.

Introduction

Although various transonic computational methods have been developed during the past decade, viable computational results for more general flow conditions can only be achieved by incorporating the viscous effects. The transonic shock wave-boundary layer interaction that weakens the shock wave relative to the inviscid case usually significantly alters the loading over the entire airfoil. The shock penetrates the boundary layer and smears as it reaches the sonic line in the lower part of the boundary layer. The shock pressure rise is attenuated through the subsonic portion of the viscous region.

The problem of transonic shock wave-boundary layer interaction is best described using the Navier-Stokes equations. Because

---

The present work was supported by the United States Naval Air Systems Command (AIR-310D, AIRTASK WR 023-02).

*Senior Research Scientist, Aircraft Division, Aviation and Surface Effects Department.

Turbulent Shear Layer/Shock Wave Interactions
IUTAM Symposium Palaiseau 1985
Editor: J. Délery

of extensive computer capacity requirements, proper turbulence modeling, and other numerical problems, aerodynamicists will continue to rely on other procedures to adequately account for the viscous effects. The purpose of the present paper is to distinguish the strong from the weak shock wave-boundary layer interaction and to describe the systems of equations to illustrate the concept of the two formulations. Results based on the two systems are compared and discussed.

Here, "Strong Interaction" implies direct communication between the outer inviscid flow and the inner viscous flow. On the other hand, "Weak Interaction" implies that the viscous effect is accounted for, but communication between the two flows takes place indirectly through specified quantities rather than by direct coupling.

Strong Interaction System

The direct communication between the outer inviscid flow and the inner viscous flow in a Strong Interaction is represented by (a) a significant flow entrainment from the outer stream to the inner region and (b) a thickening boundary layer that produces a continuous rise in pressure, causing further thickening of the boundary layer. These features are schematically shown in Fig. 1.

To accommodate these features, the viscous and inviscid equations are coupled through a common variable to allow direct inflow of the outer stream to the inner region and propagation of the positive pressure disturbance from the embedded shock wave upstream through the subsonic portion of the boundary layer. The boundary between the inviscid and viscous flows changes dynamically and, in turn, alters the surface pressure. The influence of the boundary layer growth on the surface pressure is determined by using the viscous equations, which are local and, therefore, are more consistent with the physics of the flow. This idea was adopted earlier by the author [1] to the transonic laminar viscous-inviscid interaction problem from a supersonic free interaction method [2]. Application of the concept to the laminar flow control at transonic speed has been carried out recently by Ram and Harvey [3,4] at the NASA

Langley Research Center. Very good agreement between the calculated results and the experimental measurements has been reported [3,4].

To apply the concept to the turbulent boundary layer-shock wave interaction, the two-dimensional, turbulent boundary layer integral equations are written as follows:

Continuity:

$$\frac{d}{dx}(\delta-\delta^*) + \frac{\delta-\delta^*}{U_e}\frac{dU_e}{dx} = E \qquad (1)$$

Momentum:

$$\frac{d\Theta}{dx} + (2+H)\frac{\Theta}{U_e}\frac{dU_e}{dx} = \frac{C_f}{2} \qquad (2)$$

where E is the entrainment coefficient, and other symbols retain their usual meaning. The coordinate systems are shown in Fig. 2.

The simple power-law velocity profile is used with the classical relations for the entrainment and the skin friction coefficients [5] for illustration purposes. The simplicity of the present approach voids the singularity problem in an earlier method [6].

$$\frac{u}{U_e} = \left(\frac{y}{\delta}\right)^{\frac{H-1}{2}} \qquad (3)$$

$$E = 0.025H - 0.022 \qquad (4)$$

$$\left(\frac{C_f}{C_{f_o}} + 0.5\right)\left(\frac{H}{H_o} - 0.4\right) = 0.9 \qquad (5a)$$

where C_{f_o} and H are functions of Reynolds number based on the boundary layer momentum thickness.

$$C_{f_o} = \frac{0.01013}{\ell n Re_\Theta - 1.02} - 0.00075 \qquad (5b)$$

$$H_o = \frac{1}{1-6.55\sqrt{0.5\, C_{f_o}}} \qquad (5c)$$

The transformed compressible shape parameter H may be taken as the equivalent of the conventional shape parameter in an incompressible flow [7].

Equations for the Strong Interaction system are derived as:

$$\frac{dU_e}{dx} = \frac{A_1C_4 - A_2C_2}{C_1C_4 - C_2C_3} \tag{6}$$

$$\frac{dH}{dx} = \frac{A_2C_1 - A_1C_3}{C_1C_4 - C_2C_3} \tag{7}$$

where H is the shape factor, $H = \delta^*/\Theta$. The right-side quantities are:

$$A_1 = \frac{C_f}{2} - \frac{1}{H}\frac{d\delta^*}{dx} \tag{8a}$$

$$A_2 = E - \frac{2}{H-1}\frac{d\delta^*}{dx} \tag{8b}$$

$$C_1 = (2+H)\frac{\Theta}{H};\; C_2 = -\frac{\Theta}{H} \tag{8c,d}$$

$$C_3 = \frac{2H\Theta}{(H-1)U_e};\; C_4 = -\frac{2H\Theta}{(H-1)^2U_e} \tag{8e,f}$$

Equations (6) and (7) are valid for attached flow only because of the restrictions imposed by Eqs. (3) through (5).

The viscous system is linked to the inviscid system through a common variable ψ, by which the exchange of flow properties of the outer flow may be transmitted to the inner flow or vice versa:

$$\Psi = \sin^{-1}\left[\frac{V_{e(\text{inviscid})}}{U_{e(\text{viscid})}}\right] \tag{9}$$

where v_e is calculated by the inviscid system (to be presented later), and the velocity U_e is obtained by Eq. (6). Since the two flows are connected by the flow angle, the boundary layer grows as a result of mass transfer between the outer and the inner flow:

$$\frac{d\delta^*}{dx} = \tan\psi - \tan\psi_b \tag{10}$$

where ψ_b is the surface inclination. Here, the usual procedure of using the boundary layer displacement thickness as the "effective surface" is adapted. The values of $d\delta^*/dx$ from Eq. (10) are the sources of Eqs. (6) and (7) for calculating U_e and H.

Weak Interaction System

The basic equation of the Weak Interaction system consists of the usual boundary layer equations with specified inputs. Since the inputs are "specified," the solution is dictated accordingly. The popular procedure known as the Boundary Layer Correction method is employed to allow a direct comparison with the Strong Interaction formulation.

Starting from Eqs. (1) and (2), with the aid of Eqs. (3) through (5), equations for the Weak Interaction system can be written:

$$\frac{d\Theta}{dx} = B_1 \qquad (11)$$

$$\frac{dH}{dx} = \frac{B_2 - B_1 D_1}{D_2} \qquad (12)$$

where Θ is the boundary layer momentum thickness, and the right-side quantities are:

$$B_1 = \frac{C_f}{2} - (2+H)\,\frac{\Theta}{U_e}\,\frac{dU_e}{dx} \qquad (13a)$$

$$B_2 = E - \frac{2H\Theta}{(H-1)U_e} \qquad (13b)$$

$$D_1 = \frac{2H}{H-1};\; D_2 = -\,\frac{2\Theta}{(H-1)^2} \qquad (13c,d)$$

The values of the velocity gradients, dU_e/dx, from the outer inviscid solution are employed as the sources of Eqs. (11) and (12) for calculating Θ and H. The problem is solved in an iterative fashion by treating the inviscid and boundary layer flows as separated problems. The final solution is reached iteratively when both the inviscid pressure and the viscous displacement converge to a specified tolerance. The pressure rise due to the shock wave-boundary layer interaction inside the viscous region is accounted for by subsequent corrections of the "effective" surface. The system works when the interaction is presumably weak [8].

Other procedures classified as Weak Interactions are:

Instant Boundary Layer Correction. In this procedure, the boundary layer correction is implemented in each relaxation interval in the inviscid solution process by the finite difference scheme. The global iterative process between the inviscid and the viscous solutions is eliminated [9], but the nature of the boundary layer correction remains unchanged.

Inverse Boundary Layer Method for Separated Flow. A Strong interaction can cause flow separation; however, a separated flow, which usually causes a "strong" viscous effect, is *not* always associated with a strong interaction. In fact, the flow near the trailing edge is usually separated with weak viscous-inviscid interactions. In the case of separated flow, the boundary layer equations must be solved by inversely specifying the displacements instead of the usual inviscid pressure distribution [10]. Again, the nature of the boundary layer correction remains unchanged.

Other Inviscid System

The ordinary differential equations for the outer inviscid flow, reduced by means of the N-2-strip integration scheme associated with the hybrid method [11], are:

$$\frac{du_e}{d\bar{x}} = G_1 \tag{14}$$

$$\frac{dv_e}{d\bar{x}} = G_2 \tag{15}$$

where G_1 and G_2 are from Ref. [11].

The advantage of using the hybrid method is that the inviscid external system can be solved simultaneously with the viscous system. This capability is especially useful in the case of the Strong Interaction because of the direct coupling requirement.

Discussion of Results

Equations (14) and (15) are coupled with Eqs. (6) and (7) for Strong Interaction or with Eqs. (11) and (12) for Weak Interaction, and integrated by the fourth-order Runge-Kutta scheme. For a direct comparison of results, both the Strong and the Weak Interaction systems have been applied to the case of supercritical flow over a Korn-Garabedian airfoil at $M_\infty = 0.78$ and $\alpha = 0$.

Values of the boundary-layer displacement thickness and the velocity at the edge of the boundary layer are shown in Figs. 3 and 4, respectively. The effects of the interplay between these two variables are clearly depicted. In the case of the Strong Interaction, a continuous distribution of decelerating velocity at the edge of the boundary layer has been achieved based on the variation of the normal velocity of the outer inviscid flow via the boundary layer displacement thickness through Eq. (9). The boundary layer thickens more rapidly before the sonic line in the Strong Interaction system than that obtained with the Weak Interaction formulation.

The resulting pressure distributions over the airfoil using both systems are compared in Fig. 5, along with the original finite difference inviscid solution and the experimental data. An immediate pressure rise in front of the shock wave is observed in the Strong Interaction procedure as a consequence of the decelerating velocity shown in Fig. 4. The pressure increases continuously throughout the shock region towards the trailing edge. The computation is terminated when the flow separates at $\bar{x}/c = 0.92$. The power-law profile becomes inadequate when $H > 2$ and is totally invalid aft the separation point ($H = 2.55$). Nonetheless, the overall comparison between the theoretical and the measured pressures is surprisingly good. The basic feature of a smearing shock due to viscosity is captured. The Weak Interaction system, on the other hand, always converges to an abrupt pressure jump at the discontinuity. An oblique shock with $\beta = 70$ deg that partially simulates the viscous smearing was assumed in the computation. Generally, the result reflects some viscous effects, since better agreement is observed with the experiment than with the potential flow solution.

Concluding Remarks

A simple boundary-layer integral method has been used to describe the Strong and the Weak Interaction systems for transonic shock wave-turbulent boundary-layer interaction. The Strong Interaction system that gives continuous decelerating flow throughout the shock wave region captures the basic feature of a smearing shock in the interaction process. The Weak Interaction formulation, on the other hand, always converges to an abrupt pressure jump at the discontinuity, although some viscous effects are accounted for.

References

1. Tai, T.C.: Transonic Laminar Viscous-Inviscid Interaction Over Airfoils. *AIAA Journal*, Vol. 13, No. 8, Aug 1975, pp. 1065-1072; also NSRDC Report 4362, Jun 1974.

2. Klineberg, J.M.; Lees, L.: Theory of Laminar Viscous-Inviscid Interactions in Supersonic Flow. *AIAA Journal*, Vol. 17, Dec 1969, pp. 2211-2221.

3. Ram, B.R.; Harvey. W.D.: Hybrid Approach to Steady Transonic Normal Shock-Compressible Laminar Boundary Layer Interactions over Airfoils. AIAA Paper 84-1561, Jun 1984.

4. Ram, B.R.; Harvey, W.D.: Hybrid Approach to Steady Transonic Normal Shock-Compressible Laminar Boundary Layer Interaction over Airfoils with Suction. AIAA Paper 85-0522, Mar 1985.

5. Head, M.R.: Entrainment in the Turbulent Boundary Layer. Aeronautical Research Council Report R&M 3152, 1958.

6. Tai, T.C.: Transonic Turbulent Viscous-Inviscid Interaction over Airfoils. AIAA Paper 75-78, Jan 1975.

7. Green, J.E.: Application of Head's Entrainment Method to the Prediction of Turbulent Boundary Layers and Wakes in Compressible Flow. Royal Aeronautical Establishment Technical Report 72079, 1972.

8. Collyer, M.; Lock, R.: Prediction of Viscous Effects in Steady Transonic Flow Past an Airfoil. *Aeronautical Quarterly*, No. 30, p. 485, 1979.

9. Streett, C.L.: Viscous-Inviscid Interaction for Transonic Wing-Body Configurations Including Wake Effects. *AIAA Journal*, Vol. 20, No. 7, Jul 1982, pp. 915-923.

10. Carter, J.E.: A New Boundary-Layer Inviscid Iteration Technique for Separated Flow. *Proceedings of AIAA 4th Computational Fluid Dynamics Conference*, 1979.

11. Tai, T.C.: Hybrid Approach to Transonic Inviscid Flow with Moderate to Strong Shock Wave. in *Transonic Aerodynamics*, edited by D. Nixon, Progress in Astronautics and Aeronautics, Vol. 81, New York, 1982.

12. Kacprzynski, J.J.; Ohman, L.H.; Garabedian, P.R.; Korn, D.G.: Analysis of the Flow Past a Shockless Lifting Airfoil in Design and Off-Design Conditions. National Research Council of Canada, Ottawa, Canada, Aero Report LR-544, Nov 1971.

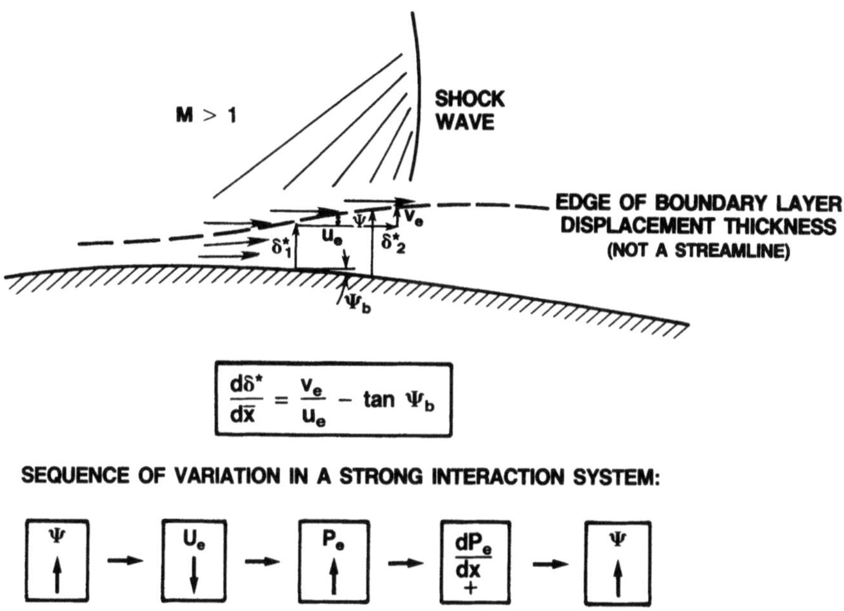

Figure 1 - Strong Interaction System for Transonic Shock Wave-Boundary Layer Interaction

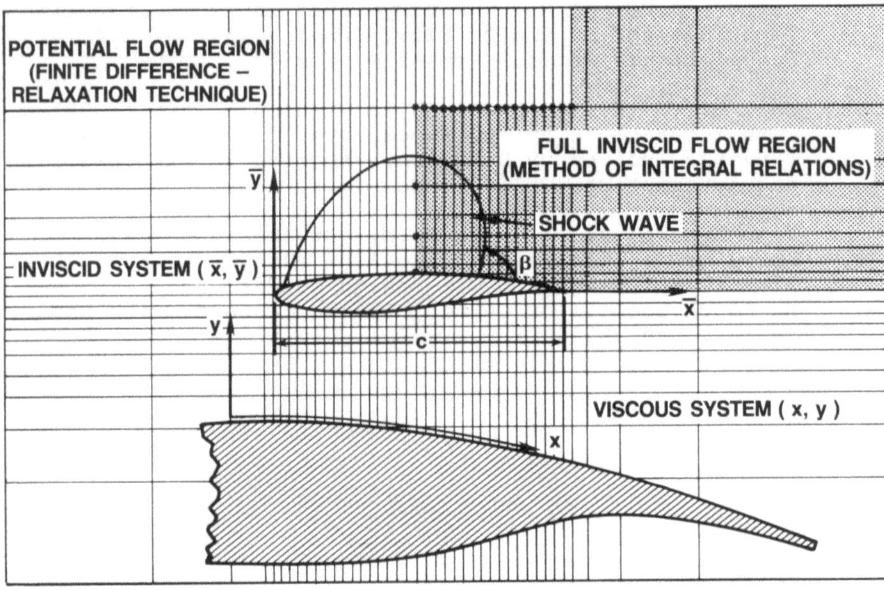

Figure 2 - Coordinate Systems for Inviscid and Viscous Flow

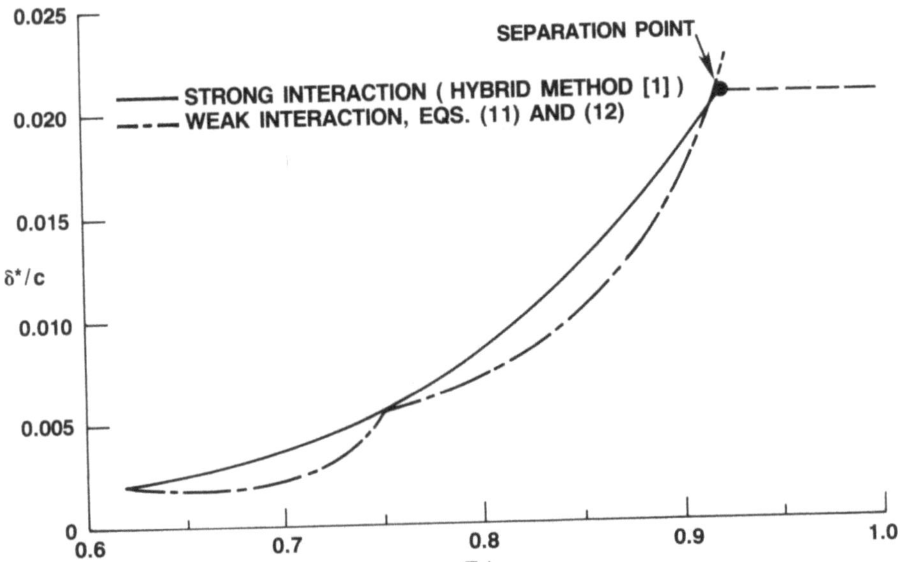

Figure 3 - Boundary-Layer Displacment Thickness over a Korn-Garabedian Airfoil at M_∞ = 0.78 and α = 0

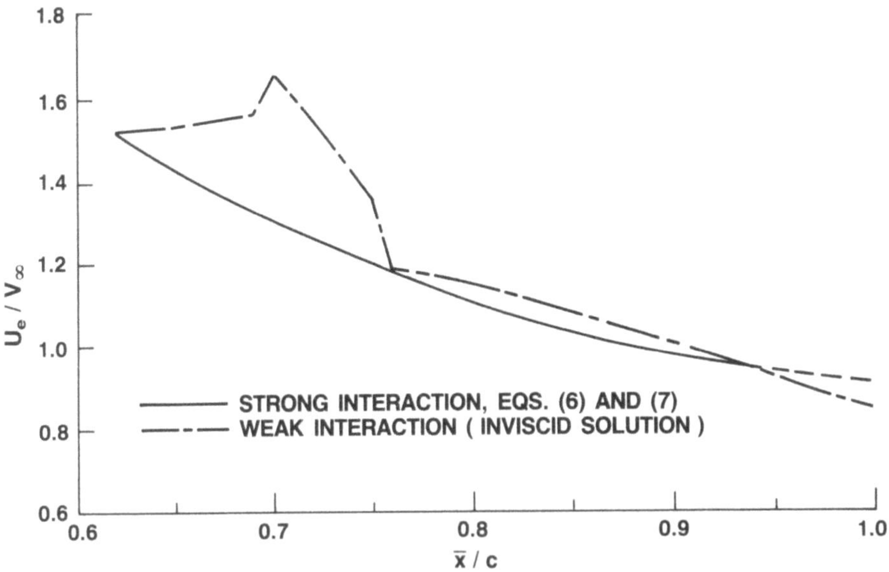

Figure 4 - Velocity at Edge of Boundary-Layer Displacement Thickness over a Korn-Garabedian Airfoil at $M_\infty = 0.78$ and $\alpha = 0$

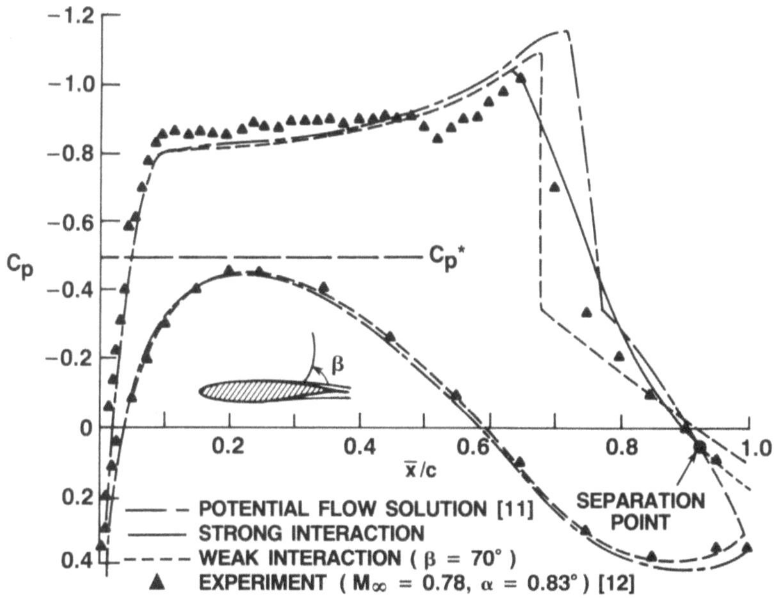

Figure 5 - Pressure Distribution around a Korn-Garabedian Airfoil at $M_\infty = 0.78$ and $\alpha = 0$

Application of Interacting Boundary-Layer Theory in the Analysis of Transonic Shock Induced Separation

DAVID E. EDWARDS*, DAVID L. WHITFIELD** AND JAMES E. CARTER*

*United Technologies Research Center
East Hartford, CT 06108

**Department of Aerospace Engineering
Mississippi State University
Mississippi State, MS 39762

1. Introduction

Accurate prediction of the flow field in a transonic shock-wave boundary-layer interaction is an important aerodynamic problem since this flow phenomenon has a large influence on the overall forces on various aerodynamic components in both internal and external flow applications. The intersection of a shock wave with a turbulent boundary layer in transonic flow produces a complex flow in which disturbances created by the impingement of the shock wave on the viscous layer propagate upstream in the subsonic portion of the viscous layer resulting in a rapid increase of the viscous layer thickness. If the upstream Mach number has a value of about 1.3 or greater, the flow will separate resulting in a complex flow structure, as shown in Fig. 1, where there is a bifurcation of the shock wave into a front and rear leg near the wall with a vortex sheet emanating from the triple point and a separation bubble at the foot of the shock wave in the lower portion of the viscous layer. The overall objective of this paper is to present a reliable and accurate procedure for the prediction of the flow which occurs when a transonic normal shock wave, of sufficient strength to cause separation, impinges on a turbulent boundary layer.

Numerous efforts have been made in recent years to develop techniques to analyze the transonic shock-wave boundary-layer interaction phenomenon. These approaches range from asymptotic analyses [1,2,3] to a more general treatment through the numerical solution of the Reynolds averaged Navier-Stokes equations [4,5,6]. A different approach has been taken in the present

Turbulent Shear Layer/Shock Wave Interactions
IUTAM Symposium Palaiseau 1985
Editor: J. Délery

work which is based on Interacting Boundary Layer Theory (IBLT). This approach assumes that for high Reynolds numbers, the flow field can be represented as regions of viscous and inviscid flow with the regions coupled through the viscous displacement thickness. It has been demonstrated by LeBalleur [7], Meauze and Delery [8], Edwards, et. al. [9] and Om and Childs [10] that IBLT can be applied in the analysis of transonic shock induced turbulent separated flow. The principal focus of the earlier work of Edwards, et. al. [9] was to extend IBLT to include normal pressure gradients in the viscous region. It was concluded in that study that displacement thickness interaction is the dominant effect which must be included in a viscous-inviscid interaction analysis of transonic shock induced separated flow with normal pressure gradients having a higher order effect.

In the present study, which is a continuation of earlier work [9], the boundary-layer equations, expressed in either finite difference or integral form, are iteratively solved through displacement thickness coupling with a stream function-vorticity representation of the inviscid flow in which entropy corrections are applied in the inviscid analysis to give the proper jump conditions at the shock wave. Normal pressure gradients in the viscous region are neglected. The viscous equations are coupled to the inviscid equations using either Carter's semi-inverse method [11] or Edwards' and Carter's extension [12] of Veldman's quasi-simultaneous method [13]. The axisymmetric experimental data of Om [14] has been used in this study to assess the accuracy of this approach. In particular, the impact of including entropy corrections and a vortex sheet, due to the shock wave, in the inviscid analysis is evaluated. Several conclusions are drawn from this study; the principal one is that Interacting Boundary Layer Theory is capable of accurately resolving many of the details of transonic shock-induced separated flow.

2. Analysis

The procedure used in this study of transonic shock induced separated flow is Interacting Boundary Layer Theory (IBLT) in which the entire flow field is represented as regions of viscous and inviscid flow with the two regions coupled through the viscous displacement thickness. The viscous region is assumed to be governed by Prandtl's boundary-layer equations. In this study, either a finite difference or an integral representation of the boundary-layer equations is applied in the IBLT analysis. The finite difference boundary-layer method, which is described in Ref. 15, is a compressible direct (pressure specified) or inverse (displacement thickness specified) method that solves the continuity, x-momentum and energy equations for two-dimensional or axisymmetric flow using an implicit finite difference procedure that is first-order accurate in the streamwise direction and second-order accurate in the normal direction. This analysis uses a modified Cebeci-Smith turbulence model [16] in which the incompressible momentum thickness is used as the length scale in the outer region of the boundary layer instead of the usual treatment which is based on the incompressible displacement thickness. This length scale modification was made based on observations of the data of Simpson, et. al. [17] for low speed separated flow. It was found in Ref. 9 that use of this modified algebraic turbulence model improved the agreement between an IBLT analysis and Kooi's data [18] for transonic shock induced separated flow.

The integral boundary-layer method was developed by Whitfield, et. al. and is described in Ref. 19. This method is a compressible turbulent inverse method that solves the x-momentum and mean-flow kinetic energy integral equations for two-dimensional or axisymmetric adiabatic flow. The representation of the velocity profiles is that developed by Swafford [20] for attached and separated turbulent flow. The turbulence modeling used for the dissipation integral is based on that developed by Thomas [21] for separated flow. The equations are solved using a fourth-order Runge-Kutta integration scheme in the streamwise direction.

The compressible stream function-vorticity method of Hafez and Lovell [22] is used to represent the inviscid flow field and is described in Ref. 23. This approach has a major advantage over that based on a potential flow approximation since it is equivalent to a solution of the Euler equations provided the entropy and vorticity produced by a curved shock wave are taken into account. Previous work by Carter [24] had used an isentropic inviscid flow model in an IBLT procedure for transonic shock-induced separated flow in which good comparison between experimental data and the analysis was obtained. In this case the local Mach number upstream of the shock wave was about 1.3; for $M_\infty > 1.3$, an isentropic inviscid flow model significantly overpredicts the shock strength. Three different inviscid flow models are examined in this study to assess the effect of entropy treatment due to the shock wave in the inviscid analysis. The first inviscid flow model assumes the inviscid flow to be isentropic (no change in entropy, $\Delta s = 0$). The second model assumes the entropy increases at the shock wave based on the Rankine-Hugoniot relations in the inviscid flow ($\Delta s \neq 0$ at the shock); this inviscid flow model is what Hafez and Lovell [25] refer to as an irrotational, non-isentropic flow. The third inviscid flow model assumes the entropy increases at the shock wave with a vortex sheet placed downstream of the shock wave. The location of the vortex sheet is determined from Om's [12] experimental total pressure profiles; an entropy increase is imposed across the vortex sheet as shown in Fig. 1 with the vorticity computed from Crocco's theorem.

The viscous and inviscid flow fields are coupled using either Carter's semi-inverse method [11] or Edwards' and Carter's [12] finite difference extension of Veldman's quasi-simultaneous method [13]. Both methods are described in detail in Ref. 12. The semi-inverse method is an explicit coupling of the viscous and inviscid flow fields whereby the displacement thickness is simultaneously specified for a direct (shape specified) solution of the inviscid equations and for an inverse solution of the boundary-layer equations. The displacement thickness is then updated based upon the mismatch between the inviscid surface tangential velocity and the viscous edge velocity. This procedure is repeated until the two velocities are within a specified tolerance. In general, use of this semi-inverse scheme in a wide range of problems has indicated that under-relaxation must be applied as the extent of the separation region increases.

The second coupling method, the quasi-simultaneous method, has been extended to compressible flow in this study. This procedure solves the boundary-layer equations for both the pressure and displacement thickness subject to a localized relation between these two quantities which is obtained from the inviscid analysis. The displacement thickness obtained from the viscous solution is then specified in the inviscid solution. This procedure is repeated until the inviscid surface tangential velocity and viscous edge velocity are within a specified tolerance of each other. In this study the quasi-simultaneous technique was applied only for the IBLT method that used the finite-difference boundary-layer analysis.

3. Results and Discussion

Results are presented for the analysis of axisymmetric transonic normal shock wave, turbulent boundary-layer interaction. The experimental data of Om [14] has been extensively used in this investigation to evaluate the present IBLT method. Om performed a detailed investigation of an axisymmetric transonic normal shock-wave boundary-layer interaction in a nozzle with a diameter of 51.9 mm at free stream Mach numbers of 1.28, 1.37 and 1.48 and a constant free stream Reynolds number per meter of 4.92×10^6. A normal shock wave was established in the nozzle by placing a blunt cone as a centerbody downstream of the test section. A summary of the experimental test conditions and incoming turbulent boundary-layer parameters upstream of the normal shock wave is given in Table 1. Two series of calculations using the present method were made for these experimental flows. In the first series of calculations the effect of including or excluding the entropy rise at the shock and the vortex sheet in the IBLT analysis for the $M_\infty = 1.48$ flow is examined. In the second series of calculations the present IBLT method using either a finite difference or integral boundary-layer analysis is evaluated through comparison with Om's data for all three experimental flows. When comparing numerical results to experimental data, all quantities have been non-dimensionalized using Om's upstream flow parameters of pressure (p_o), boundary-layer thickness (δ_o), displacement thickness (δ_o^*), and skin friction (C_{f_o}).

For all calculations, the inviscid mesh consisted of 201 evenly distributed grid points in the x-direction over the region -10 mm $<x<$ 110 mm and 21 points in the y-direction from the wall (y = 25.95 mm) to the centerline (y = 0) distributed non-uniformly such that the smallest step occurs at the wall. The start of interaction is assumed to be at x = 0 and the normal shock wave is imposed on the centerline boundary at x = 11 mm (Om indicated that the normal shock wave occurred at about 5 upstream boundary-layer thicknesses from the start of interaction). The x-stepsize used in the inviscid analysis corresponds to approximately 1/4 of the incoming experimental boundary-layer thickness for each of the calculations. Both viscous analyses used the same streamwise mesh as the inviscid analysis to avoid interpolation between the two analyses. The finite-difference boundary-layer analysis used 51 nonuniformly distributed points across the boundary layer in the normal direction. Each calculation was made with 100 global iterations between the inviscid analysis and the viscous analysis. In this study, all calculations made with the quasi-simultaneous coupling technique used a relaxation factor of ω = 1. and all calculations using the semi-inverse technique used a relaxation factor of ω = .2.

3.1 Effect of Entropy Treatment on IBLT Results

A series of calculations were made to examine the effect of the entropy rise and hence the total pressure loss due to a normal shock wave at a free stream Mach number of 1.48 on the computed results of the present viscous-inviscid analysis. The three inviscid flow models discussed previously were applied in this series of calculations. For the third inviscid flow model, the vortex sheet was located based on Om's experimental results at 15.45 mm above the centerline of the nozzle. Calculations in this series were made using the finite difference boundary-layer analysis. Figure 2 shows a comparison of the results of the

IBLT calculations with the experimental data of Om for wall pressure and displacement thickness. Figure 2(a) shows that the results of the analyses using all three inviscid flow models are in good agreement with the experimentally measured surface pressure with a mild discrepancy downstream of the shock wave occurring between the experiment and the computed results of the analysis which used the irrotational, non-isentropic inviscid flow model. It is observed in Fig. 2(b) that the predicted displacement thickness using the third inviscid flow model which includes the effect of the vortex sheet is in good agreement with the experimental displacement thickness. Further, it is observed that there are major discrepancies between the interaction results using the other two inviscid flow models and the experimental data. The analysis using the isentropic inviscid flow model significantly overpredicts the displacement thickness which is probably due to the overprediction of the actual shock strength when the flow is assumed to be isentropic. This stronger shock wave results in a more severe adverse pressure gradient region at the wall which leads to a larger overall growth in the displacement thickness. It is interesting to note that the overall pressure rise remains the same since the larger displacement thickness causes the downstream subsonic flow to accelerate thereby droping the pressure level. The analysis using the irrotational, non-isentropic inviscid flow severely under-predicts the displacement thickness downstream of the shock wave indicating a less severe wall pressure gradient across the shock than the experimental distribution.

An additional IBLT calculation was performed using the third inviscid flow model. In this calculation, the inviscid grid in the normal direction was increased from 21 points to 81 points in order to see if the results of the IBLT analysis were grid dependent due to the specification of the vortex sheet behind the shock wave. The results of this very fine grid calculation and the coarse grid calculation differed less than 5% in terms of the surface pressure and the displacement thickness. Mach number contour lines obtained from this calculation are shown in Fig. 3. In this figure, one can see rapid Mach number variation in the shock wave and the vortex sheet downstream of the shock wave.

Figure 4 shows convergence histories for the IBLT method using both the semi-inverse and quasi-simultaneous coupling methods for the analysis employing the isentropic inviscid flow and the analysis which modeled the inviscid flow with a entropy rise at the shock wave and a vortex sheet. For both of these flow cases the quasi-simultaneous procedure is converging at a faster rate than the semi-inverse technique. However, before a conclusion can be made about the efficiency of one coupling technique compared to another for transonic flow calculations, a more complete numerical study should be made.

This series of calculations demonstrates that modeling the vortex sheet behind the shock wave in the inviscid analysis has a significant impact on the calculated IBLT results for transonic shock-induced separated flow at a high Mach number. For the remainder of the numerical calculations presented in this paper, the inviscid model with a entropy rise at the shock wave and a vortex sheet behind the shock wave is used as the inviscid flow model in the present IBLT approach.

3.2 Effect of Boundary-Layer Analysis on IBLT Results

Comparisons are presented between the results of the present IBLT method and the data of Om [14] for transonic shock induced separated flow in an axisymmetric nozzle. A series of calculations was made to assess the impact on the numerical results when applying either Whitfield's integral boundary-layer analysis or a finite difference boundary-layer analysis in the IBLT method. Figure 5 shows a comparison of the results from the IBLT computations and the experimental data of Om for the wall pressure, displacement thickness and skin friction at M_∞ = 1.28, 1.37 and 1.48. Figures 5(a) and 5(b) show that the predicted wall pressure and displacement thickness from the present IBLT method using either of the viscous analyses are in good agreement with the experimental data. The distributions of the skin friction coefficient in Fig. 5(c) show that ahead of the shock wave the results of both analyses are in good agreement with the experiment. Downstream of the shock wave, the skin friction results from the integral boundary-layer analysis are in good agreement overall but do not predict separation in any of the cases while the skin friction results from the calculations using the finite-difference boundary-layer analysis are not recovering as rapidly as the experiment (lower skin friction, longer extent of separation). The lack of agreement for the skin friction in the finite-difference boundary-layer analysis indicates that further work is required to improve the accuracy of this IBLT method. Specifically, work is required in modeling the turbulence near the wall. This is further demonstrated in Fig. 6 where profiles for the u-component of velocity at various streamwise locations upstream and downstream of the shock wave from the finite difference boundary-layer analysis are compared to the measurements of Om at $M_\infty = 1.48$. Linear streamwise interpolation was used in the calculation to obtain the computed velocity profiles at the same x-location as the experiment. The edge of the boundary-layer used in the calculation was assumed to be at the y-location where $u = .995\ u_e$. The agreement is good in the wake region of the boundary-layer, but near the wall downstream of the shock wave, the calculation underpredicts the velocity recovery as expected based upon the skin friction results. Overall, the results of the IBLT method using either of the two viscous analyses are encouraging.

4. Concluding Remarks

Results from an IBLT method which uses a stream function-vorticity representation of the inviscid flow and either a finite difference or integral representation of the boundary-layer equations have been presented. The method uses either a semi-inverse technique or a quasi-simultaneous technique to couple the two flows. Favorable comparisons which have been obtained with the axisymmetric transonic normal shock wave boundary-layer interaction data of Om ($M_\infty = 1.28$, 1.37 and 1.48) demonstrate that the present procedure, using either a finite difference or integral boundary-layer analysis, is capable of accurately resolving many of the details of transonic shock-induced separated flow. In this investigation, it was shown that the inviscid modeling of the entropy rise due to the shock wave and the resulting vortex sheet behind the shock have significant impact on the results of an viscous-inviscid interaction for transonic shock-induced separated flow at high Mach numbers and hence should not be neglected. This suggests that an Euler analysis which has adequate grid resolution to capture both the shock wave and the resulting vortex sheet emanating from the triple point should be applied in the IBLT analysis. Further, this suggests that for the Navier-Stokes procedures, the numerical grid must be carefully selected not only to capture the shock wave and viscous region but also the vortex sheet in order to obtain accurate solutions for transonic shock-wave, boundary-layer interaction at high Mach numbers.

Acknowledgements

This work was supported by the compressor group at Pratt and Whitney of the United Technologies Corporation and by the Office of Naval Research under Contract N00014-81-0381.

References

1. Melnik, R. E. and Grossman, B.: AIAA Paper No. 74-598, 1974.
2. Adamson, T.C., Jr. and Feo, A.: SIAM J. Appl. Math., Vol. 29, 1975, pp. 121-145.
3. Messiter, A. F.: NASA CR-3194, 1979.
4. Viegas, J. R. and Horstman, C. C.: AIAA J., Vol. 17, August 1979, pp.811-820.
5. Shea, J. R., III: AIAA Paper No. 78-1170, 1978.
6. Om, D., Childs, M.E. and Viegas, J.R.: AIAA Paper No. 82-0990, 1982.
7. LeBalleur, J.C. and Blaise, D.: ONERA TP-1984-152, 1984.
8. Meauze, G. and Delery, J.: ONERA TP-1983-62, 1983.
9. Edwards, D. E., Carter, J. E. and Hafez, M. M.: AIAA Paper No. 85-0371,1985.
10. Om, D. and Childs, M. E.: AIAA Paper No. 83-1402, 1982.
11. Carter, J. E.: AIAA Paper No. 79-1450, 1979.

12. Edwards, D. E. and Carter, J. E.: Presented at Third Symposium on Numerical and Physical Aspects of Aerodynamic Flows, Long Beach, CA., January 1985.
13. Veldman, A. E. P.: NLR TR7 9023 U, 1979.
14. Om, D.: A Study of Transonic Normal Shock Wave Turbulent Boundary Layer Interaction in Axisymmetric Internal Flow, Ph.D Thesis, University of Washington, 1982.
15. Carter, J. E. : NASA TP-1208, September 1978.
16. Cebeci, T. and Smith, A. M. O.: Analysis of Turbulent Boundary Layers, Academic Press, 1974.
17. Simpson, R. L. , Chew, T. T. and Shivaparsad, B. V.: J. Fluid Mechanics, Vol. 113, 1981, pp. 23-51.
18. Kooi, J. W.: AGARD CP-1168, 1975.
19. Whitfield, D. L., Swafford, T. W. and Donegan, T. L.: Recent Contributions to Fluid Mechanics, Ed. A. Haase, Springer-Verlag, 1982, pp. 294,302.
20. Swafford, T. W.: AIAA J., Vol. 21, No. 6, June 1983.
21. Thomas, J. L. : AIAA Paper No. 84-1615, 1984.
22. Hafez, M. and Lovell D.: AIAA J., Vol. 21, No. 3, March 1983.
23. Carter, J. E., Edwards, D. E. and Hafez, M. M.: AFOSR TR-83-1283, 1983.
24. Carter, J. E.: AIAA Paper No. 81-1241, 1981.
25. Hafez, M. and Lovell, D.: AIAA Paper No. 83-1926, 1983.

Table 1. Om's Experimental flow conditions and upstream boundary-layer parameters.

M_∞	p_o, atm	δ_o, mm	δ_o^*, mm	C_{f_o}
1.48	.3325	2.10	0.438	.003725
1.37	.3264	2.34	0.455	.003647
1.28	.3239	2.62	0.509	.003734

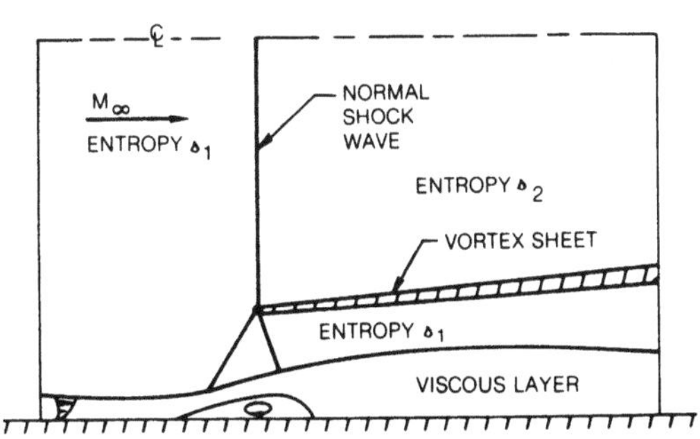

Fig. 1. Normal shock wave impinging on a boundary layer.

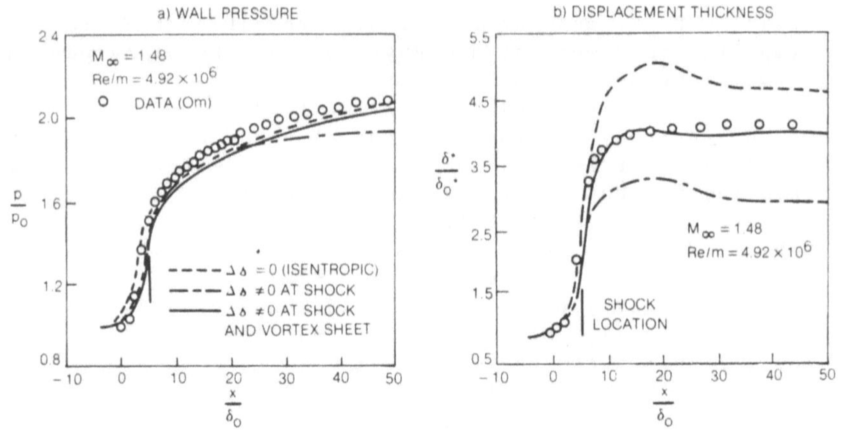

Fig. 2. Effect of entropy treatment on predicted results for transonic shock induced separated flow.

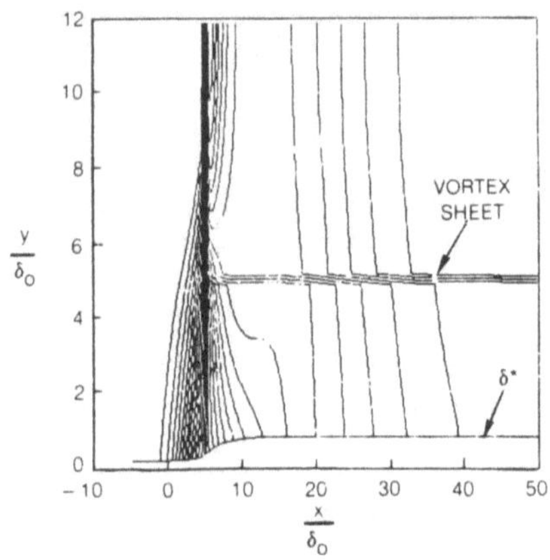

Fig. 3. Computed Mach number contours for Om's experimental flow, $M_\infty = 1.48$.

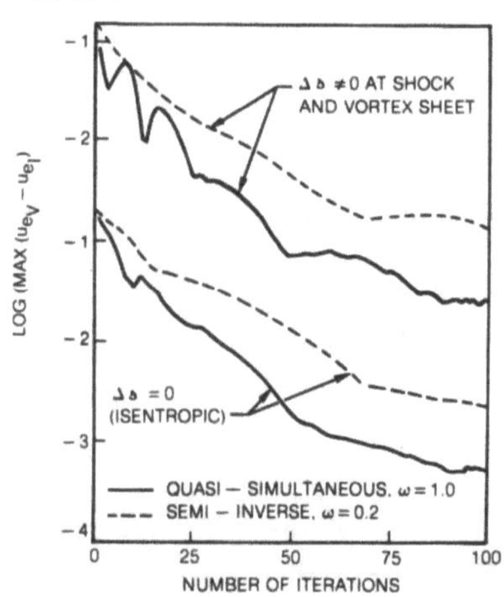

Fig. 4. IBLT convergence histories for Om's experimental flow, $M_\infty = 1.48$.

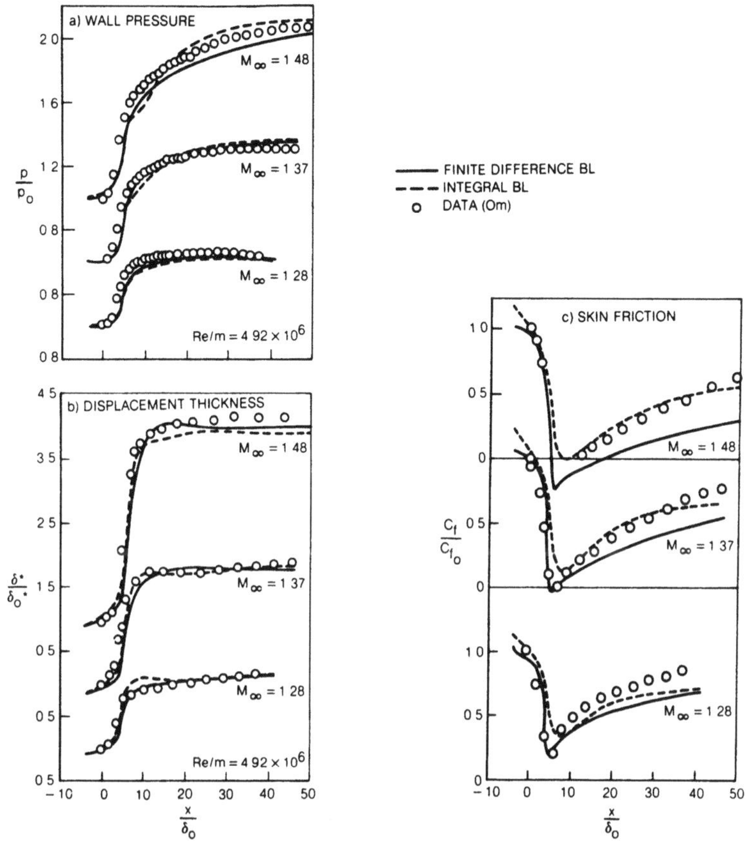

Fig. 5. Comparison of IBLT results with Om's experimental data.

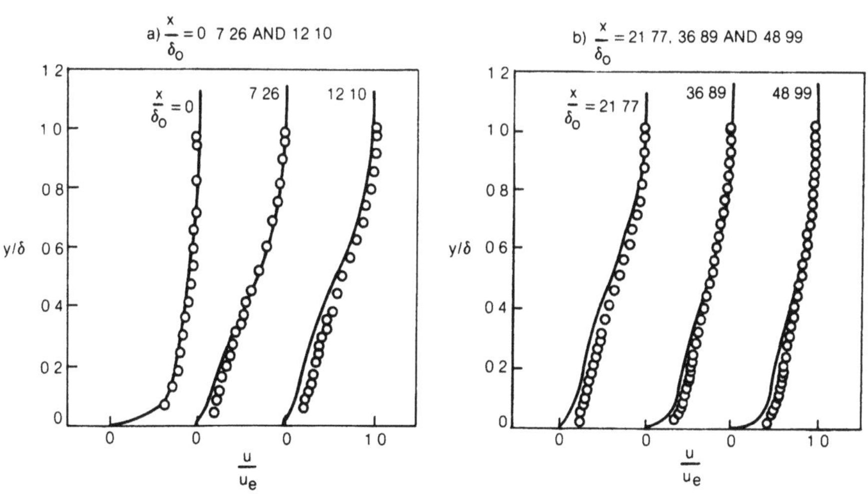

Fig. 6. Comparison of velocity profiles from IBLT with Om's experimental data for $M_\infty = 1.48$.

The Shock-Wave/Turbulent Boundary-Layer Interaction on Curved Surface at Transonic Speed

X. LIU and L. C. SQUIRE

Cambridge University Engineering Department

Summary

Detailed experimental measurements have been made of the shock-wave/turbulent boundary-layer interaction on curved surface at transonic speed. The flow field is calculated by a wide range of numerical methods including two boundary-layer methods, two shock/boundary-layer interaction methods and one Euler-solver inviscid method.

1. Introduction

Shock/boundary-layer interaction occurs in many practical situations such as the flow around transonic aerofoils and the flow in turbomachines. In most practical situations the interaction occurs on curved surfaces at transonic speeds, but understanding of this flow is still far from perfect. Due to its complexity, the flow field of shock/boundary-layer interaction provides a very severe test case for the numerical predictions. In the present research the flow field of shock-wave/turbulent boundary-layer interaction on curved surface at transonic speed is measured in detail and is calculated by a wide range of numerical methods in order to test and compare their performances.

2. The experiment

The experiment was performed in an intermittent blow-down wind tunnel with a two-dimensional test section of 114 mm in width and 62.5 mm in height. The shock waves were generated above two-dimensional circular-arc models or bumps mounted on the floor of the wind tunnel test section as shown in Fig.1. A second throat aerofoil was used to stabilize the shock wave and to control its strength. Upstream of the shock the boundary layers developed in a favourable pressure gradient and were fully turbulent. Five different models

Turbulent Shear Layer/Shock Wave Interactions
IUTAM Symposium Palaiseau 1985
Editor: J. Délery

were used. The ratio of the boundary-layer thickness in front of the shock to the radius of the surface curvature ranged from 0.015 to 0.068, while for measurements on the top flat surface this ratio was zero. For each model a series of shock waves with different strengths was produced as shown from the surface pressure distributions in Fig.2, where M_p is the peak surface Mach number; R is the radius of the bump; c is the chord length of the bump and X is the streamwise coordinate measured from the mid-chord point of the bump. In the figure the separation position "S" and the reattachment position "R" obtained from the oil flow test are shown by the arrows. The peak Mach number in front of the shock ranged from 1.00 to 1.82. This covered cases where there was no separation and cases where the boundary layer was separated extensively. The Reynolds number based on the model chord length was about 1.6 million.

The flow field was measured by laser holographic interferometry, schlieren visualization, oil flow tests, surface pressure surveys and extensive boundary layer traverses with a pitot probe. The surface Mach number was calculated from the surface pressure and the stagnation pressure using the isentropic relationship, and the boundary layer velocity profile was calculated from the measured pitot pressure profile and surface static pressure using Crocco's temperature-velocity relationship.

The flow was steady according to the schlieren visualization and surface pressure trace. The oil flow test and surface pressure measurement showed that good two-dimensionality was achieved in all test cases except some very extensively separated cases. The full experimental results will be published elsewhere.

3. The numerical methods

The flow field of some typical measured cases were calculated by five different numerical methods. They include the "time-marching" Euler-solver inviscid method of Denton (1983), the "eddy-viscosity" boundary-layer method of Verma(1971); the "lag-entrainment" integral boundary layer method of Green et al. (1972); the "control volume" normal shock-turbulent boundary layer interaction method of Baker (1980) and the "small disturbance, multi-region" normal shock-turbulent boundary layer interaction method of Inger and Mason

(1976). In the last two methods there is no assumption of zero normal pressure gradient as there is in an ordinary boundary-layer method.

4. Results and discussions

4.1 The boundary layers

Results from the boundary layer programs of Verma and Green and from the shock/boundary -layer interaction programs of Inger and Baker are presented here for an attached and a separated case. The main inputs to Verma's program are the initial velocity profile, the initial temperature profile, and the Mach number distribution. The main inputs to Green's program are the measured surface pressure coefficient distribution and the initial boundary-layer momentum thickness and the shape factor H. In Baker's and Inger's program the conditions at the peak surface Mach number position were taken as the upstream conditions in the input.

Fig.3 shows the surface pressure distribution and the boundary layer traverse positions of the case R=401mm and M_p=1.09 on the top flat surface. The corresponding oil flow test shows that in this case the boundary layer remained attached throughout. The calculated velocity profiles U/U_e together with the measurements are shown in Fig.4, and the calculated surface pressure distribution across the interaction region by Inger's method is compared with the measurement in Fig.5 where P is the static pressure and P_0 is the stagnation pressure. As can be seen, both the predicted velocity profile and the predicted pressure distribution agree with the experiment very well. For this attached case other calculated boundary layer parameters also agree with the experiment very well.

Fig.6 shows the pressure distribution and the boundary layer traverse positions of the case R=163mm and M_p=1.27 on the bump surface. In this case the oil flow showed that the boundary layer was separated. Fig.7 compares the boundary layer velocity profiles predicted by Verma's program with the measured ones. At X/0.5c=0.0, the agreement between the experiment and the prediction is very good. At this station the program has already marched 15 initial boundary layer thicknesses. This shows that the program perform well under a favourable pressure gradient. The prediction agrees with the measured profile up to X/0.5c=0.5, which is around the actual separation point. After the actual reattachment, Verma's program predicts fuller profiles than the measured ones. This shows that the method is inadequate in crossing the actual boundary layer separation. In fact it did not predict separation at all, as will be seen later. The discrepancy gets smaller as the calculation marches downstream. At the last station, X/0.5c=5.4, which is 32 initial boundary-layer thickness downstream of the shock, Verma's prediction and the measured profile are again in close agreement.

Figs.8 to 10 show the results for the boundary layer parameters. As can be seen in Fig.8, the large increase of the displacement thickness δ^* across the shock disturbance is not predicted by any of the four methods. This is because in the real flow the boundary layer separated under the shock foot which considerably thickened the boundary layer. However, this separation is not predicted by any of the methods as will be seen later. Before the shock and at the initial stage of the shock compression, i.e. up to the third measuring station,all the four methods perform equally well and the agreement with experiment is good. Green's prediction shows that the displacement thickness increases to a peak value in crossing the shock, then decreases and finally levels off. This process should be expected in crossing the shock. As is seen, the measured displacement thickness decreases after the shock. However, the prediction of Verma's program does not show this decrease, and instead the predicted displacement thickness grows all way through with no peak values. Before the peak value point of displacement thickness in Green's prediction, the results of the two boundary layer methods are quite close. Inger's shock/boundary-layer interaction method predicts a sharper increase of the displacement thickness, an earlier finish of the shock disturbance,and a lower downstream displacement thickness value than the two boundary layer methods. In Baker's method the predicted slope of the displacement thickness increase up to the sonic point agrees well with the experiment. (In Baker's program the downstream point is the sonic point if the predicted boundary layer is not separated.) Baker's program is able to deal with separation but it did not predict separation for this separated case. Presumably this is because the separation in this test case is partly caused by the surface curvature which is not taken into account in this program. It is seen that the two shock/boundary-layer interaction methods did not show any advantages over the ordinary boundary-layer methods.

Fig.9 shows the development of the shape factor H. It is interesting to notice that the two boundary-layer methods predict two peaks at the same streamwise locations. As can be seen from the skin friction in Fig.10, the first peak of H is around the separation point corresponding to the initial shock compression where the surface pressure has a sharp rise, while the second peak is around the reattachment point. The predicted location of the first peak of H agrees with the measurement very well. The shape factor values at the first peak predicted by the two different boundary-layer methods are quite similar but they are all slightly lower than the measured value. Because of the difficulty in measuring the separated and reattaching boundary layer with an ordinary pitot tube, there are no measurements to show the existence of the second peak. Downstream of the second peak, the predicted values of H by the two methods are all higher than the measured values. Baker's method predicts the first peak.

Fig.10 shows the results for the skin-friction coefficient C_f. The experimental skin friction was calculated by fitting the measured velocity profile to the law of the wall and the zero skin friction was obtained from the oil flow test. It is seen that the boundary layer separated around X/0.5c=0.5 and reattached around X/0.5c=1.4. Upstream of the separation, the results from all the methods are similar and agree with the experiment reasonably well, except that Inger's method predicts smaller skin friction values and quicker recovery than other methods. None of the methods predicted separation. After the actual reattachment, the predicted skin friction coefficients are all lower than the measured values. Green's prediction shows that the skin friction increases as the boundary layer develops downstream of the reattachment, which agrees with the experiment, but the results from other methods do not show this increase.

Fig.11 shows the static pressure distribution on the bump surface across the interaction region, where the solid line is the prediction by Inger's program and the symbols are the measured values. The initial compression and the downstream pressure are well predicted, but the "plateau" corresponding to the separated region is not predicted at all because Inger's method cannot deal with separated flow. Attempt was also made to calculate the flow with the "turbulent kinetic energy" boundary layer method of Bradshaw, but the calculation result was peculiar under the shock disturbance where the pressure gradient was very high and no way was found to overcome this problem.

4.2 The whole flow field

The whole flow field was calculated by combining the inviscid main flow calculation with the boundary layer calculation. The inviscid flow was calculated by Denton's "time-marching" Euler-solver method. Since Denton's method has been developed for the turbomachinery cascade flow calculation, in the present calculation the flow in the wind tunnel test section as shown in Fig.1 was treated as equivalent to the flow passing through a single passage of an infinite cascade with each blade in the cascade consisting of the circular-arc model mounted on one side of a zero-thickness plate extending upstream and downstream, thus the existence of the tunnel top roof was automatically taken into account. The main inputs to Denton's program are the wall geometry, the stagnation pressure and temperature, and the static pressure at the exit of the computational passage. Green's "lag-entrainment" program was used to calculate the displacement thickness of the boundary layers on both bottom and top surfaces **using the measured pressure distributions** as input. These calculated boundary-layer displacement thicknesses were added to the bottom and the top surfaces to form the modified surfaces. Denton's program was run with these modified surfaces as the boundaries.

Fig.12(a) is the result of such calculation for the case R=163mm and Mp on bump =1.27 and shows the calculated density contours of the flow field. Note that the gap between the contours and the wall surfaces corresponds to the calculated boundary layer displacement thickness. These calculated density contours can be compared with the measurement shown by the interferogram of Fig.12(c) in which the interference fringes are the density contours of the actual flow field. The interferogram corresponds to the region marked by the rectangle in the calculation result. The black square on the interferogram shows the scale with each side 10mm long. The fringes on the interferogram turn upstream close to the wall surfaces, corresponding to the edge of the boundary layer. Of course this turning is not predicted by the inviscid calculation in Fig.12(a). It is seen from the interferogram of Fig.12(c) that the shock is very thin in the main stream while it is spread out into a compression fan as the wall surface is approached, which is typical of the shock/boundary-layer interaction. This feature is not predicted by the inviscid calculation in Fig.12(a) and the predicted shock location is quite a distance downstream of the actual shock location shown on the interferogram.

The actual boundary layer was separated but this separation was not predicted as was shown in the previous section. Since no method capable of predicting the flow in the separated region was available, it was decided to obtain a displacement surface, representing the effect of the separation bubble, from the interferogram as was done by Bryanston-Cross and Denton (1982). The displacement so found was added to the solid surface together with the boundary layer displacement thickness calculated by Green's method. The result is shown in Figs.12(b) where the dotted line is the displacement caused by the separation bubble obtained from the interferogram and adjusted by the method of trial and error. After several tries the configuration shown in Fig.12(b) was found to give the best results. The displacement has a straight edge starting around the separation point on the bump surface and extending to the exit of the computation passage where this displacement is still finite. The reason why the displacement is extended beyond the reattachment point is that the large increase in the displacement thickness caused by the separation bubble continues downstream of reattachment as was shown in Fig.8. Compared with Fig.12(a), a considerable improvement is observed, especially in the predicted shock position, and the predicted density contour pattern agrees with the interferogram quite well. However, the predicted shock is straight whereas the actual shock turns towards the upstream direction near the bump surface due to the shock/boundary-layer interaction. Thus if the predicted shock location is correct at the bottom, the shock will be too forward on the top surface. As a compromise, the displacement caused by separation was chosen in such a way that the predicted shock location is somewhat downstream on the bottom and somewhat upstream on the top compared with the measurement as is shown in the surface pressure distribution on Fig.13. It is seen from Fig.13 that upstream of the shock and downstream of the interaction region the predicted surface pressure agrees with the experiment very well. With the good agreement of surface pressures, it can be inferred that the density value on a contour line in Fig.12(b) agrees well with the density value on the corresponding fringe in Fig.12(c).

The calculations described above relies on the use of a bubble shape from the interferogram and, therefore, it is not a complete prediction. Nevertheless it demonstrates that the inviscid flow calculation can give good predictions as long as the boundary layers are correctly incorporated. In view of this, a fully interactive method based on Denton's "time-marching" program and the "lag-entrainment" method of Green et al. has been developed as described in ref.6.

5. Conclusions

From all the calculations made it was found that all prediction methods perform well for attached flow with a weak shock but for the well separated flow none of the methods is adequate. In particular the large increase in the displacement thickness across the shock disturbance is underpredicted, also none of the methods predicted separation when the actual flow was separated. Being an integral method Green's method is the fastest and its prediction is no worse than any of the other methods used. Verma's method performs well in predicting the velocity profiles for the attached boundary layer. Baker's method predicts the correct trend of the boundary layer development across the shock, but it did not predict separation. Inger's method predicts well the static pressure distribution across the shock when the boundary layer is not separated. However for the displacement thickness and skin friction, Inger's method predicts too early an end to the shock disturbance and underestimates the changes which occur through the interaction region. The shock/boundary-layer interaction methods give more detailed information than the boundary-layer methods, but in predicting the boundary-layer parameters, which the ordinary boundary-layer methods can also predict, the former methods do not show any advantage over the latter methods.

The overall features of the flow field are predicted by Denton's inviscid method, but the details of the prediction are not satisfactory due to the viscous effect. It is shown that the inviscid calculation can be considerably improved if the effect of separation bubble is incorporated. In other words, the inviscid calculation can give good predictions as long as the boundary layers are correctly incorporated.

References

1 BAKER C.J. (1980) The prediction of boundary layer development through a normal shock wave turbulent boundary-layer interaction, Cambridge University, CUED/A-AERO/TR.10(1980)

2 BRYANSTON-CROSS P.J. and DENTON J. D. (1982) Comparison of interferometric measurements and computed flow around a wedge profile in the transonic region, ASME paper 82-GT-258

3 DENTON J.D. (1983) An improved time-marching method for turbomachinery flow calculation, Transactions of the ASME, Journal of Engineering for power, Vol. 105, July 1983, pp514-521

4 GREEN J.E., WEEKS D.J. and BROOMAN J.W.F. (1972) Prediction of turbulent boundary layers and wakes in compressible flow by a lag-entrainment method, R.A.E. T.R.72231

5 INGER G.R. and MASON W.H.(1976) Analytical theory of transonic normal shock-turbulent boundary-layer interaction, AIAA J. vol.14, No.9, p.1266

6 LIU X. (1985) Shock/boundary-layer interaction on curved surfaces at transonic speeds, Ph.D. Dissertation, Cambridge University Engineering Department,1985

7 VERMA V.K. (1971) A method of calculation for two-dimensional and axisymmetric boundary layers, CUED/A-AERO/TR3 (1971)

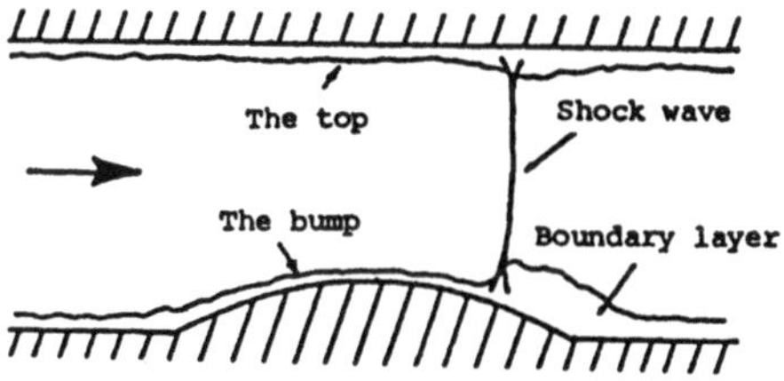

Fig.1 The shock/boundary-layer interaction

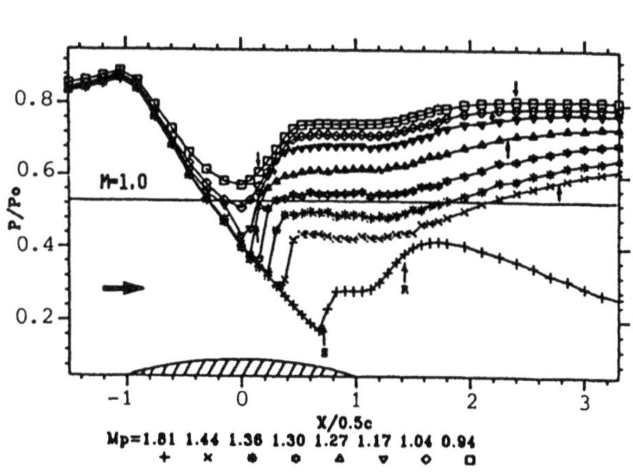

Fig.2 The surface pressure distributions with different shock strengths on the R = 73 mm bump

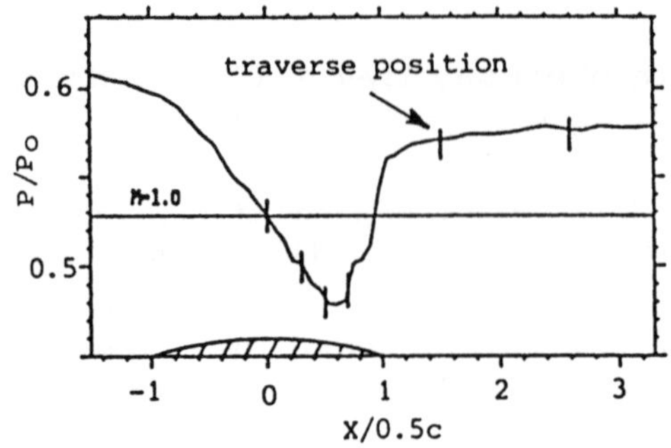

Fig.3 The pressure distribution and the traverse position for the case R=401mm, M_p=1.09, on top

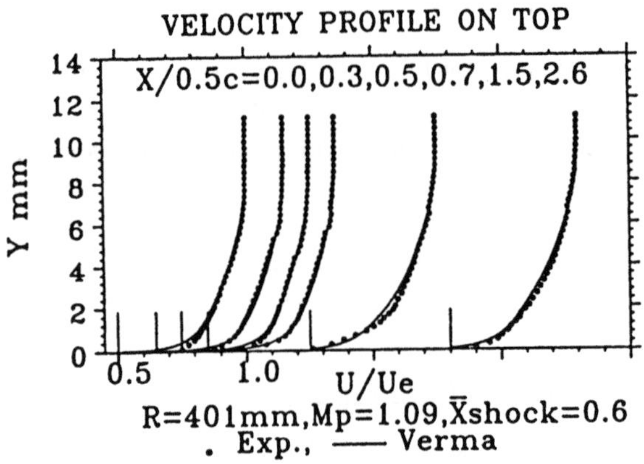

Fig.4 Comparison with experiment of the velocity profiles calculated by Verma's method

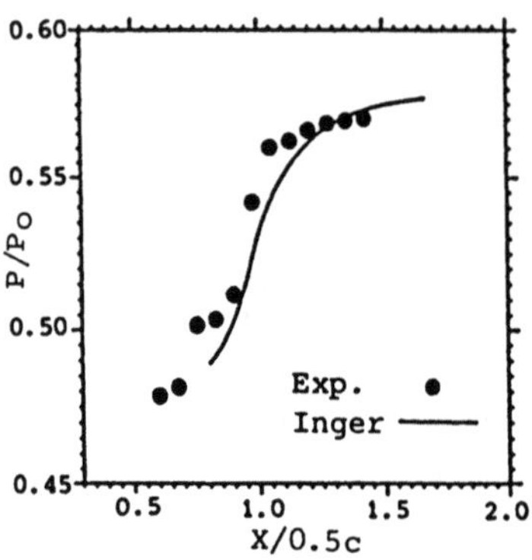

Fig.5 Pressure distribution for the case R=401mm,M_p=1.09 on top

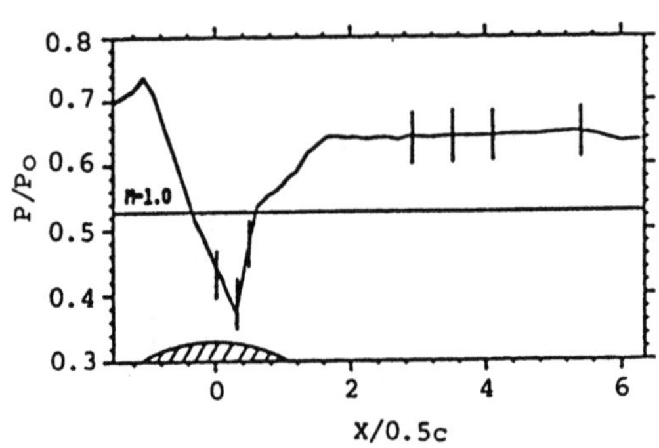

Fig.6 The pressure distribution and the traverse position for the case R=163mm, M_p=1.27, on the bump surface

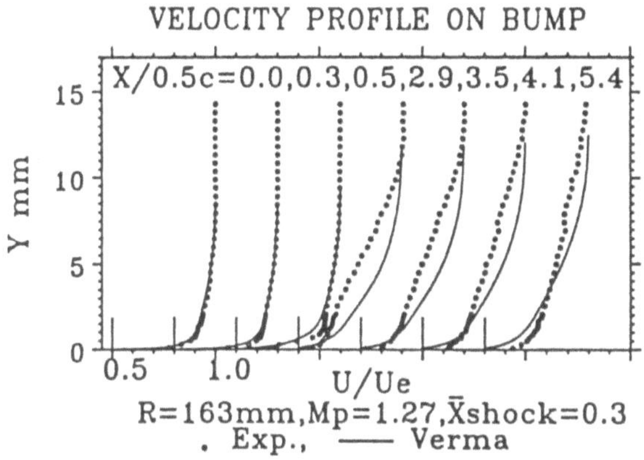

Fig.7 Comparison with experiment of the velocity profiles calculated by Verma's method

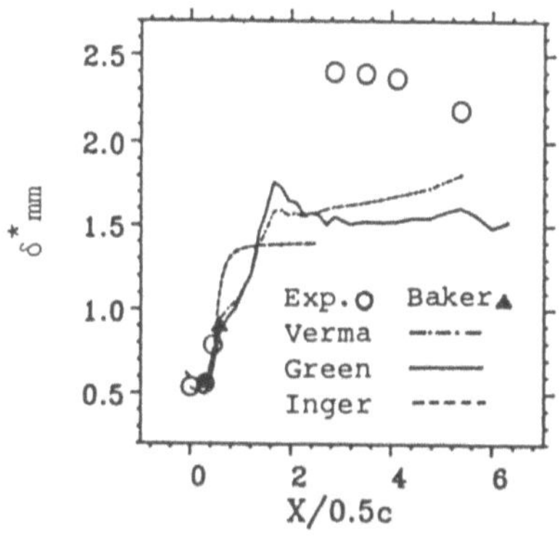

Fig.8

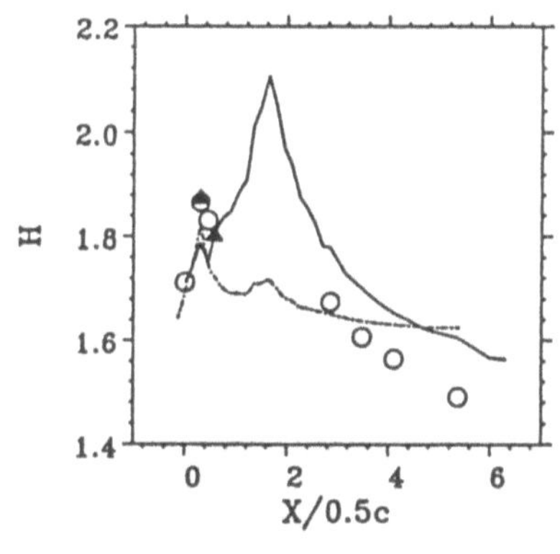

Fig.9

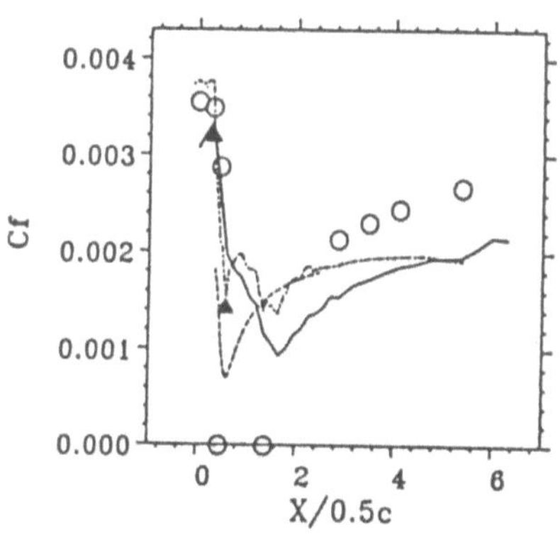

Fig.10

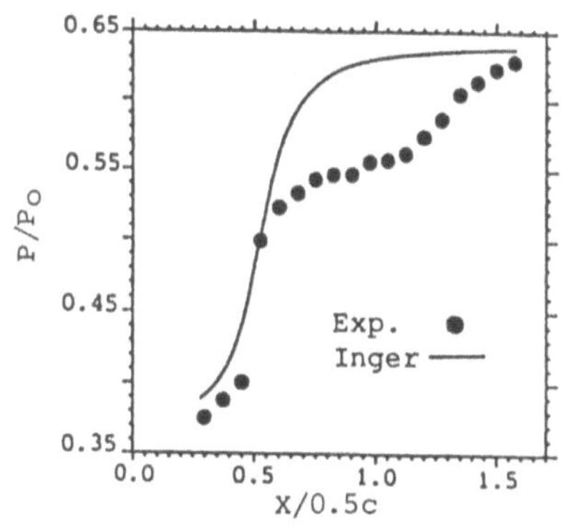

Fig.11

The results for the case R=163mm,M_p=1.27,on bump

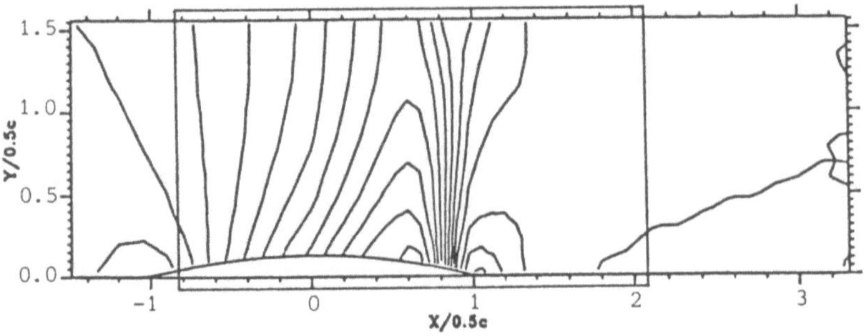

12(a) Calculated with no account of separation

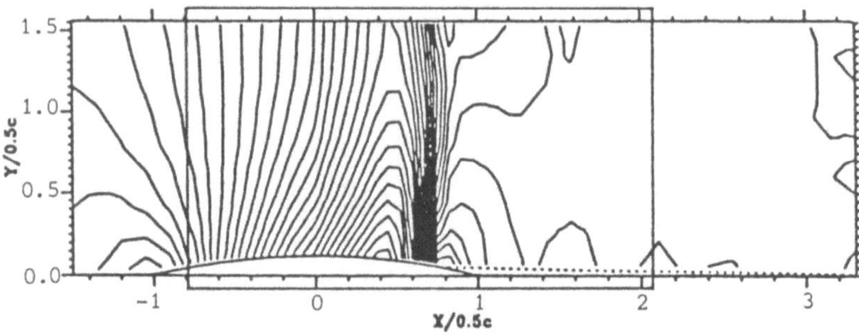

12(b) Calculated with surface modified by the separation bubble

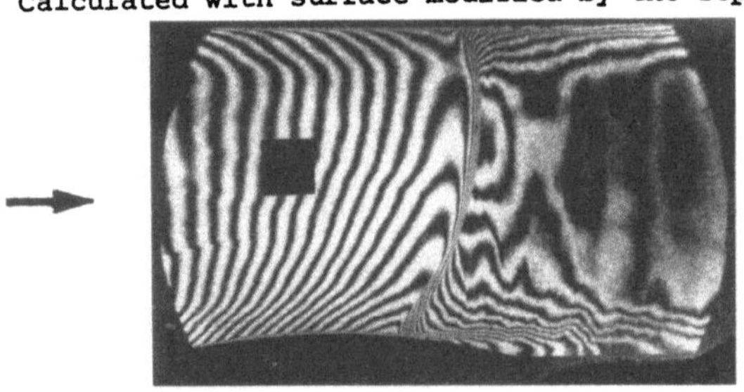

Fig.12 Calculated and measured density contours for the case R= 163 mm, M_p(on bump)=1.27

12(c) The interferogram showing the measured density contour

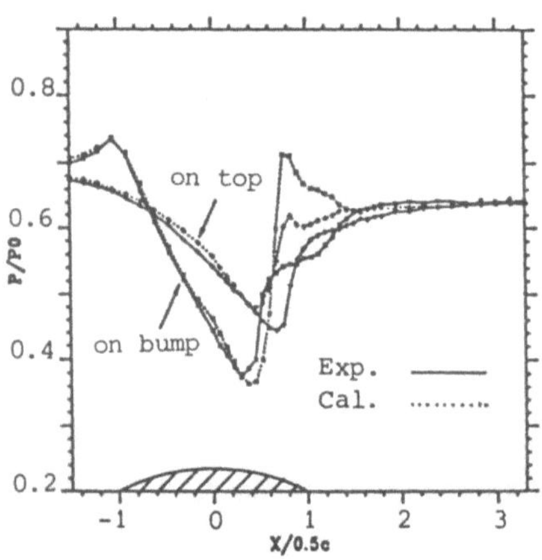

Fig.13 Comparison with experiment of the surface pressure calculated with the surface modified by the separation bubble for the case R=163mm, M_p(on bump)=1.27

Calculation of 2D Turbulent Shock/Boundary-Layer Interaction at Curved Surfaces with Suction and Blowing

R. Bohning and J. Zierep

Institut für Strömungslehre und Strömungsmaschinen
Universität (TH) Karlsruhe, FRG

Introduction

The interaction between a normal shock wave and a turbulent boundary layer is of great importance for the flow around transonic airfoils because the limits of their performance are influenced by this local occurence very strongly. Often a shock causes flow separation and thereby produces global disturbances of the flow field (e.g. rapid drag rise and buffet onset). Recently, in order to counteract the separation, endeavours were taken to weaken the shock by suction or ventilation. Some of the experimental results are summarized in [1].

In the following we give a survey of our investigations concerning the interaction between a weak normal shock and a two dimensional compressible turbulent boundary layer at a curved wall. Our local interference model was embedded in different global numerical solutions of the outer flow field coupled with boundary layer procedures [2]. Because of the more accurate treatment a better convergence of the global calculation results. Recently a new boundary condition at the wall was introduced with the aim to simulate and to calculate analytically the passive control of the interaction region by blowing and suction (ventilation) at a perforated strip. An example of results is discussed in this paper.

Flow field model and method of solution

A local supersonic region, closed by a nearly normal shock, is assumed on the upper surface of a supercritical wing (Fig. 1). In the vicinity of the wall the shock generally penetrates the boundary layer and vanishes therein because of the influence of viscosity (Fig. 2). The upstream influence through the subsonic region close to the wall causes a thickening of the boundary layer already in front of the shock (Fig. 1 and 2). Due

Turbulent Shear Layer/Shock Wave Interactions
IUTAM Symposium Palaiseau 1985
Editor: J. Délery

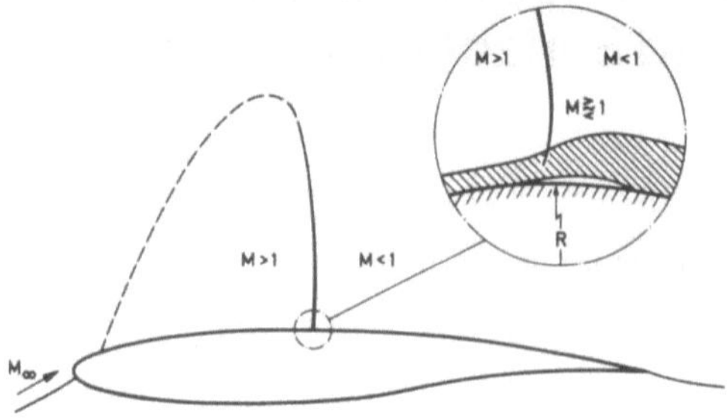

Fig. 1 Δ
Transonic airfoil flow

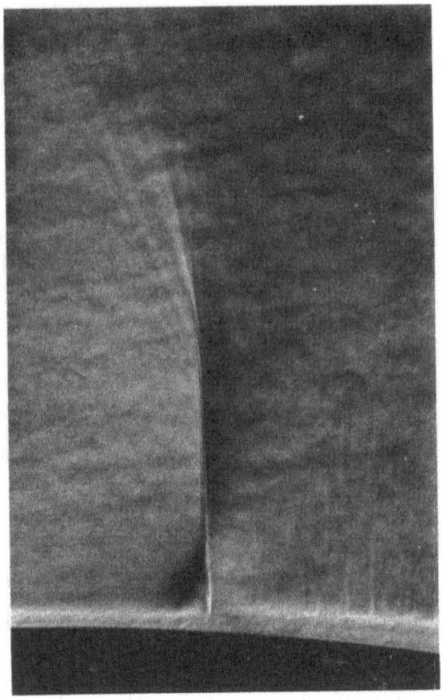

Fig. 2 ▷
Schlieren picture of a shock-boundary layer interaction (M_δ = 1.27)

to this a compression region arises in the supersonic flow and reduces the shock strength just outside the boundary layer.

Behind the shock the displacement effect on the one hand and the influence of the wall curvature on the other causes a post-shock expansion. This region influences critically the flow behaviour in the direct neighbourhood of the wall. The greater the post-shock expansion the less the tendency of flow separation.

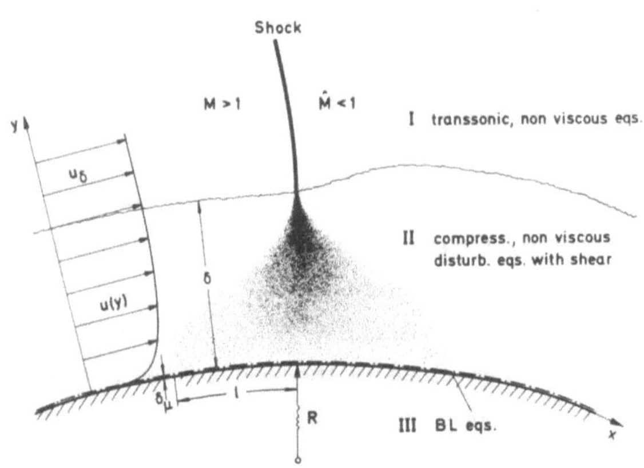

Fig. 3
Schematic flow model

Figure 3 illustrates the used model. The turbulent boundary layer at a curved wall in a transonic slightly supersonic flow (Fig. 3, left) is disturbed by a weak normal shock (Fig. 3, middle). The flow field is divided into three domains (Fig. 3, right). With regional solutions in closed form, which are coupled iteratively, we obtain explicitly the pressure and velocity distribution in the boundary layer [3] - [6].

Boundary conditions at the perforated wall

Figure 4 shows schematically the arrangement for the wall ventilation. Through a perforated wall in the touch down region of the shock a secondary flow is induced due to the pressure rise. At the downstream section of the perforation, part of the boundary layer streams into the cavity underneeth the perforation and leaves it in the region of lower pressure. In this way the pressure rise should be reduced so much that a possible shock induced flow separation is suppressed. In our model the ventilation is simulated by a certain normal velocity that is antisymmetric relativ to the shock position. This velocity distribution is used as boundary condition at the wall.

Results

Figure 5 shows a typical example of the calculated pressure distribution for different wall distances. The influence of the shock reaches over 2 - 3 boundary layer thicknesses upstream. At the edge of the boundary layer ($\tilde{y}/\delta = 1$) behind the shock the post-shock expansion is to be seen. This behaviour was first observed experimentally by Ackeret, Feldmann and Rott [7] and later found theoretically in the inviscid case by Oswatitsch and Zierep, too.

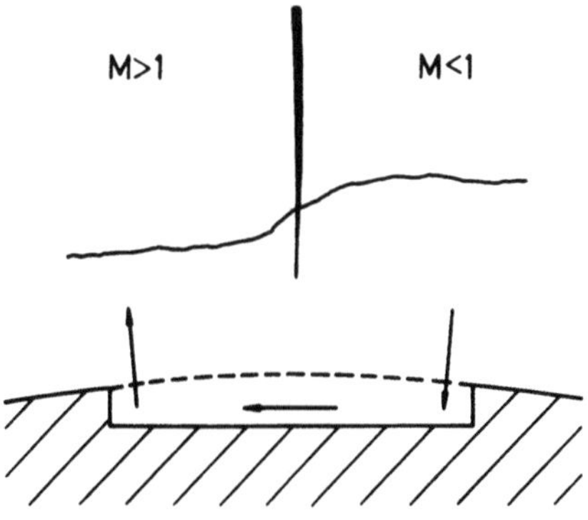

Fig. 4
Arrangement for wall ventilation

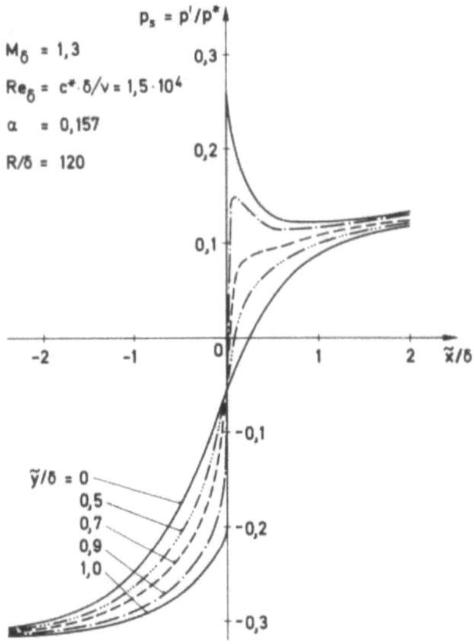

Fig. 5

Pressure distribution in the boundary layer

The structure of the flow field without and with wall ventilation is given in Figure 6 and 7a,b by lines of constant Mach number. The isolines are lifted already upstream of the shock corresponding to the pressure distribution. By blowing and suction however, as described above, this effect increases in front of the shock and decreases behind. Depending on the intensity of the ventilation a separation bubble can occur (Fig. 7a,b) which is obviously sucked off totally behind the shock. This artificial ramp

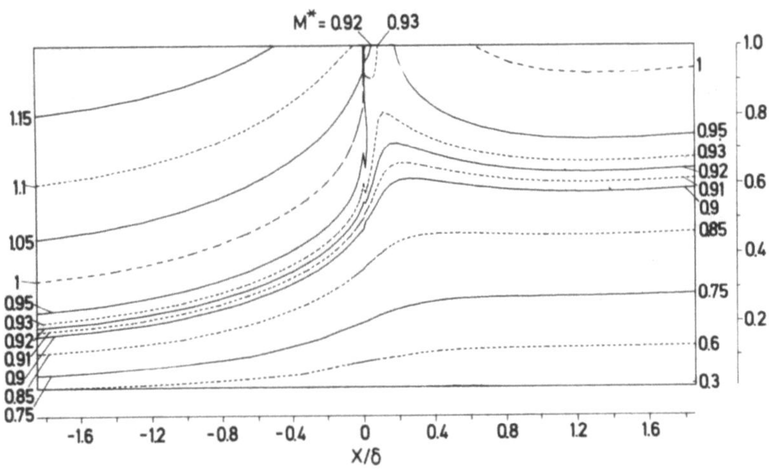

Fig. 6 Lines of constant velocity without ventilation (M_δ = 1.25, R/δ = 175, α = 1/7, $Re_\delta = 2 \cdot 10^4$)

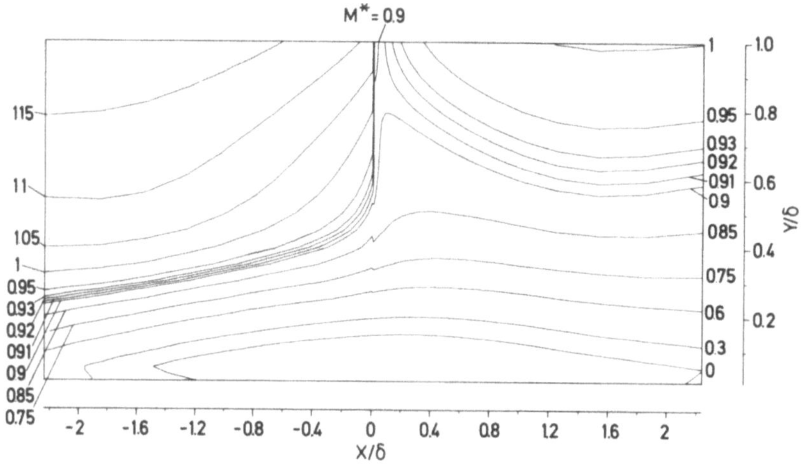

Fig. 7a Lines of constant velocity with ventilation (M_δ = 1.25, R/δ = 175, α = 1/7, $Re_\delta = 2 \cdot 10^4$)
Boundary condition at the wall:

$$V_n = - 0.05 \cdot c^* \cdot \frac{x}{\delta} \exp \frac{1}{2} (1 - x^2)$$

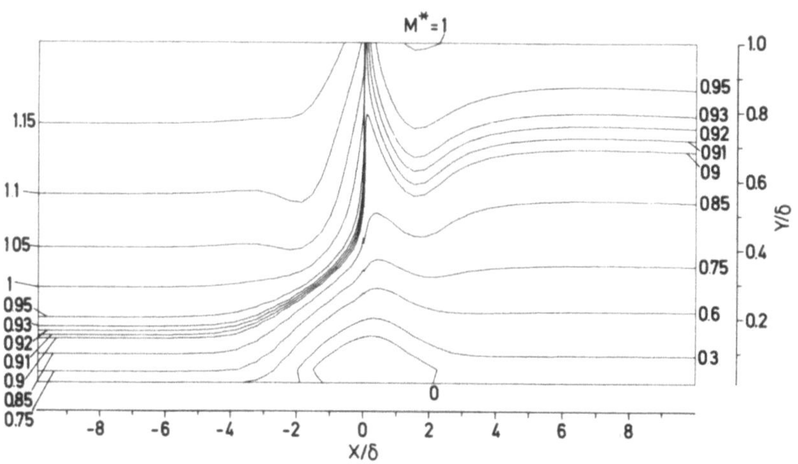

Fig. 7b as 7a, except for the scale in x-direction

may be responsible for the fact that the shock position is stabilized by ventilation. This influences conveniently the boundary layer downstream of the shock as experiments show [8].

In the solution (Fig. 6; 7a,b) only one iteration was used. Therefore the flow near the wall is well described but the influence of the ventilation on the outer flow is not yet established.

Figures 8a,b present a solution of higher order. It concerns static pressure distribution in the boundary layer; without ventilation Figure 8a and with ventilation Figure 8b. At the edge of the boundary layer (y/δ = 1) the discontinuity of the pressure distribution can be observed. It diminishes in the boundary layer due to friction. When ventilation is introduced the shock intensity is substantially reduced (Fig. 8b). In this case the static pressure distribution along the wall has a wavy shape which disappeares away from the wall.

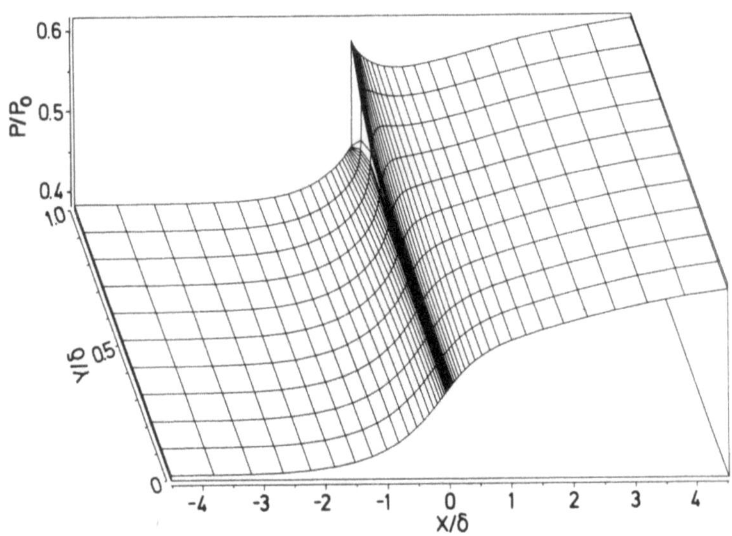

Fig. 8a Static pressure distribution in the boundary layer without ventilation

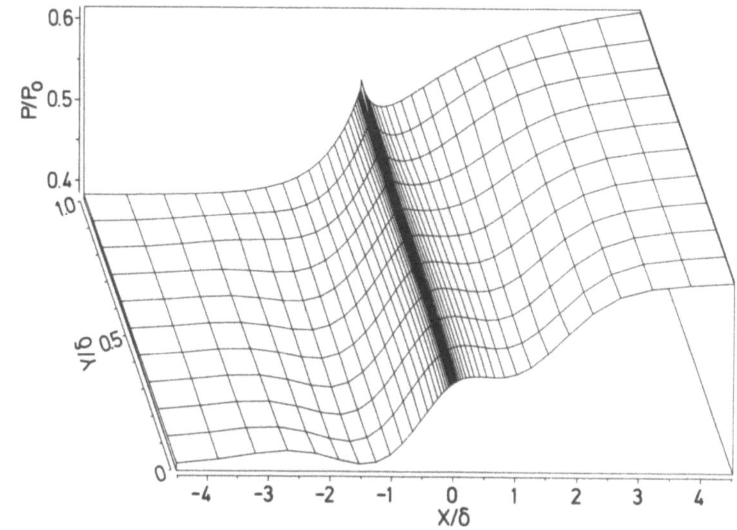

Fig. 8b Static pressure distribution in the boundary layer with ventilation (M_δ = 1.25, $R/\delta = \infty$, α = 1/7, $Re_\delta = 2 \cdot 10^4$)

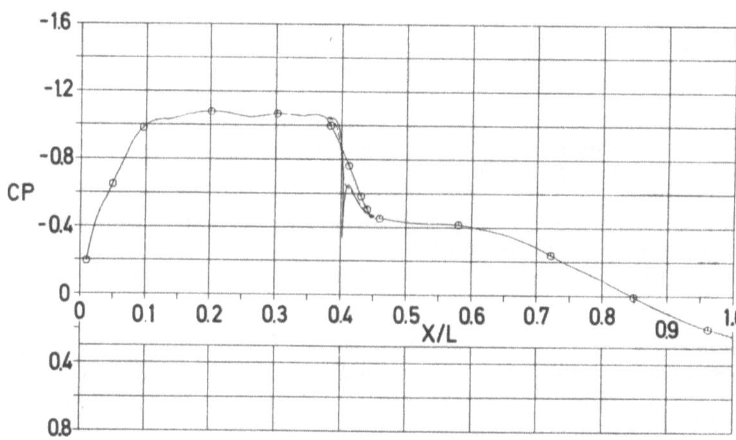

Fig. 9 Pressure distribution for the CAST7/DoA1 profil (M_δ = 1.246, R/δ = 230)

Figure 9 shows the pressure distribution (represented by the pressure coefficient c_p) for the profile CAST7/DoA1 which is calculated with an integral boundary layer procedure of Rotta [9]. In the region of shock interaction the method breaks down in general due to the rapid pressure rise. To bridge this region our local analytical solution for the shock-boundary layer interaction is embedded in the global boundary layer calculation. It is fitted again approximately four boundary layer thicknesses behind the shock to the procedure of Rotta. In Figure 9, the singular pressure distribution at the edge of the boundary layer can be seen, as well as the wall pressure, smoothed by the influence of viscosity.

Obviously the analytical interference model is suited not only for the detailed resolution of the flow structure in the interference region but also for the support of numerical global methods for calculating the flow over transonic airfoils.

References

[1] Thiede, P., Krogmann, P., Stanewsky, E.: Active and Passive Shock/Boundary Layer Interaction Control on Supercritical Airfoils. AGARD-FDP-Symposium, Brussels, May 1984, pp 34-1 - 24-13

[2] Dargel, G., Jakob, H.: Numerische Simulation viskoser transsonischer Profilumströmungen mit verbesserter Erfassung der Stoß-Grenzschicht-Interferenz. STAB-Workshop, 19. - 20.9.1984, DFVLR Köln-Porz

[3] Oswatitsch, K., Zierep, J.: Das Problem des senkrechten Stoßes an einer gekrümmten Wand. ZAMM 40, 143-144 (1960), Tagungsheft

[4] Bohning, R., Zierep, J.: Der senkrechte Verdichtungsstoß an der gekrümmten Wand unter Einfluß der Reibung. ZAMP 27, 225-240 (1976)

[5] Bohning, R., Zierep, J.: Normal Shock-Turbulent Boundary Layer Interaction at a Curved Wall. AGARD CP No. 291, Computation of Viscous-Inviscid Interactions, pp 17-1 - 17-8, 1980

[6] Bohning, R.: Die Wechselwirkung eines senkrechten Verdichtungsstoßes mit einer turbulenten Grenzschicht an einer gekrümmten Wand. Habilitationsschrift, Universität Karlsruhe, 1982

[7] Ackeret, J., Feldmann, F., Rott, N.: Untersuchungen an Verdichtungsstößen und Grenzschichten in schnell bewegten Gasen. Mitteilungen aus dem Institut für Aerodynamik, ETH Zürich, No. 10, 1946

[8] Krogmann, P., Thiede, P.: Passive Beeinflussung der Stoß-Grenzschichtinterferenz an transsonischen Tragflügelprofilen. STAB-Workshop 18. - 19. September 1984, DFVLR Köln-Porz

[9] Rotta, J.C.: Fortran IV Programm für Grenzschichten bei kompressiblen ebenen und achsensymmetrischen Strömungen. DFVLR Göttingen, Forschungsbericht 71-51 (1971)

Improvement of Transonic Airfoil Performance through Passive Shock/Boundary-Layer Interaction Control

P. THIEDE[+] and P. KROGMANN[++]

[+] Messerschmitt-Bölkow-Blohm GmbH
Unternehmensbereich Transport- und Verkehrsflugzeuge, D-2800 Bremen

[++] Deutsche Forschungs- und Versuchsanstalt für Luft- und Raumfahrt e.V.
Institut für Experimentelle Strömungsmechanik, D-3400 Göttingen

Summary

Experimental investigations are presented, aiming at the improvement of transonic airfoil performance through passive control of the shock/boundary layer interaction (SBLI) by boundary layer ventilation in the shock region.

The experiments were carried out in the DFVLR 1m x 1m Transonic Windtunnel Göttingen (TWG), using the supercritical MBB airfoil VA-2 designed to have a relatively small shock shifting with Mach number. The basic model was equipped with an exchangeable insert in the shock region to allow measurements with ventilation through slotted or perforated strips and without control. The effectiveness of the different SBLI control methods was evaluated from pressure distribution, boundary layer and wake measurements as well as Schlieren observations.

It is shown that boundary layer ventilation in the shock region mainly weakens the shock wave and stabilizes the shock position, resulting in substantial improvements of the transonic airfoil performance at off-design conditions.

Introduction

The aerodynamic performance of supercritical airfoils at transonic speeds is strongly influenced by compression shocks, occuring at off-design conditions behind supersonic flow fields and interacting with the turbulent boundary layer, Fig. 1.

Not only the off-design characteristics suffer from this phenomenon (e.g. additional wave drag and higher viscous drag), but also buffet and drag rise boundaries are a consequence of shock-induced separations.

Because of the key role of shock/boundary layer interaction (SBLI) in the transonic airfoil performance, it is suggested to look for practicable SBLI control methods on supercritical airfoils with the aim to weaken

Turbulent Shear Layer/Shock Wave Interactions
IUTAM Symposium Palaiseau 1985
Editor: J. Délery

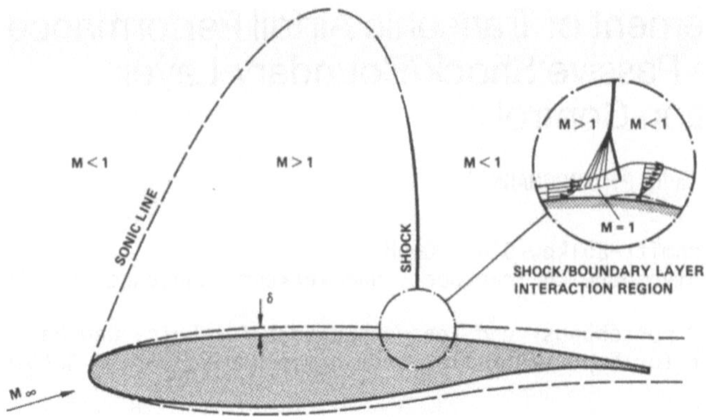

Fig. 1 Shock/boundary layer interaction at transonic airfoil flow

the shock wave and to stabilize the shock position. In order to reach this objective, methods have been investigated to control actively SBLI on supercritical airfoils by tangential blowing (Ref. 1) or boundary layer suction in the shock region (Refs. 2, 3). However, the easiest way SBLI control can be accomplished is passively by boundary layer ventilation in the shock region through a permeable surface with a cavity located underneath (passive SBLI control, Refs. 4 to 6).

Passive control mechanism

The passive effect is reached by a partial pressure compensation between the sub- and supersonic flow fields through a cavity underneath the shock, Fig. 2.

At the top, the well-known shock structures without control are shown: a) a weak shock configuration with an isentropic recompression and a rear normal shock near the wall at low Mach numbers in front of the shock $M_\delta \lesssim 1.3$ and b) a strong λ-shock configuration consisting of an oblique front shock and a normal rear shock with a shock-induced separation at higher Mach numbers $M_\delta \gtrsim 1.3$. The λ- shock configuration leads to a strong drag rise and buffet onset.

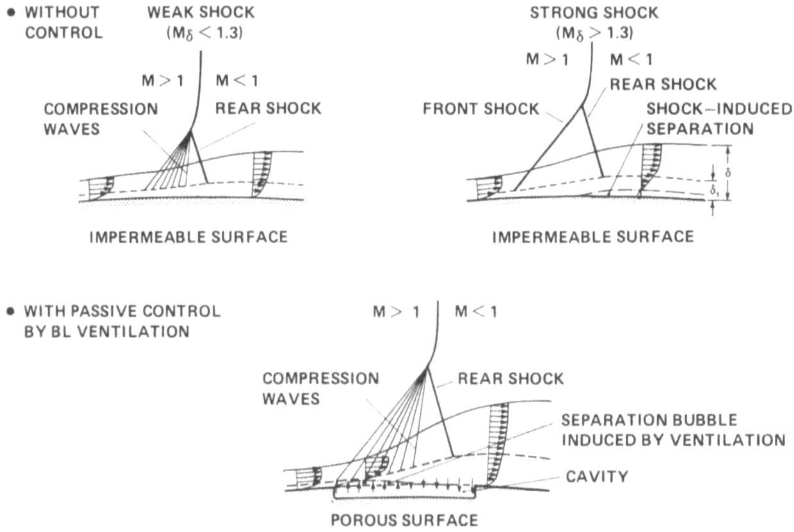

Fig. 2 Passive SBLI control mechanism

However, with a porous surface in the shock region and a cavity underneath as shown at the bottom, a secondary flow develops due to the pressure rise in the shock region, leading to blowing upstream and sucking downstream of the shock. As a consequence of the shock-induced ventilation, the boundary layer thickening in the shock region is increased, forming a steeper compression ramp for the shock wave and changing the flow structure near the surface to a weak shock with an isentropic recompression. This favourable passive effect is increased by the strong influence of blowing on the turbulence structure.

The ventilation process strongly intensifies the capability of the boundary layer to weaken the shock wave and to stabilize the shock position. As a result of passive control, wave and shock-induced viscous drag are drastically reduced and the onset of shock-induced flow separation is delayed, see also Refs. 7 and 8.

Experimental investigations

Experimental investigations were conducted to study the effects of passive SBLI control by boundary layer ventilation in the shock region on the flow characteristics of a modern supercritical airfoil with the aim to improve its off-design performance.

The experiments were carried out in the DFVLR 1m x 1m Transonic Windtunnel Göttingen (TWG), using the two-dimensional supercritical MBB airfoil VA-2 (Ref. 9). This airfoil was chosen because of its relatively small shock shifting with Mach number at off-design conditions (Ref. 10).

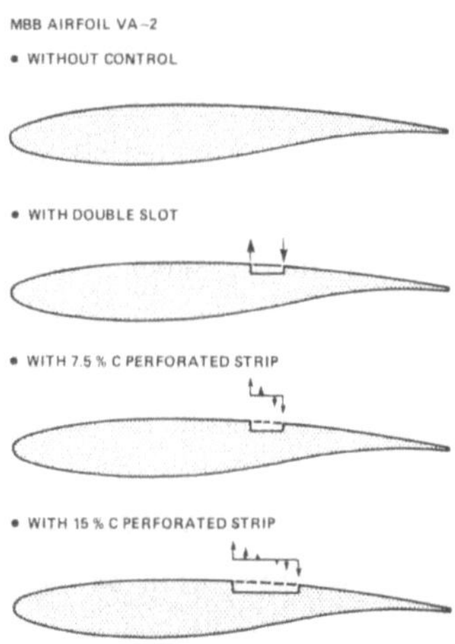

Fig. 3 **Model configurations tested with passive SBLI control**

The model was equipped with an exchangeable insert in the shock region to allow measurements with ventilation through slotted or perforated surface strips and without control. Fig. 3 shows the model configurations tested with passive SBLI control. A double slot insert (slot width s = 0.6 mm, slot distance d = 7 %c) and two perforated strip inserts with same opening (8.16 %) but different width (s = 7,5 %c and 15 %c) were investigated.

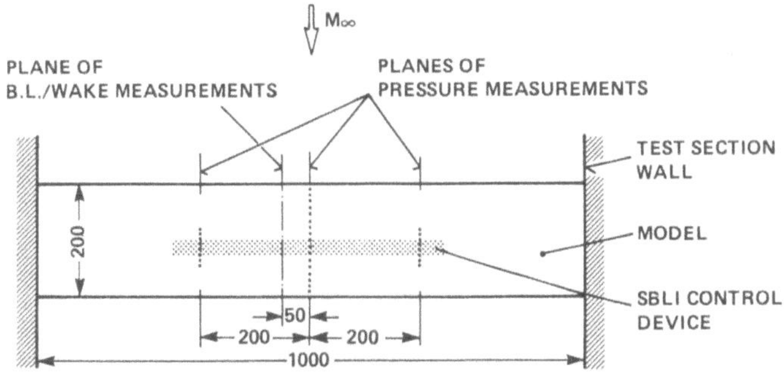

Fig. 4 Test set-up for pressure/b.l./wake measurements and Schlieren observations

The TWG of the DFVLR described in Ref. 11 is a pressurized closed-circuit tunnel, allowing independent Reynolds and Mach number variations. Fig. 4 shows the general arrangement of the 2-D model (chord length c = 200 mm, span b = 1 m) in the test section. The experimental investigations include surface pressure distribution, boundary layer and wake measurements as well as Schlieren observations.

Experimental results

Before discussing the effects of passive SBLI control on the airfoil performance, the influence of boundary layer ventilation in the shock region on the complex transonic airfoil flow development will be analysed from typical pressure distribution measurements, concentrating on 15 %c perforated strip results.

Fig. 5 shows the passive effect of the perforated strip on the pressure distribution of the VA-2 airfoil. At small incidences, the measured pressure distributions differ only in the shock region, where the pressure rise is definitely flattened by ventilation. But at higher incidences, the upper surface pressure distributions are quite different in both cases: without control the data show the buffet boundary clearly exceeded with the shock located far upstream and a shock-induced separation reaching downstream to the trailing edge, while in the ventilation case buffeting did not occur, showing a pressure distribution with the shock stabilized immediately upstream of the perforated strip and the flow attached

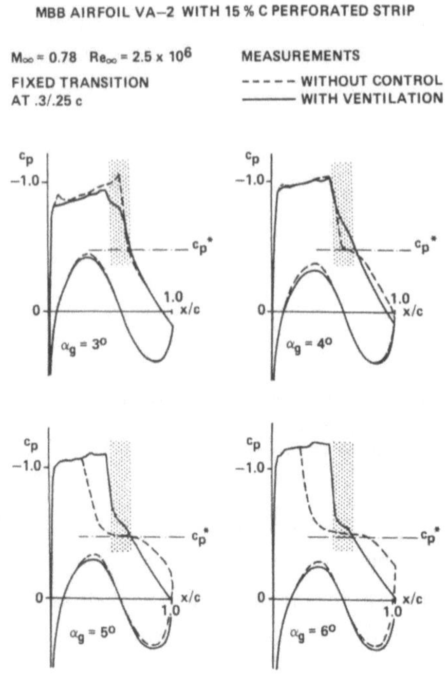

Fig. 5 Pressure distribution with/without passive SBLI control

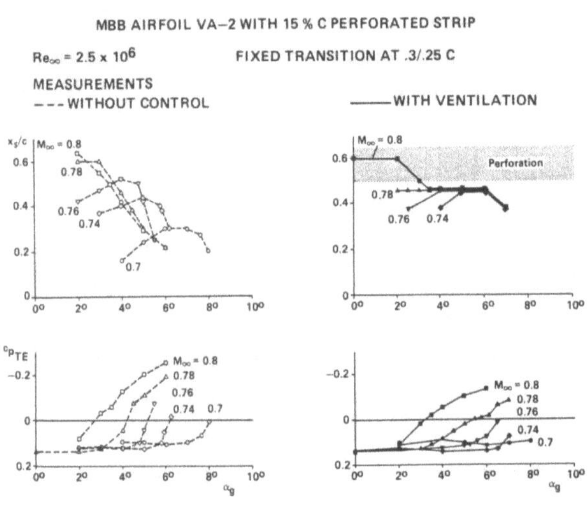

Fig. 6 Shock position and trailing edge pressure with/without passive SBLI control

downstream. These ventilation effects were also indicated by Schlieren observations not presented here. Boundary layer data, showing the increased boundary layer thickening in the shock region in the ventilation case, are not yet available for the 15 %c perforated strip configuration but published in Refs. 12 to 14 for the double slot and 7,5 %c perforated strip configurations.

Fig. 6 summarizes the influence of the 15 %c perforated strip on shock position and trailing edge pressure as a function of Mach number and incidence. It is obvious, that passive SBLI control by boundary layer ventilation in the shock region mainly causes the shock structure to change near the wall, leading to a weaker shock wave and a more stable shock position up to higher incidences or/and Mach numbers.

As a result of passive SBLI control, wave and shock-induced viscous drag are drastically reduced and the onset of shock-induced separation is delayed, leading to considerable lift increases at high incidences as well as to drastic off-design drag reductions.

In Fig. 7 lift polars of the model equipped with the double slot and both the perforated strip inserts are compared with the basic model results without ventilation. For the double slot and the wide perforated strip configurations a favourable passive effect was obtained at all Mach numbers tested showing no distinct lift maxima. On the contrary, the lift maxima of the small perforated strip configuration are increased only at high Mach numbers due to the shock shifting with Mach number.

Fig. 8 demonstrates, that passive SBLI control offers a large potential for off-design drag reductions. Even at low incidences, the drag coefficents are slightly decreased in comparison with the basic model. However, at higher incidences with strong shocks drag reductions up to 45 % are obtained for all ventilated models.

The large influence of passive SBLI control on the lift/drag polars of the investigated airfoil, using both the double slot and the 15 %c perforated strip inserts, is shown in Fig. 9. It is remarkable, that not only the off-design performance with strong shocks but also the L/D optima

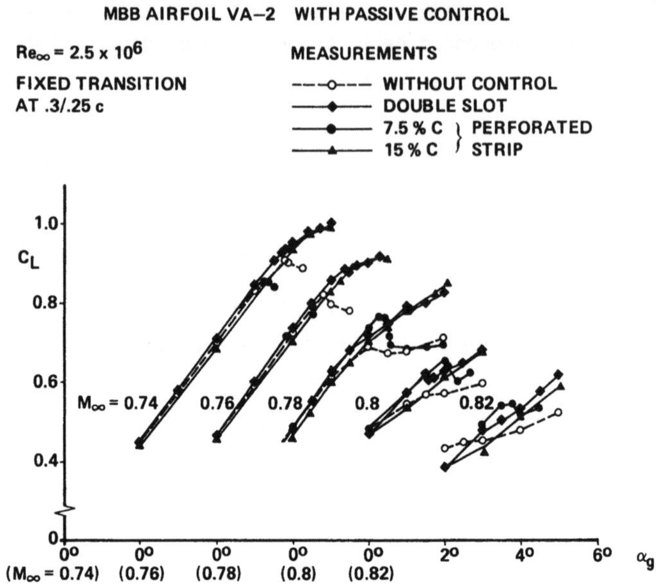

Fig. 7 Lift increases by passive SBLI control

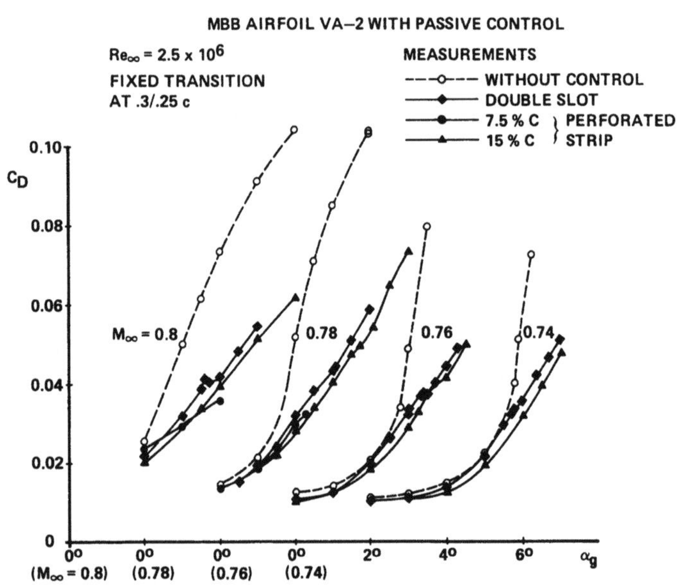

Fig. 8 Drag reductions by passive SBLI control

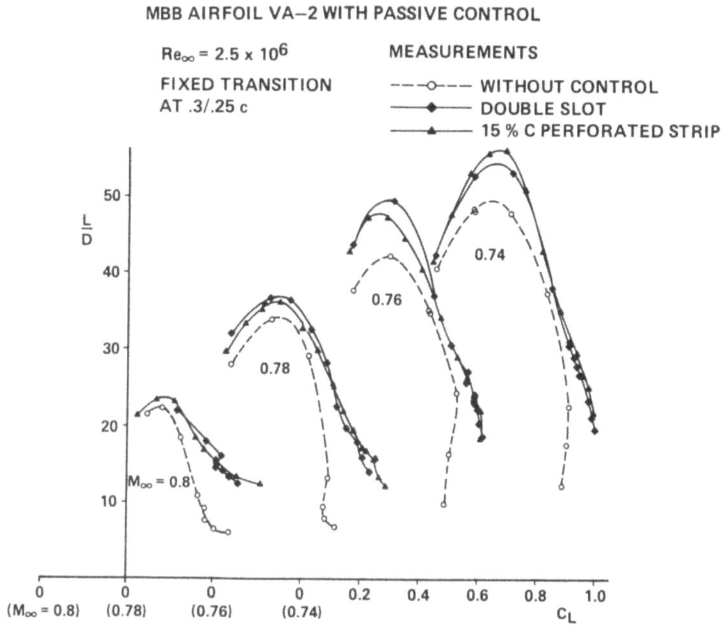

Fig. 9 Lift/drag improvements by passive SBLI control

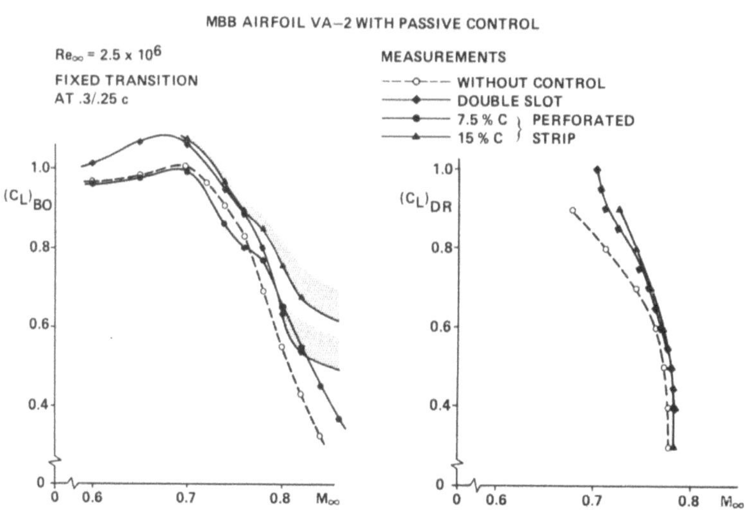

Fig. 10 Extension of buffet and drag rise boundaries by passive SBLI control

could be improved in the entire Mach number range tested. On the contrary, the smaller perforated strip was not effective in the whole Mach number range due to the shock shifting with Mach number.

Besides the gain in lift and drag, a favourable passive effect on the buffet and drag rise boundaries was obtained by boundary layer ventilation in the shock region, Fig. 10. However, the most striking result of the ventilation experiments was, that beyond the buffet boundary neither shock oscillations could be detected nor severe buffeting was indicated. Moreover, in some high Mach number cases buffeting was completely ceased, although the flow was separated downstream of the shock. It seems that the strong coupling between shock and trailing edge flow behaviour is largely suppressed by boundary layer ventilation.

Conclusions

The present investigations have generally confirmed that the transonic airfoil performance can be considerably improved through passive SBLI control.

The results show that passive control by boundary layer ventilation in the shock region mainly weakens the shock wave and stabilizes the shock position, leading to lift increases at high incidence and drastic off-design drag reductions as well as to an extension of the buffet and drag rise boundaries. Beyond the buffet boundary, passive control prevents severe buffeting and can suppress buffeting completely.

So far the understanding of the passive SBLI control mechanism is by no means complete. The present measurements did not allow a detailed investigation of the flow field structure in the shock region. More detailed investigations are necessary, using sophisticated measurering techniques, to gain a deeper insight into this complex flow phenomenon.

References

1. K. Finke: Stoßschwingungen in schallnahen Strömungen. VDI-Forschungsheft 580 (1977).

2. P. Thiede: Supercritical airfoil flow control by slot suction in the shock region. Proc. of the 5th USAF/FRG DEA-Meeting on "Viscous and Interacting Flow Field Effects", Annapolis, Md., AFFDL-TR-80-3088 (1980).

3. E. Stanewsky, P. Krogmann, P. Thiede, G. Dargel: Stoß-Grenzschicht-Interferenz und ihre Beeinflussung durch Grenzschichtabsaugung. 3. DGLR-Symposium "Strömungen mit Ablösung", Stuttgart (1981).

4. P. Krogmann, P. Thiede: Aktive und passive Beeinflussung der Stoß-Grenzschicht-Interferenz an überkritischen Tragflügeln. 4. DGLR-Symposium "Strömungen mit Ablösung", Göttingen (1983).

5. L. Bahi, J.M. Ross, H.T. Nagamatsu: Passive shock wave/boundary layer control for transonic airfoil drag reduction. AIAA 21st Aerospace Science Meeting, Reno, Nev., Paper 83-0137 (1983).

6. G. Savu, O. Trifu, L.Z. Dumitrescu: Suppression of shocks on transonic airfoils. Proc. of the 14th International Symposium on "Shock Tubes and Shock Waves", Sydney (1983).

7. R. Bohning, J. Zierep: Calculation of 2-D turbulent shock/boundary layer interaction at curved surfaces with suction and blowing. IUTAM-Symposium on "Turbulent Shear Layer/Shock Wave Interactions", Palaiseau, France (1985).

8. A. Bertelrud: Passive shock modification in flight. EUROMECH 181, Saltsjöbaden, Sweden (1984).

9. K. D. Klevenhusen, R. Hilbig, W. Burgsmüller: Entwicklung von superkritischen Profilen und Voruntersuchungen zum superkritischen Flügelentwurf, Teil 1. BMFT-Bericht 523-8891-LFF 34, VFW-Fokker, Bremen (1976).

10. R.D. Boehe: Transsonikmessungen am VFW-F-Profil VA-2 bei der DFVLR Göttingen. ZKP-Flügelsektion-Bericht 8, Bremen (1976).

11. T. Hottner, W. Lorenz-Meyer: Der Transsonische Windkanal der Aerodynamischen Versuchsanstalt Göttingen. DGLR-Jahrbuch (1968).

12. P. Krogmann, E. Stanewsky, P. Thiede: Effects of local boundary layer suction on shock/boundary layer interaction and shock-induced separation. AIAA 22th Aerospace Science Meeting, Reno, Nev., Paper 84-0098 (1984).

13. P. Thiede, P. Krogmann, E. Stanewsky: Active and passive shock/boundary layer interaction control on supercritical airfoils. AGARD-CP-365, Brussels (1984).

14. P. Krogmann, E. Stanewsky, P. Thiede: Transonic shock/boundary layer interaction control. ICAS-Paper 84-2.3.2, Toulouse (1984).

Interactions in Two-Dimensional Steady Flows

Physical Analysis of Turbulent Boundary-Layer/ Shock-Wave Interactions Using Second Order Closure Predictions

D. Vandromme(*) and H. Ha Minh(**)

Université des Sciences et Techniques de Lille, U.E.R. de Mathématiques, Service de Mécanique,
59655 Villeneuve d'Asq Cédex (France)*

Institut de Mécanique des Fluides, Laboratoire associé au C.N.R.S. LA 0005
Avenue du Professeur Camille Soula, 31400 Toulouse (France)**

Summary

The present contribution describes the Reynolds stress behaviour through a supersonic compression cornerflow. After a comparison between increasing complexity turbulence models (algebraîc-two equation and second order) for mean flow quantities in the 16° angle case, a more detailed analysis of the Reynolds stress anisotropy is made for the 20° angle flow.

1. INTRODUCTION

Considerable progress has been made in recent years in the development of computationally efficient numerical methods for solving non-linear partial differential equations. Together with improved high speed large memory electronic computers, these developments have allowed solutions of very complex fluid mechanical problems such as turbulent boundary-layer / shock wave interactions.

But when it is time to choose an adequate turbulence model for complex turbulent flow computations, it is often necessary to compromise between performance and cost. Accordingly, in most of complex flow calculations, eddy-viscosity concept models can often be preferred to second order closure models for simplicity reasons. In most turbulent shear flows, eddy-viscosity turbulence models (for example, the popular two-equation k, ε model) are generally able to yield results as good as those given by second order closure models. More surprisingly in some circumstances, for example, for equilibrium boundary layers, mixing length theory seems to give more satisfactory results than the second order closure models.

In fact, numerical considerations enforce the previous choice, not only because much computing time may be saved by the use of zero-equation or two-equation models, but also the main reason of this success is more numerical. When eddy-viscosity concept models are used, the local Peclet number (convective effects / diffusive effects) is actually small or moderate. Therefore numerical schemes performed for low or moderate Reynolds number flows are

Turbulent Shear Layer/Shock Wave Interactions
IUTAM Symposium Palaiseau 1985
Editor: J. Délery

well adapted for the calculation of turbulent flows using eddy-viscosity models although the nominal Reynolds numbers of these flows are generally very high. When second order closure models are used, this convenient arrangement is no longer available : the main momentum fluxes (the Reynolds stresses) appear as pressure terms and not as gradient fluxes like in viscous laminar flows. Accordingly, the Reynolds equation character is more hyperbolic like the Euler equation one than elliptic like the classical Navier-Stokes viscous equation. This remark emphasizes the necessity of performing numerical methods more adapted to very high Reynolds number flows like those used for Euler equations and then more adapted to second order closure models. Following a precedent published work [1] devoted to the numerical study of the 12° supersonic expansion [2], the present investigation deals with the Reynolds stress behaviour through a supersonic compression corner flow. After a comparison between increasing complexity turbulence models (algebraîc, two-equation and second order) for mean flow quantities in the 16° angle case, a more detailed analysis of the Reynolds stress anisotropy is made for the 20° angle flow.

2. THE MATHEMATICAL MODEL

Using mass-weighted averaging, neglecting buoyancy effects and all viscosity and thermal conductivity fluctuations, the open equations for mass, momentum and energy conservation are :

$$\frac{\partial \bar{\rho}}{\partial t} + \frac{\partial \bar{\rho}\widetilde{U}_k}{\partial x_k} = 0 \tag{1}$$

$$\frac{\partial \bar{\rho}\widetilde{U}_i}{\partial t} + \frac{\partial}{\partial x_k}\left(\bar{\rho}\widetilde{U}_k\widetilde{U}_i + \bar{\rho}\widetilde{u_i u_k} - \bar{\sigma}_{ik}\right) = 0 \tag{2}$$

$$\begin{aligned}\frac{\partial \bar{\rho}\widetilde{E}}{\partial t} + \frac{\partial}{\partial x_k}\Bigg[&\left(\bar{\rho}\widetilde{E} + \bar{p}\right)\widetilde{U}_k + \bar{\rho}\widetilde{u_i u_k}\widetilde{U}_i + \bar{\rho}\frac{\widetilde{u_i u_i u_k}}{2} \\ &+ \gamma C_v \bar{\rho}\widetilde{\theta u_k} - \lambda\frac{\partial \widetilde{T}}{\partial x_k} - \mu\left(\widetilde{S_{ik}}\widetilde{U}_i + \overline{s_{ik}}\widetilde{U}_i + \overline{s_{ik}u_i} + \widetilde{S_{ik}}\overline{u_i}\right)\Bigg] = 0\end{aligned} \tag{3}$$

where small symbol like u_k indicates fluctuating part in Favrian sense and where :

$$p = (\gamma - 1)\,\rho e \qquad \overline{\sigma}_{ik} = -\overline{p}\delta_{ik} + \overline{\mu S_{ik}}$$

and

$$S_{ik} = \left(\frac{\partial U_i}{\partial x_k} + \frac{\partial U_k}{\partial x_i}\right) - \frac{2}{3}\delta_{ik}\frac{\partial U_l}{\partial x_l}$$

It is not worth reminding here different turbulence model levels already given in ref. [3]. Only the second order closure model is related here. The Navier-Stokes equations for compressible turbulent flows may be written in a vector form in cartesian coordinates as (for 2-D problems) :

$$\frac{\partial U}{\partial t} + \frac{\partial F}{\partial x} + \frac{\partial G}{\partial y} = H \tag{4}$$

where U, F and G represent respectively the mean flow variables and the corresponding flux vectors in the streamwise and normal directions. U vector can be written as :

$$U = \left(\bar{\rho}\ ,\ \bar{\rho}\widetilde{U}\ ,\ \bar{\rho}\widetilde{V}\ ,\ \bar{\rho}\widetilde{E}\ ,\ \bar{\rho}\widetilde{u^2},\ \bar{\rho}\widetilde{v^2},\ \bar{\rho}\widetilde{w^2},\ \bar{\rho}\widetilde{uv}\ ,\ \bar{\rho}\widetilde{\varepsilon}\right)^T \tag{5}$$

in which $\bar{\rho}$, $\widetilde{U}$, $\widetilde{V}$, $\widetilde{E}$, $\widetilde{u^2}$, $\widetilde{v^2}$, $\widetilde{w^2}$, $\widetilde{uv}$ and $\widetilde{\varepsilon}$ stand respectively for the mean density, the two mean velocity components, the total energy per unit mass, the Reynolds normal stresses, the turbulent shear stress and the turbulent dissipation rate. The vector H contains all the source terms (non-divergent terms).

The Reynolds stresses can be obtained from transport equation :

$$\begin{aligned}
\frac{\partial}{\partial t}\left(\bar{\rho}\widetilde{u_iu_j}\right) &+ \frac{\partial}{\partial x_k}\left(\bar{\rho}\widetilde{u_iu_j}\widetilde{U}_k - \mu\frac{\partial}{\partial x_k}\widetilde{u_iu_j} - C_s'\frac{\widetilde{k}}{\widetilde{\epsilon}}\widetilde{u_ku_l}\frac{\partial}{\partial x_l}\widetilde{u_iu_j}\right) \\
&= P_{ij} - \frac{C_2+8}{11}\left(P_{ij} - \frac{2}{3}\delta_{ij}P\right) - \frac{8C_2-2}{11}\left(D_{ij} - \frac{2}{3}\delta_{ij}D\right) \\
&- \frac{30C_2-2}{55}\bar{\rho}\widetilde{k}\widetilde{S_{ij}} - C_1\bar{\rho}f_1\frac{\widetilde{\epsilon}}{\widetilde{k}}\left(\widetilde{u_iu_j} - \frac{2}{3}\delta_{ij}\widetilde{k}\right) \\
&+ \left(C_3\bar{\rho}\frac{\widetilde{\epsilon}}{\widetilde{k}}\left(\widetilde{u_iu_j} - \frac{2}{3}\delta_{ij}\widetilde{k}\right) + C_4\left(P_{ij} - D_{ij}\right) + C_5\bar{\rho}\widetilde{k}\widetilde{S_{ij}}\right)\frac{\widetilde{k}^{\frac{3}{2}}}{\widetilde{\epsilon}x_m} \\
&- \bar{\rho}\frac{\widetilde{\epsilon}}{\widetilde{k}}\left(\widetilde{u_iu_j}f_s + (1-f_s)\frac{2}{3}\delta_{ij}\widetilde{k}\right) - \overline{u_i}\frac{\partial\bar{p}}{\partial x_j} - \overline{u_j}\frac{\partial\bar{p}}{\partial x_i}
\end{aligned} \tag{6}$$

$$P = D = -\bar{\rho}\widetilde{u_iu_j}\frac{\partial\widetilde{U_i}}{\partial x_j} \quad ; \quad \overline{u_i} = \frac{\widetilde{U_k}\widetilde{u_iu_k}}{(n-1)C_p\widetilde{T}}$$

where

$$P_{ij} = -\bar{\rho}\left(\widetilde{u_iu_k}\frac{\partial\widetilde{U_j}}{\partial x_k} + \widetilde{u_ju_k}\frac{\partial\widetilde{U_i}}{\partial x_k}\right) \quad ; \quad D_{ij} = -\bar{\rho}\left(\widetilde{u_iu_k}\frac{\partial\widetilde{U_k}}{\partial x_j} + \widetilde{u_ju_k}\frac{\partial\widetilde{U_k}}{\partial x_i}\right)$$

The dissipation rate is calculated from equation :

$$\frac{\partial}{\partial t}(\bar{\rho}\tilde{\epsilon}) + \frac{\partial}{\partial x_k}\left(\bar{\rho}\tilde{\epsilon}\widetilde{U}_k - (\bar{\mu}\delta_{kl} + C_\epsilon\bar{\rho}\frac{\widetilde{k}}{\widetilde{\epsilon}}\widetilde{u_k u_l})\frac{\partial \epsilon}{\partial x_l}\right)$$
$$= -C_{\epsilon 1}\bar{\rho}\frac{\widetilde{\epsilon}}{\widetilde{k}}\widetilde{u_i u_k}\frac{\partial \widetilde{U}_i}{\partial x_k} - C_{\epsilon 2}\bar{\rho}\frac{\widetilde{\epsilon}}{\widetilde{k}}f_\epsilon\left(\widetilde{\epsilon} - 2\nu\left(\frac{\partial\sqrt{\widetilde{k}}}{\partial x_m}\right)^2\right)$$
$$- C_{\epsilon 3}\bar{\mu}\frac{\widetilde{k}}{\widetilde{\epsilon}}\widetilde{u_j u_k}\left(\frac{\partial^2 \widetilde{U}_i}{\partial x_j \partial x_l}\right)\left(\frac{\partial^2 \widetilde{U}_i}{\partial x_k \partial x_l}\right) - C_{\epsilon 4}\frac{\widetilde{\epsilon}}{\widetilde{k}}\overline{u_i}\frac{\partial \bar{p}}{\partial x_i} \tag{7}$$

and the total energy equation is modelled as :

$$\frac{\partial \bar{\rho}\widetilde{E}}{\partial t} + \frac{\partial}{\partial x_k}\left[\left(\bar{\rho}\widetilde{E} + \bar{p} + \frac{2}{3}\mu\frac{\partial \widetilde{U}_l}{\partial x_l}\right)\widetilde{U}_k + \left(\bar{\rho}\widetilde{u_i u_k} - \mu\frac{\partial \widetilde{U}_i}{\partial x_k}\right)\widetilde{U}_i\right.$$
$$- \mu\frac{\partial}{\partial x_k}\left(\frac{\widetilde{U}_i\widetilde{U}_i}{2} + \frac{\widetilde{u_i u_i}}{2}\right) - 0.5C'_s\frac{\widetilde{k}}{\widetilde{\epsilon}}\widetilde{u_k u_l}\frac{\partial}{\partial x_l}\widetilde{u_i u_i}$$
$$\left. - \left(C_{se}C_v\gamma\bar{\rho}\frac{\widetilde{k}}{\widetilde{\epsilon}}\widetilde{u_k u_l} + \delta_{lk}\lambda\right)\frac{\partial \widetilde{T}}{\partial x_l}\right] = 0 \tag{8}$$

The equation (4) represents a strongly non-linear system of non-conservative second order partial differential equations.

Note, when equation (4) is written for non orthogonal meshes, $\tilde{U}$, $\tilde{V}$ and their fluctuations remain in the x and y directions respectively. This avoids having the enormous complexity of transforming the turbulence field equations into a generalized coordinate system.

The numerical method used in this study is the explicit-implicit scheme of Mac CORMACK [4]. This method represents an implicit extension of an explicit method also developed by Mac CORMACK [5]. This method contains two stages. In the first stage, Mac CORMACK's explicit predictor-corrector scheme is used. Here, the finite difference equations approximate the governing fluid flow equations to second order accuracy in space and time but are subject to restrictive explicit stability criteria. The second stage removes these stability conditions by numerically creating an implicit form of the finite difference equations. This method is unconditionally stable and can be made second order accurate for most flows.

3. APPLICATION TO THE SUPERSONIC COMPRESSION FLOW

3.1. Flow description

An equilibrium supersonic boundary layer is submitted to an abrupt compression (16 and 20 degrees)(Fig. 1). These flows have been investigated by Princeton Group [6], [7]. The free stream Mach numbers, prior to the interaction region are respectively equal to 2.98 and 2.90. As experimentally the incoming boundary layer was developed on a rather long distance, it was necessary to generate numerically the input profiles from a boundary layer calculation, using the same second order turbulence model as for the Navier-Stokes codes [8]. The use of this preliminary calculation allows to supply the Navier-Stokes flow field with realistic Reynolds stress anisotropy in the wall vicinity. As the incoming flow field is supersonic almost everywhere, the inlet boundary conditions can easily be handled by using frozen quantities.

With these significant flow deflection angles(16° and 20°), the interaction zone is characterized by a rather complex structure. The recirculation zone is preceded by a separation oblique shock, upstream the apex. The reattachment is related to a second weak shock. Both shock waves coalesce the interaction to produce the classical inviscid shock wave of the supersonic compression. It is worth noticing here that experimental observations yield some reserves about the flow stability. In fact, it is conjectued that the unsteady effects are significant in the interaction zone.

3.2. The 16 degree compression flow : comparison between different turbulence models :

Calculations were performed using the zero equation (mixing length) model of CEBECI, SMITH and MOSINSKI [3], the two equation model of JONES and LAUNDER [3] and a second order closure (Reynolds stress equation) model. The influence of the positive pressure gradient on the skin friction distribution is given on the Figure 2 where the results are obtained from the three previous turbulence model. It is shown that there is a recirculating zone in the 16 degree compression corner flow, experimental results being obtained from MUCK et al [6]. As well prior to the first shock as behind the reattachment shock, the second order closure model seems to be better than the zero equation model and the two equation one, these two last models being not able to recover the right level of skin friction in the relaxation zone after separation.

3.3. The 20 degree compression flow : behaviour of the Reynolds stress anisotropy

Fig. 3 (a, b, c and d) present the evolution of $\widetilde{u^2}$, $\widetilde{v^2}$, $\widetilde{w^2}$ and $\widetilde{uv}$ along the pressure gradient zone at different position inside the wall layer. The stresses are followed along four given streamlines identified with their initial y+ values (respectively 10, 50, 300 and 2,000). Obviously, along the pressure gradient zone, the y^+ does not remain constant. Note in the figures the Reynolds stresses are in cartesian axes aligned with the free stream. Figures 3 show that the compression effects on the Reynods stress productions are more important in the outer part than in the wall zone. These results display a turbulent stress behaviour completely different from the evolution of the same quantities in the supersonic expansion flow [1]. To emphasize the relative importance of each Reynolds stress component with respect to the others, the corresponding terms of the anisotropy tensor $b_{ij} = \widetilde{u_i u_j}/\tilde{k} - 2/3$, are plotted for the same streamlines. Here, the values of the anisotropy tensor are sensitive to the direction to which the Reynolds stresses are referred. Figures 4 show that the anisotropy levels increase about 60 % through the first compression shock. The results are not as clear for the second shock due to the presence of some expansion waves and the beginning of rotation effects. In spite of the rotation effects which minimize the anisotropy character after the corner, this anisotropy increases plainly through the whole compression corner shock. What is seen in Figures 4 is the rate at which the flow goes towards anisotropy. This is clearly dependent on the distance from the surface and is likely dependent on the extent of the shock system.

4. CONCLUDING REMARKS

The major difference between results obtained with the eddy-viscosity concept models and the Reynolds stress model is the information about the way the turbulent energy is distributed among its normal components. If the large changes in the anisotropy of the turbulence seems to have minor effect on the mean quantities in some flows (an expansion corner [9], for example), the information regarding turbulence anisotropy, however, may be useful in itself for other complex flows, the transonic flow over a bump, for example [9] where the successive pressure gradients in shock should be correctly described only by the use of a second order closure model.

V. ACKNOWLEDGMENTS

The authors are indebted for the continuous collaboration in this work of

Dr. M.W. RUBESIN and Dr. J.R. VIEGAS of NASA Ames Research Center. Part of the computing time has been provided by the Centre de Calcul Vectoriel pour la Recherche. The authors are also grateful to Mrs. NGUYEN-TRISTANI for typing the manuscript so efficiently.

6. REFERENCES

1. Ha Minh, H. ; Kollmann, W. and Vandromme, D.D. : The behaviour of turbulence anisotropy through shock waves and expansions. 5th T.S.F. Symp. Cornell Ithaca, New-York (1985).

2. Dussauge, J.P. : Evolution de transferts turbulents dans une détente rapide en écoulements supersoniques. Thesis for Doctor of Physical Sciences, University of Marseille II, France (1981).

3. Ha Minh, H. and Vandromme, D.D. : Modelling of compressible turbulent flows : present possibilities and perspectives. I.U.T.A.M. Symp. on Turbulent shear layer / shock-wave interactions, Paris, sept. (1985).

4. Mac Cormack, R.W. : Numerical solution of the interaction of a shock wave with a laminar boundary layer. Lecture notes in Physics, Springer Verlag, 8, 151 (1971).

5. Mac Cormack, R.W. : A numerical method for solving the equations of compressible viscous flow. A.I.A.A., 81-0110, A.I.A.A. 19th Aerospace Sciences Meeting, St-Louis, Missouri, Jan. 12-15 (1981).

6. Muck, K.C ; Hayakawa, K. and Smits, A.E. : Compilation of turbulence date for a 16 degree compression corner at Mach 2.9. Princeton Gas Dynamics Lab. MAE Report 1619 (1983).

7. Muck, K.C. ; Hayakawa, K. and Smits, A.E. : Compilation of turbulence data for a 20° degree compression corner at Mach 2.9. Report MAE-1620, Princeton Univ. (1983).

8. Vandromme, D.D. : Contribution à la modélisation et à la prédiction d'écoulements turbulents à masse volumique variable. Thèse de Docteur ès Sciences, Univ. des Sciences et Techniques de Lille, (France) (1983).

9. Ha Minh, H. ; Rubesin, M.W. ; Vandromme, D.D. and Viegas, J.R. : On the use of second order closure modelling for the prediction of turbulent boundary layer / shock-wave interactions : Physical and numerical aspects. I.S.C.F.D., Tokyo, Sept. (1985).

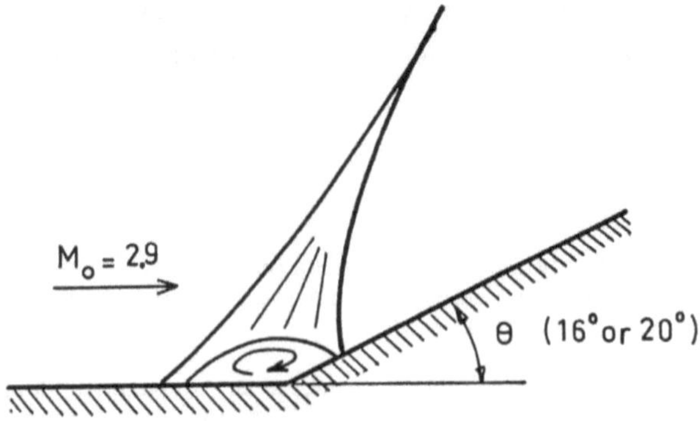

Figure 1 : Compression corner pattern

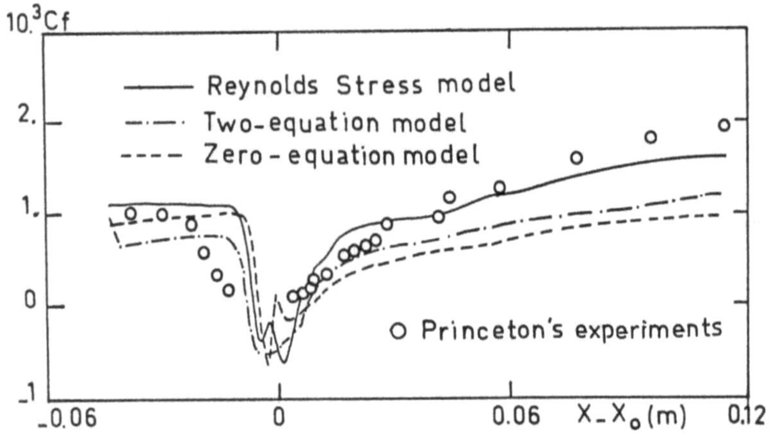

Figure 2 : Skin friction distribution (θ = 16°)

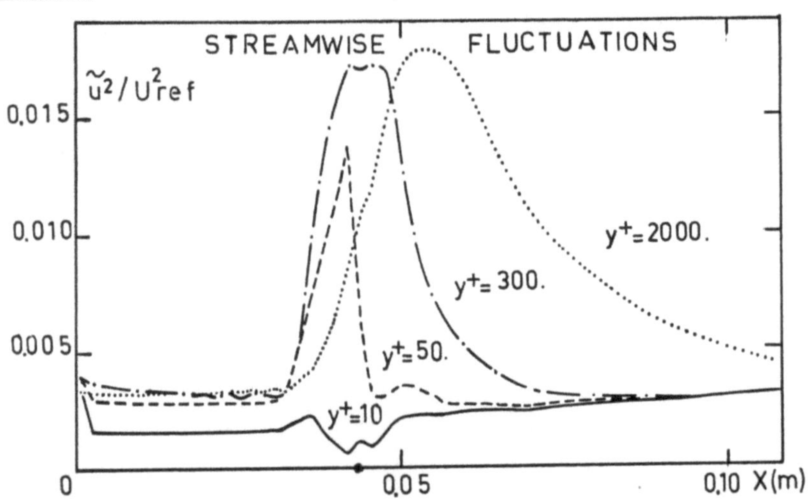

Figure 3a : Evolution along streamlines of $\widetilde{u^2}$

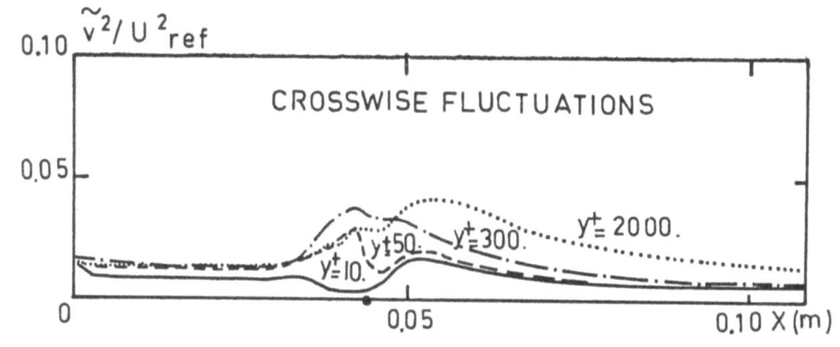

Figure 3b : Evolution along streamlines of v

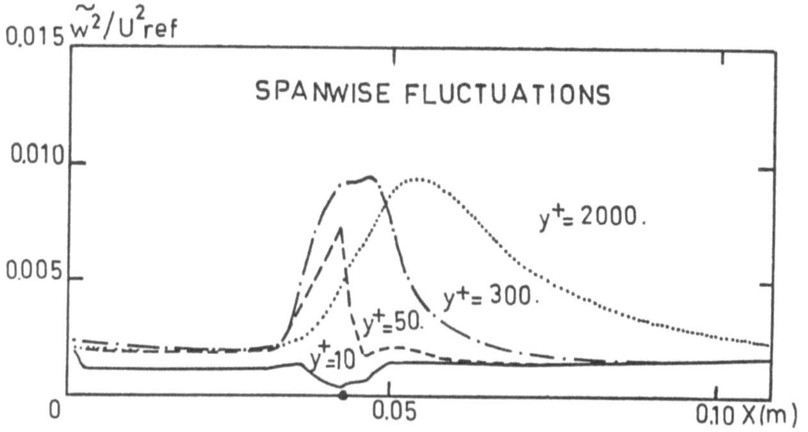

Figure 3c : Evolution along streamlines of w

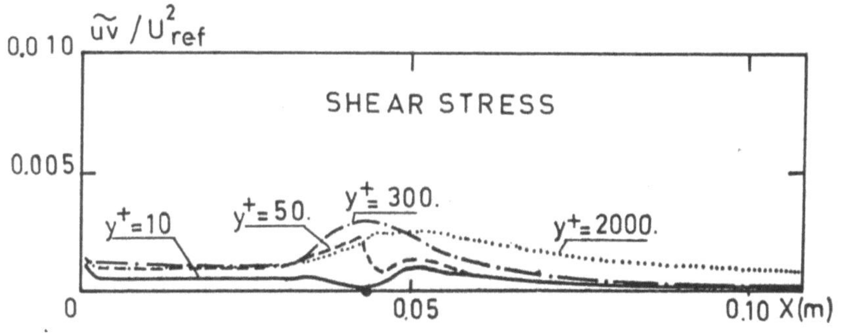

Figure 3d : Evolution along streamlines of w

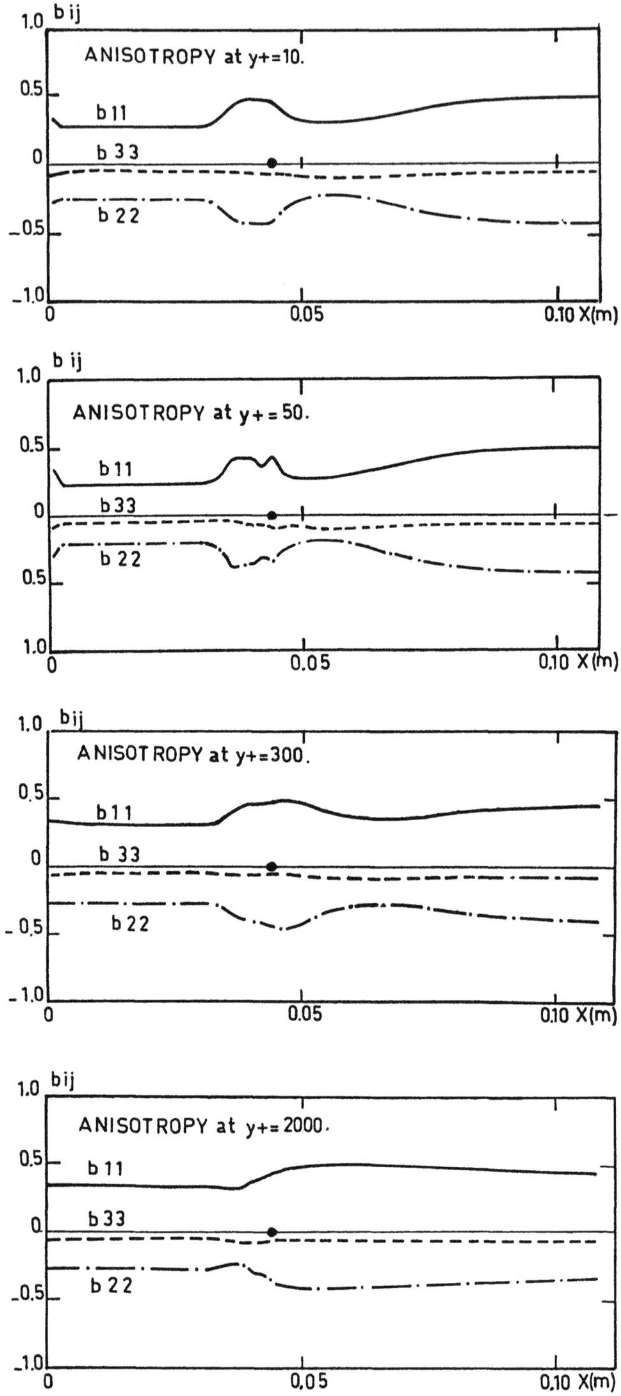

a

b

c

d

Figure 4 : Anisotropy evolutions along streamlines

Turbulent Properties of Axisymmetric Shock-Wave/Boundary-Layer Interaction Flows

J. L. BROWN, M. I. KUSSOY, AND T. J. COAKLEY

NASA, Ames Research Center
Moffett Field, California

Summary

A combined experimental and computational investigation of an axisymmetric turbulent shock-wave boundary-layer interaction flow is presented. Experimental measurements include both mean and fluctuating data obtained by LDV techniques and identify large scale unsteady motions associated with shock induced separation. Computations using the compressible Navier-Stokes equations, and a two-equation turbulence model are in relatively good agreement with experimental measurements. It is found that the large scale unsteady motions do not appear to have a critical impact on the ability to compute the mean properties of the flows investigated in this paper.

Introduction

Currently there is considerable interest within the aerodynamic community in high speed compressible turbulent boundary layer interaction flows.[1-3] This interest arises from the dominant role these interactions play in determining the flow fields about modern flight vehicles. The ability to predict and control these flows is an important element in the design of aeronautical systems. It is complicated, however, by non-equilibrium effects arising from shock induced flow separation and large scale unsteadiness in shock motions which may pose special problems in flow prediction methods.[4]

The basic objective of this paper is to advance the understanding and prediction of compressible turbulent boundary layer interaction flows by a coordinated experimental and computational investigation of a series of representative flows. The flows are supersonic and consist of axisymmetric cylinder-cone-cylinder geometries (Fig. 1) in which both shock boundary-layer and expansion-fan boundary-layer flow interactions

Turbulent Shear Layer/Shock Wave Interactions
IUTAM Symposium Palaiseau 1985
Editor: J. Délery

studied within the same experimental context. By varying the cone angle, the strength of the interactions can be varied from unseparated weak interactions to strong separated interactions. The present investigation is part of an overall program designed to document and model complex turbulent compressible 2-D and 3-D flows.[5]

Model and Instrumentation

The experimental configuration consists of a 5.08 cm diameter cylinder with interchangeable conical flare and cylindrical afterbody assemblies (see Fig. 1). Results of experiments and computations corresponding to cone angles of 12.5° and 30° are reported here. These two cases correspond to both fully attached (12.5°) and fully separated (30°) flow at the cylinder-cone intersection. The experiments were conducted in the 25.4 by 38.1 cm test section of the High Reynolds Number Channel I at the Ames Research Center.

The nominal free stream Mach number was 2.85 with a total temperature of 270°K. The compression corner was placed 1 meter from the cusped nose and the total pressure was 1.7 atm resulting in a Reynolds number of 18×10^{6} per meter. The initial boundary layer thickness just upstream of the compression corner was 1.1 cm (for both cases). The present study focuses on mean and turbulent flow measurements which were obtained by a laser Doppler velocimeter (see Refs. 6 and 7). Seed material consisted of 0.5 micron latex particles, as verified by lag measurements. Supplemental information obtained from dynamic pressure cells and holographic interferometry is also included. More detailed results obtained with holographic interferometry are reported in Ref. 6. Oil flow measurements were made to confirm flow symmetry and to establish the extent of the separated flow region.

Numerical Computations and Turbulence Modeling

The numerical computations were performed using the computer code described in Ref. 8. This code solves the compressible Navier-Stokes equations using a time marching implicit upwind differencing method and utilizes several turbulence models. The turbulence model used in the present investigation is the two-equation q-ω model which is described in Ref. 9. Other turbulence models were also tested but the results of these studies will be reported elsewhere. The numerical grid used in the computations was a 95x95 mesh with grid points finely spaced in the interaction and shock regions. The complete experimental flow field was solved including the region from the cylinder tip to the region aft of the expansion corner. Experimental and computed boundary-layer profiles upstream of the interaction were in good agreement.

Results and Discussion

Flow field holographic interferograms for the unseparated and separated cases are shown in Figs. 2a and 2b. In these interferograms horizontal finite fringe reconstructions were used. For the 12.5^{o} case a simple oblique shock wave can be seen originating slightly upstream of the compression corner. For the 30° case the flow is considerably more complicated and consists of a separation shock wave emanating from a location substantially ahead of the compression corner, and merging with a second, reattachment, shock wave just ahead of the expansion corner. The separation shock wave is caused by flow separation in the corner, and is intrinsically an unsteady process. Interferograms taken at different times as well as shadowgraph movies indicate appreciable movement in the separation shock wave location with a time scale of about 1 msec.

Mean surface pressure and displacement thickness measurements are compared with computational results in Figs. 3a and 3b. For the 12.5^{o} unseparated case, surface pressure computations are in good agreement with measurements. For the 30° separated case, however, although qualitative features of the flow are well predicted (eg, the plateau pressure) there are quantitative differences between measurement and computation. These differences are caused primarily by a reduced length of separation in the computation compared with measurements. Although not shown here, measurements of surface pressure fluctuations are substantially higher in the separated regions than in outlying regions and are consistent with unsteady shock motions observed in shadowgraph movies. Locations of separation and reattachment as determined by surface oil flow measurements are also shown in Fig. 3.

Considering next the compressible displacement thickness distributions of Fig. 3b, it is found that although the computations qualitatively agree with measurements, there are quantitative differences. These differences are characterized by computed displacement thicknesses which decrease less rapidly than measured thicknesses downstream of the compression corner. This feature appears to be a deficiency of the turbulence model.

Experimental and computed u-velocity contours for the 30° case are shown in Figs. 4a and 4b. The velocities in this case are nondimensionalized by the free stream sonic throat speed. These figures generally show good agreement between computation and experiment except near the compression and expansion corners. Near the compression corner, the computation predicts less upstream influence than experiment while near the expansion corner it indicates a somewhat thicker boundary-layer than experiment. These results are consistent with the earlier results discussed on surface pressure and displacement thickness distributions.

Experimental and computed density contours are compared for the 30° case in Figs. 5a and 5b. In this case, densities were derived from experimental velocities by using (inviscid) shock-expansion theory to relate static pressures to measured flow angles. In the inner part of the boundary layer, a zero normal pressure gradient was assumed, ie $\partial p/\partial n = 0$. A Crocco relation was also used in relating static and total temperatures. The experimental density contours agree with contours obtained using holography.6

Comparing experimental and computed densities in Figs. 5a and 5b, it is found that differences between the two are similar to those observed with the velocity contour and displacement thickness distributions. Additional differences are: (1) the slightly greater stand off distance of the merged reattachment shock wave compared with experiment (which is caused by the shorter computed separation bubble) and (2) differences in density contour shapes close to the cone surface. These latter differences could be due either to turbulence modeling deficiencies or to inaccuracies in the data reduction technique.

Mean turbulence properties of the two flows are shown in Figs. 6 and 7. In Figs. 6a-6d, experimental and computational contours of turbulent kinetic energy (TKE) are compared for both the 12.5^o and 30° cases. In these cases, the experimental turbulent kinetic energy was computed by assuming $w'^2 = (u'^2 + v'^2)/2$. From these figures it is apparent that the overall features of the experimental TKE distributions are well predicted by the computations. In particular, the peak kinetic energies and their locations are well predicted. There are, however, some differences in detail in the regions very close to the cone surface which are probably due to turbulence modeling effects.

Experimental TKE production contours are shown in Figs. 7a and 7b corresponding to the 12.5^o and 30° cases. Velocity derivatives in the formula for production were computed by using a two-dimensional least squares technique to fit a smooth surface through measured velocity points. Results for the 12.5^o case indicate a peak production occurring just downstream of the compression shock wave and corner. Results for the 30° case indicate peaks in production both at separation and in regions downstream of reattachment (X=1 cm). Production decays in the downstream direction but not as rapidly as the TKE so that the turbulent dissipation rate remains relatively high over the cone surface. Although not shown here, for a short distance beyond the expansion corner, the TKE production actually becomes negative possibly associated with large dilatation effects.[3] This event combined with a relatively large preexisting dissipation rate results in a rapid collapse of the TKE and a near relaminarization of the boundary layer. Further downstream the turbulent production becomes positive and the TKE returns to normal values. This behavior was in part predicted by the numerical computation. It is anticipated that these measurements of TKE production will be useful in understanding and developing turbulence modeling concepts.

Experimental results dealing with unsteady aspects of the separated 3Ø^{o} case are shown in Figs. 8 and 9. Figure 8 shows contours of intermittency in the neighborhood of separation. The intermittency, or Γ^{+}, represents the percentage of the time that the u-velocity is positive. Separation usually is taken to correspond to values of Γ^{+} less than Ø.5. In Fig. 8 it is seen that the separation contour extends from X=-1.3 cm to X=Ø.3 cm and is roughly Ø.3 cm in height. This separation contour may be compared with experimental and computed zero velocity contours in Figs. 4a and 4b. It is found that the zero velocity contour of the experiment is closer to the surface than the Γ^{+}=Ø.5 contour of Fig. 8. In contrast, the zero velocity contour of the computation is approximately twice the length and height of the Γ^{+}=Ø.5 contour of Fig. 8. This computed result is in closer agreement with experimental separation and reattachment points obtained by oil flow measurements which are shown in Fig. 4a. These differences result in part from strong skewness in the probability density function for u and point out difficulties in interpretation of measurements of unsteady separation that will be addressed in future studies.

A typical joint probability distribution function (JPDF) of velocities is shown in Fig. 9. The JPDF was obtained at a location in the outer boundary layer and slightly downstream of the mean position of the separation shock wave. The bimodality of the distribution is particularly evident and is strongly indicative of unsteady shock wave motion. The two peaks, labeled s_1 and s_2, are representative of velocity states upstream and downstream of the separation shock wave. To each of these states is associated a total probability of occurrence (p_1 and p_2 = 1-p_1), mean velocities (u_1, u_2 and v_1,

v_2), turbulence normal stresses, and turbulent shear stress. A straightforward analysis reveals that the difference in mean values (u_2 - u_1) and (v_2 - v_1) for the two states contributes to the stresses. The shear stress contribution due to unsteadiness is $-p_1p_2(u_1-u_2)(v_1-v_2)$ and is 75 percent of the total shear stress. (However, at locations more distant from the shock wave, the shear stress contribution due to shock unsteadiness falls off rapidly.) This contribution to the total Reynolds stresses is due to an organized or coherent motion of the shock rather than to incoherent or dissipative turbulence, and complicates interpretation of data in the vicinity of the shock. Conditional sampling techniques may be required to more fully analyze the effect of shock unsteadiness.

An important observation concerning numerical predictions of shock induced separation can be made here. The computations appear to be capable of predicting important mean flow characteristics in the vicinity of separation in spite of large scale flow unsteadiness and difficulties in interpreting experimental data. It is felt that further improvements leading to even better agreement between computation and experiment can be achieved through improvements in turbulence modeling.

Concluding Remarks

In the preceding sections, we have described a combined experimental and computational investigation of an axisymmetric shock-wave turbulent boundary-layer interaction flow. Experimental measurements included both mean and fluctuating turbulence data and drew attention to some of the large scale unsteady motions associated with separated shock-wave boundary-layer interactions. Experimental data analysis methods were described which (1) determine the experimental mean density field from velocity measurements and (2) simplify the determination of turbulent kinetic energy production. Comparison of computational and experimental results showed good overall agreement although some important quantitative differences exist. We believe that further improvements in turbulence modeling should substantially reduce these differences. We also believe that the significant unsteady effects which were observed in the separated case do not appear to have a crucial impact on turbulence modeling or the ability to compute the mean properties of these flows.

References

1. Ardonceau, P.L.: The structure of turbulence in a supersonic shock wave/boundary layer interaction. AIAA Journal, Vol. 22, (1984) 1254-1262.

2. Settles, G.S.; Fitzpatrick, T.J.; and Bogdonoff, S.M.: Detailed study of attached and separated compression corner flow fields in high Reynolds number supersonic flow. AIAA Journal, Vol. 17 (1975) 573-585.

3. Dussauge, J.P.; Gaviglio, J.: Bulk dilatation effects on Reynolds stresses in the rapid expansion of a turbulent boundary layer at supersonic speeds. Third Symposium on Turbulent Shear Flows. Univ. of Calif., Davis, California, (1981).

4. Dolling, D.S.; Or, C.T.: Unsteadiness of the shock wave structure in attached and separated compression ramp flow fields. AIAA Paper 83-1715 (1983).

5. Marvin, J.G.: Turbulence modeling for computational aerodynamics. AIAA Journal, Vol. 21, No. 7, pp. 941-955 (1983).

6. Dunnagan, S.E.; Brown, J.L.; Miles, J. B.: A holographic interferometric study of an axisymmetric shock-wave boundary-layer strong interaction flow. AIAA Paper 85-1564 (1985).

7. Brown, J. L.; Viswanath, P. R.: An experimental study of a supercritical trailing-edge flow. AIAA Paper 84-2187 (1984).

8. Coakley, T. J.: A compressible Navier-Stokes code for turbulent flow modeling. NASA TM 85899 (1984).

9. Coakley, T. J.; Hsieh, T.: A comparison between implicit and hybrid methods for the calculation of steady and unsteady inlet flows. AIAA Paper 85-1125 (1985).

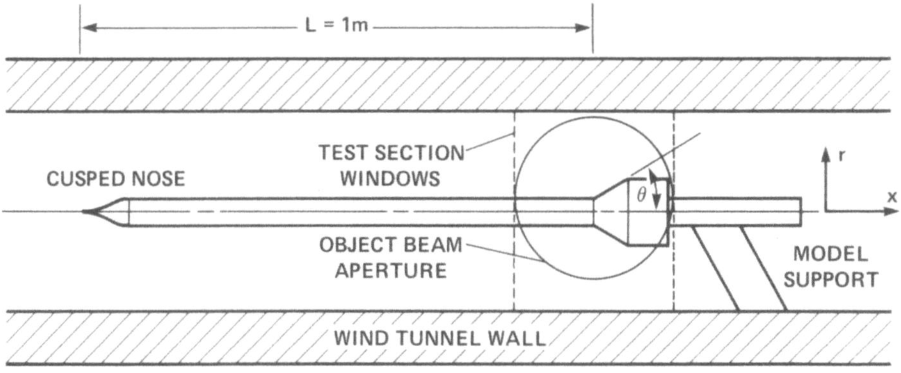

Fig. 1 Axisymmetric model geometry.

a)

b)

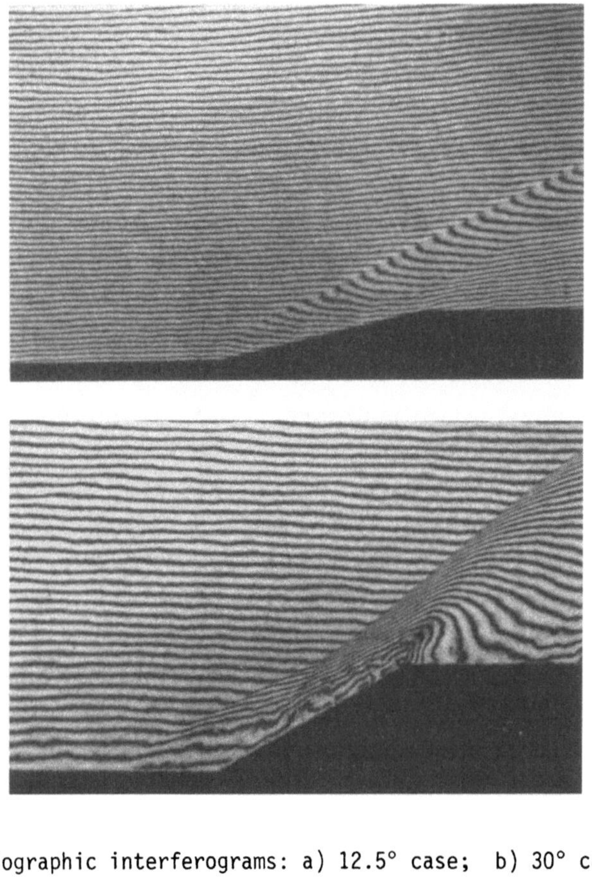

Fig. 2 Holographic interferograms: a) 12.5° case; b) 30° case.

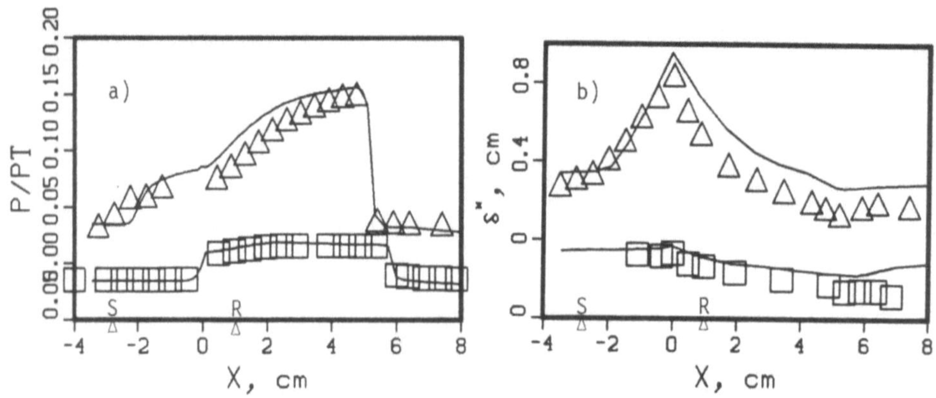

Fig. 3 Surface pressure and displacement thickness distributions: △ - experiment; 30° case; □ - experiment, 12.5° case; --- computation; a) surface pressure; b) displacement thickness.

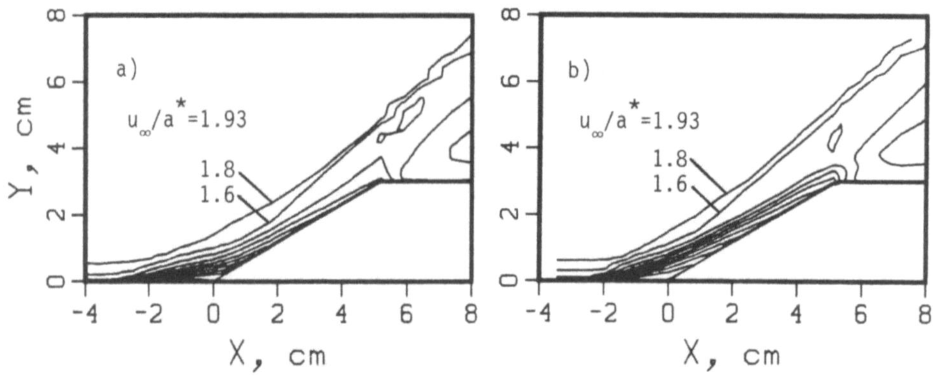

Fig. 4 u-velocity contours, u/a^*=1.8, 1.6, ..., 0.., 30° case: a) experiment, b) computation.

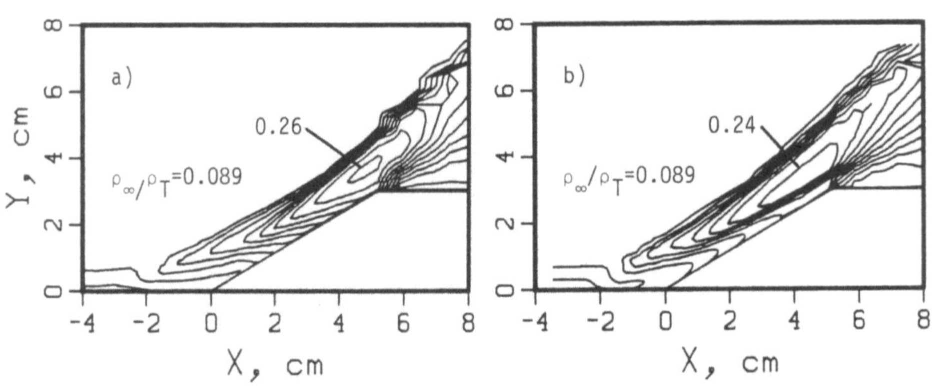

Fig. 5 Density contours, ρ/ρ_T=.26, .24, ..., 06, 30° case: a) experiment; b) computation.

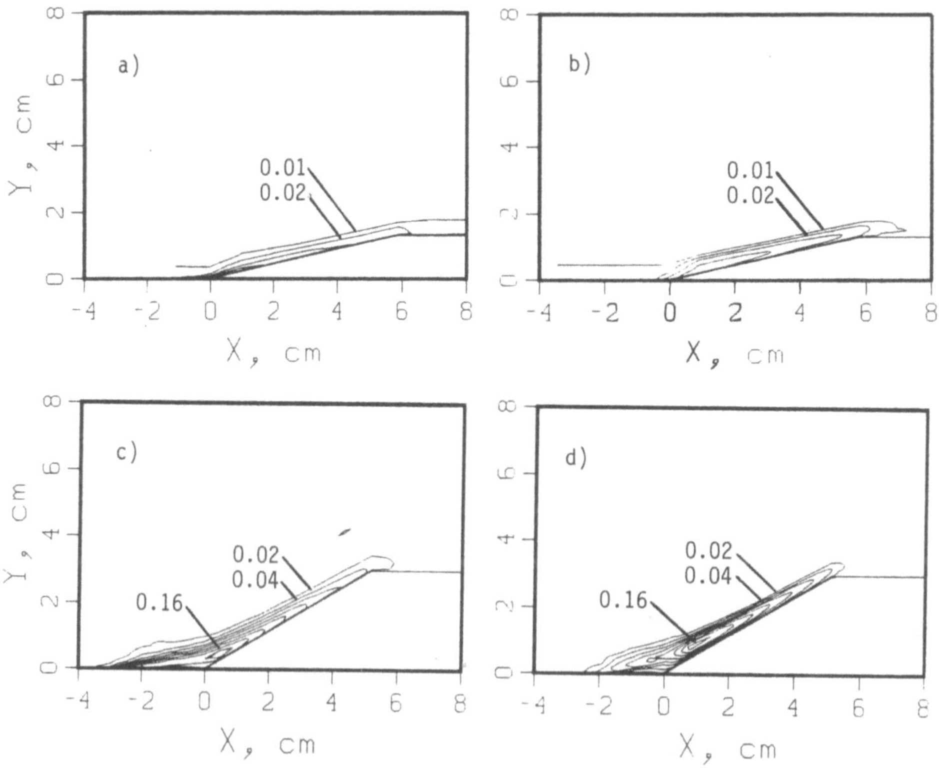

Fig. 6 Turbulent kinetic energy contours, k/a^{*2}=.16, .14, ..., .Ø2 : a) experiment, 12.5°; b) computation, 12.5°; c) experiment, 3Ø°; d) computation, 3Ø°.

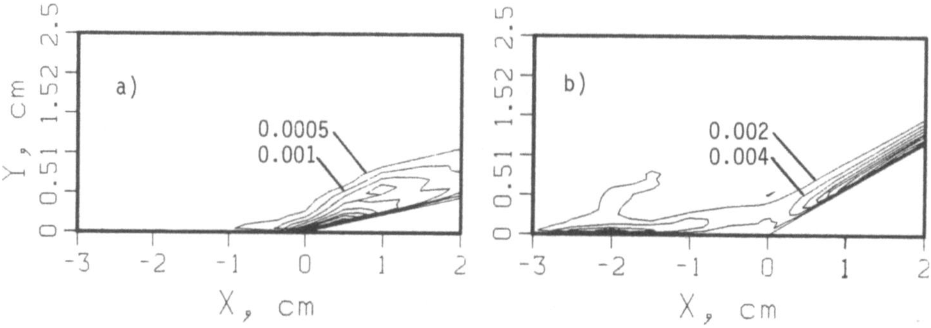

Fig. 7 Experimental turbulent kinetic energy production contours $P_k/\rho_T a^{*3}$(cm^{-1}): a) 12.5° case; b) 3Ø° case.

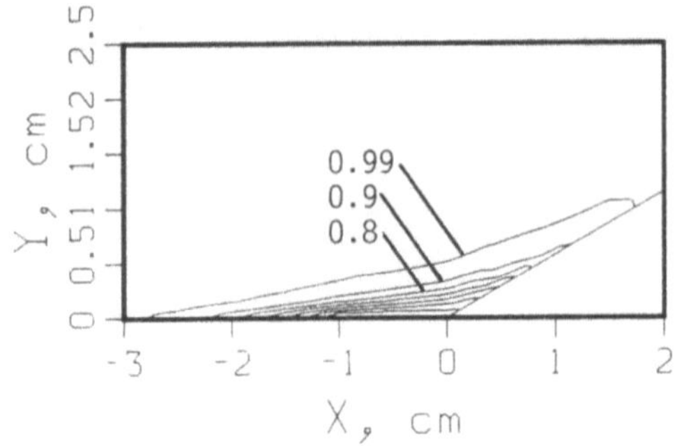

Fig. 8 Experimental intermittancy contours, Γ^+=, 99, .9, .8, ..., .4, 30° case.

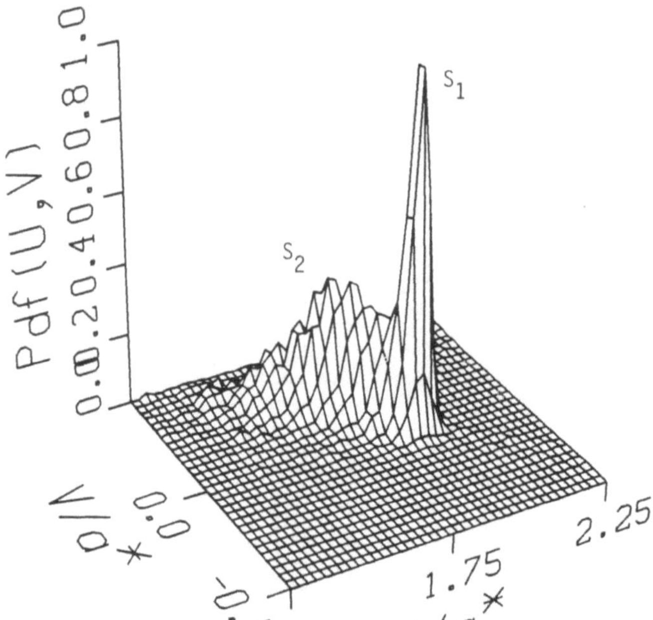

Fig. 9 Joint probability distribution function for u and v, 30° case, x=-1.5 cm, y=0.83 cm.

An Asymptotic Theory of Supersonic Turbulent Interactions in a Compression Corner

R. E. MELNIK*, R. L. CUSIC** M. J. SICLARI*
* Grumman Corporation
** University of Colorado

1. Introduction

In this paper we present a theoretical analysis of the interaction of an unseparated supersonic turbulent boundary layer with a shallow two-dimensional compression corner. In these flows the pressure rises in two stages: an abrupt, nearly discontinuous jump, followed by a more gradual rise to the inviscid wedge value far downstream. The downstream flow in the gradual part of the pressure rise is now fairly well understood. In 1969, Roshko and Thomke [1] and later Elfstrom [2] showed that this could be accurately predicted with a simple inviscid rotational flow model, provided that a suitable cut-off or slip velocity was applied to the initial velocity profile and adjusted to match the steep part of the pressure rise. Later, in 1984, Agrawal and Messiter [3] developed a rational asymptotic theory for this problem based on the limits of small wedge angle and large Reynolds number. Their approach is closely related to asymptotic theories previously developed for other turbulent interaction problems in Refs. [4-11]. Agrawal and Messiter [3] showed that their theory could predict the gradual part of the pressure rise without the need to introduce an empirical slip velocity. Unfortunately, their theory was singular at the corner and could not describe the flow in the steep part of the pressure rise. In the present paper we briefly describe a new asymptotic theory that applies to the near corner region covering the steep part of the pressure rise.

Further details of our work can be found in the forthcoming thesis of Cusic [14] and in an expanded report version of the present paper [15].

2. Formulation

The flow field under study is sketched in Fig. 1. In this report, we follow Agrawal and Messiter's [3] notation, except that here we employ θ_w for the turning angle and ε for the

Turbulent Shear Layer/Shock Wave Interactions
IUTAM Symposium Palaiseau 1985
Editor: J. Délery

basic small parameter scaling the Reynolds number effects. The flow is characterized by the value of the three basic dimensionless parameters: freestream Mach number, M_∞, friction velocity, $u_\tau \equiv \sqrt{\tau_w} \equiv \sqrt{c_{fo}/2}$, and turning angle, θ_w. As an approximation, we assume the initial velocity profile is given in the form of a Van Driest transformed law of the wall - law of the wake (e.g., see [3]). We also assume an adiabatic wall, a perfect gas with a constant specific heat ratio, γ, and an approximate uniform total temperature distribution. Then, following Agrawal and Messiter, the velocity and Mach number profiles can be written as,

$$U_0(y) = \Gamma \sin (\Gamma^{-1} \bar{U}) \quad (\Gamma = \sqrt{T_w/T_w - 1}) \tag{1}$$

$$M_0(y) = \sqrt{\frac{2}{\gamma-1}} \tan (\Gamma^{-1} \bar{U}) \tag{2}$$

$$\bar{U} = \Gamma \sin^{-1} (\Gamma^{-1}) + u_\tau T_w^{1/2} \kappa^{-1} u_0(y) \tag{3}$$

where $u_0(y)$ is the incompressible law of the wall - law of the wake form,

$$u_0(y) = \ln y - \tilde{\pi} (1 + \cos \pi y) , \tag{4}$$

$\kappa = 0.41$ is the von Karman constant, $\tilde{\pi} = 0.55$ is the Coles wake parameter and T_w is the upstream wall temperature. The problem formulation is completed by the mass and momentum conservative equations. These we do not write out here but simply note that all the terms obtained in the present report are contained within the thin-layer version of the Reynolds averaged Navier-Stokes equations. We should also note that a turbulent closure model is not required to determine the first few terms of the outer inviscid rotational flow solution.

3. The Agrawal-Messiter Solution

The Agrawal-Messiter solution is developed in a two-parameter, asymptotic expansion in terms of the friction velocity u_τ and the corner angle θ_w with $M_\infty \neq 1$. The asymptotic flow structure that arises in this limit is sketched in Fig. 2. The flow field is divided into three principle regions, which in order of their relative distances from the corner, are called the inner, intermediate and outer regions. Each of these regions, in turn, divide into the characteristic three-layer structure that arises in other turbulent interaction problems [4-11]. The wall layer is the same in all three regions and is identical to the wall layer that appears in noninteracting boundary layers. In addition, the blending layer solution has the same structure in all three regions. Because of the thinness of the sublayers, it follows that the displacement effects of the sublayers are so small that they do not influence the lowest order solution in the inviscid outer regions.

The small θ_w, u_τ limit leads to a standard linearization of the flow about the uniform supersonic external flow. The solution in the outer layer is governed by linearized supersonic shear layer equations. The equations are solved subject to the imposed flow deflection at the wedge surface and to the jump conditions at the shock wave. There are no disturbances propagated upstream of the wedge apex to all orders within this theory. For brevity we present just the solution for the pressure distribution:

$$p = 1 + [\theta_w p_{11} + \theta_w^2 p_{12} + \ldots] + u_\tau[\theta_w p_{21}(x,y) + \ldots] + \ldots \qquad (5)$$

where p_{11} and p_{12} are constants determined from the linearized and second order inviscid supersonic solution for wedges. The solution for p_{21}, which provides for the first effect of the viscid-inviscid interaction, is

$$\frac{p_{21}}{\gamma M_\infty^2} = \frac{M_\infty^2 - 2}{2B_\infty^2}\left(1 + \frac{\gamma-1}{2} M_\infty^2\right)\left\{u_0\left(\frac{x}{2B_\infty} - \frac{y}{2}\right) + u_0\left(\frac{x}{2B_\infty} + \frac{y}{2}\right)\right\} \qquad (6)$$

where $B_\infty \equiv \sqrt{M_\infty^2 - 1}$. Logarithmic singularities appear in the above solution for the pressure along the shock wave, $y_s = x/B_\infty$, and at the vertex x, y = 0. The singularities along the shock wave are eliminated in a composite supersonic/hypersonic solution developed by Agrawal and Messiter but remain at the apex. The singularities arise from the logarithmic behavior in the initial velocity profile at $y = 0$.

As a result of the singularities, the composite supersonic/hypersonic solution in the outer region is not uniformly valid at the corner. In order to extend the theory closer to the corner, Agrawal and Messiter developed a separate intermediate expansion for this region in which the solution is expressed as a linearized perturbation about the local value (in y) of the undisturbed shear layer upstream of the shock wave. The solution was developed in the following similarity form:

$$p = 1 + P_1(\sigma,\lambda;\theta_w) + u_\tau P_2(\sigma,\lambda;\theta_w) + \dots \qquad (7)$$

$$\sigma + \frac{u_\tau}{\kappa} \ln B_0(\sigma) = \frac{u_\tau}{\kappa} \ln x \quad , \quad \lambda = \frac{x}{yB_0(\sigma)} \qquad (8)$$

where $B_0 = (M_0^2-1)^{1/2}$ and $M_0(\sigma)$ is given by Eq (2) with $u_\tau u_0/\kappa$ replaced by σ. The intermediate solution describes the flow in the supersonic part of the boundary layer defined by $1 < M_0 < M_\infty$ and $\sigma^* < \sigma < 0$ where σ^* is the value of σ at sonic conditions. Agrawal and Messiter constructed more general composite solutions in [3] which extended the validity of the supersonic solutions to transonic and hypersonic speeds.

The Agrawal-Messiter theory led to lengthy but relatively simple closed form expressions for the surface pressure and also the skin friction. In the present paper we will only be concerned with the solution in the outer region for the pressure distribution. Numerical results were compared with two of the high Mach number ($M_\infty = 3.93$, 4.92) data sets of Roshko and Thomke. The results for the $M_\infty = 3.93$, $\theta_w = 14.9°$ case are repeated here in Fig. 3. The Agrawal-Messiter solution is seen to agree very well with the Roshko-Thomke computations over the smooth part of the pressure distribution. However, the asymptotic theory shows a large compressive spike that is absent from both the data and numerical results. As noted in Agrawal and Messiter, the breakdown or the theory is associated with a singular behavior of their solution for $M_0(\sigma) \to 1$ for $\sigma \to \sigma^*$,

$$(p_w - 1)/\gamma \to \frac{\theta_w}{B_0(\sigma)} + \frac{(\gamma+1)\theta_w^{\,2}}{4B_0^{\,4}(\sigma)} + \ldots + \frac{\theta_w u_\tau(\gamma+1)M_\infty(\tilde{\pi} + \frac{1}{2}\ln 2)}{\kappa B_0^3(\sigma)} + \ldots \quad (9)$$

$$B_0(\sigma) \to x \to \sqrt{(\gamma + 1)M_\infty(\sigma - \sigma^*)} \to 0 \quad (10)$$

It follows from Eq (9) that $x \to 0^+$ so that the singularity occurs as sonic conditions are approached at the corner.

In Fig. 4 we show a comparison for a lower supersonic Mach number case, $M_\infty = 1.95$, $\theta_w = 5.16°$. The comparison with data is not good in this case. As noted by Agrawal and Messiter, the reasons for this are not understood.

Aside from this low supersonic case, the Agrawal-Messiter theory agrees very well with the Roshko-Thomke data in the smooth part of the pressure rise and is completely consistent with and provides a rational basis for their inviscid rotational flow model. Unfortunately, the asymptotic solution develops a singularity which causes a breakdown of the theory in the steep part of the pressure rise.

4. Inner Region

In this section we describe a new inner solution developed by us to extend the asymptotic theory to the near corner region. The breakdown of the theory for $M_0(\sigma) \to 1$ suggests the need for an inner region with conventional transonic type scalings. The form of the initial Mach number profile for $y \to 0$ can be written in the form,

$$M_0^2(\hat{y}) = 1 + (\gamma + 1)\left(\frac{M_\infty u_\tau}{\kappa}\right) \ln \hat{y} + 0 \left(\frac{M_\infty u_\tau}{\kappa}\right)^2 \quad (11)$$

where $\hat{y} \equiv y/\delta_s$ is a stretched inner coordinate defined in terms of the height of the undisturbed sonic line, δ_s. This expansion suggests the introduction of a new small parameter, ε, in place of u_τ to serve as a measure of Reynolds number effects,

$$\varepsilon \equiv M_\infty u_\tau/\kappa \tag{12}$$

A distinguished limit arises from the requirement that the inner region be essentially transonic, $M_0^2 - 1 = 0(\theta_w^{2/3})$, from which Eq (11) leads to: $0(u_\tau) = 0(\varepsilon) = 0(\theta_w^{2/3})$. Thus, we are led to the following limit process in the inner region:

$$\begin{matrix} \varepsilon \to 0 \\ \theta_w \to 0 \end{matrix} \qquad \chi \equiv \frac{\theta_w}{(\gamma+1)^{1/2}\varepsilon^{3/2}} \quad \text{FIXED} \tag{13}$$

and stretched inner coordinates,

$$\hat{y} = y/\delta_s \qquad \hat{x} = (\gamma+1)^{-1/2}\varepsilon^{-1/2}x/\delta_s \tag{14}$$

The solution in the outer layer is represented as a perturbation expansion about sonic speed. The solution, including the initial velocity and Mach number profiles, is represented by asymptotic expansions in powers of ε, assuming χ is an order one constant. The solution for the dimensionless velocity components are sought in the form

$$u/a^* = 1 + \varepsilon\, \ell n\hat{y} + \varepsilon\, \chi[\hat{u}(\hat{x},\hat{y};\chi) + 0(\varepsilon)] \tag{15a}$$

$$v/a^* = \varepsilon^{3/2}(\gamma+1)^{1/2}\, \chi[\hat{v}(\hat{x},\hat{y};\chi) + 0(\varepsilon)] \tag{15b}$$

where a^* is the critical speed of sound in the undisturbed boundary layer. The second term in the expansion for the streamwise velocity component arises from the undisturbed nonuniform flow, and the terms proportional to χ arise from the interaction of the upstream shear flow with the inviscid wedge flow. Since the vorticity in the boundary layer is small, the perturbation velocities can be expressed as a gradient of a potential, $\hat{u}, \hat{v} = \nabla\phi$.

Substitution of the above expansion into the full Reynolds averaged equations leads to the following generalized transonic small disturbance equation for the perturbation potential:

$$[\ln \hat{y} + \chi \hat{\phi}_{\hat{x}}] \hat{\phi}_{\hat{x}\hat{x}} - \hat{\phi}_{\hat{y}\hat{y}} = 0 \tag{16}$$

This is similar to the equation derived by Melnik and Grossman [4] for the normal shock wave problem, except that here their more general profile function $u_0(y)$ is replaced by a near sonic approximation. Because of the thinness of the sublayers, the wall boundary condition takes the form

$$\phi_{\hat{y}}(o,\hat{x}) = H(\hat{x}) \tag{17}$$

where $H(\hat{x})$ is the unit step function.

The above scaling leads to an approximate similarity in which the solution near the corner for various values of the controlling parameters, $\theta_w, M_\infty, u_\tau, \gamma, \tilde{\pi}$, depends only on the value of χ. For example, the pressure coefficient can be expressed in the form

$$C_p(x,y;\theta_w,M_\infty,u_\tau,\tilde{\pi};\gamma) = \frac{\varepsilon\chi}{M_\infty^2} \hat{C}_p(\hat{x},\hat{y};\chi) \tag{18}$$

where $\hat{C}_p = 2\hat{u}$. The above asymptotic theory is valid provided the basic parameter, ε, is small enough and the dimensionless interaction parameter, χ, is $O(1)$. For the cases considered in the Roshko-Thomke experiments, ε and χ fall in the ranges $0.1 \lesssim \varepsilon \lesssim 0.2$ and $1.4 \lesssim \chi \lesssim 3.5$. Although ε is not impressively small, the range of values is such that the theory may be useful for many practical problems of interest.

To complete the solution we demonstrate the matching of the present solution to the intermediate solution of Agrawal and Messiter. We follow the standard procedure of comparing the outer limit $(\hat{x} \to \infty)$ of the inner solution to the inner limit $(x \to 0)$ of the outer (intermediate) solution. The outer limit of the inner solution is obtained by expanding the inner problem defined by Eq (16) and (17) for $\hat{x}, \hat{y} \to +\infty$. An appropriate form for the expansion can be guessed from the similarity form of the intermediate solution (Eq (4.1)-(4.4) of [3]). Thus,

$$\hat{u} = \hat{\sigma}^{-1/2}[a_0(\hat{\lambda}) + a_{1/2}(\hat{\lambda})\hat{\sigma}^{-1/2} + a_1(\hat{\lambda})\hat{\sigma}^{-1} + a_{3/2}(\hat{\lambda})\hat{\sigma}^{-3/2} + \ldots] \tag{19}$$

$$\hat{v} = b_0(\hat{\lambda}) + b_{1/2}\hat{\sigma}^{-1/2} + b_1(\hat{\lambda})\hat{\sigma}^{-1} + b_{3/2}(\hat{\lambda})\hat{\sigma}^{-3/2} + \ldots \tag{20}$$

where $\hat{\lambda}$ and $\hat{\sigma}$ are similarity variables defined by

$$\hat{\sigma} + \frac{1}{2}\,\ell n\,\hat{\sigma} = \ell n\,\hat{x}\ , \qquad \hat{\lambda} = \hat{x}/\hat{y}\ \hat{\sigma}^{1/2} \tag{21}$$

The nonlinear terms are small in the far field, and the coefficients $a_i(\hat{\lambda})$, $b_i(\hat{\lambda})$ satisfy linear ordinary differential equations that can be derived by substituting the above expansions into Eq (16) (written in terms of $\hat{u},\hat{v}$). These equations can be integrated explicitly and the solution completed by imposing the expanded form of the shock wave and tangency boundary conditions at $\hat{\lambda} = \hat{\lambda}_s(\sigma)$ and $\hat{\lambda} \to \infty$, respectively. Thus the far field behavior of the surface pressure is found to be,

$$p_{inner}(\hat{x},o) = 1+\varepsilon\chi\gamma\{\hat{\sigma}^{-1/2}\}\{1+(\tfrac{1}{2}\ \ell n\ 2)\hat{\sigma}^{-1}+(\tfrac{1}{4}\ \chi)\hat{\sigma}^{-3/2}+\ldots\} \tag{22}$$

for $\hat{x} \to \infty$. Next, introducing an inner variable, $\bar{\sigma} = (\sigma - \sigma^*)\,M_\infty/\varepsilon$, into Eq (9) and (10) leads to the following limiting expression for the Agrawal-Messiter solution for the surface pressure for $x \to 0$:

$$p_{int}(\hat{x},0) = 1+\varepsilon\chi\gamma\{\bar{\sigma}^{-1/2}\}\{1+[\tilde{\pi}+(\tfrac{1}{2}\ \ell n\ 2)\bar{\sigma}^{-1}+(\tfrac{1}{4}\ \chi)\bar{\sigma}^{-3/2}+\ldots\} \tag{23}$$

Substituting the definition of $\bar{\sigma}$ into Eq (8) and expanding the resulting relation yield the following expression for $\bar{\sigma}$ as function of the inner variable, $\hat{x}$:

$$\bar{\sigma} + \frac{1}{2}\,\ell n\,\bar{\sigma} = (\ell n\,\hat{x}) + 2\,\tilde{\pi} + O(\varepsilon\bar{\sigma}) \tag{24}$$

The two limiting expressions for the surface pressure given by Eq (22) and (23) cannot be directly compared because of the different definitions of the similarity coordinates $\hat{\sigma}$ and $\bar{\sigma}$.

From Eq (21) and (24) we see that they can be related by the following expansion for $\hat{\sigma} \to \infty$:

$$\hat{\sigma} = \bar{\sigma}\,[1 + (2\tilde{\pi})\,\bar{\sigma}^{-1} + O(\bar{\sigma}^{-2}) + O(\varepsilon\bar{\sigma})] \tag{25}$$

If we substitute this expansion into Eq (22), we find that it reduces to Eq (23) with an error term that is $o(\varepsilon\bar{\sigma})$. The two expressions for the limiting behavior of the outer and inner solutions then match term by term in an overlap domain defined by $O(1) < \bar{\sigma} < O(\varepsilon^{-1/2})$. A multiplicative type composite solution that is uniformly valid in the inner, intermediate and outer regions can be constructed in the usual way. Thus,

$$p_c = \frac{p_{inner}(\hat{x})\; p_{am}(\sigma, x)}{p_{cp}(\bar{\sigma})} \tag{26}$$

where p_{am} is the composite intermediate/outer solution given by Eq (4.12) of Agrawal-Messiter [3] and $p_{cp}(\bar{\sigma})$ is the "common part" defined by Eq (22).

5. Results

We have solved numerically the generalized, nonlinear, transonic, small disturbance equation using a conservative finite difference scheme of Murman-Cole type [12]. A sheared hyperbolic coordinate system is employed to approximately align the grid with the shock wave. We typically use a grid with 100 points along the body and 50 normal to it. The solution for the perturbation potential is set to zero at the upstream boundary while extrapolated outflow conditions are set at the upper and downstream boundaries. The scheme produces good results for the surface pressures and wiggle free shock waves that are spread over about 10-20 points. Grid clustering near shock waves has been used in some of the computations to improve the resolution of the shock waves.

Numerical results for the reduced pressure coefficient for values of χ equal to 1.0, 2.0, and 4.0 are presented in Fig. 5. The solution appears to have a weak singularity at the corner with a local maximum in the numerical solution that slowly increases as the mesh size is reduced. The analytic structure of this singularity has not yet been determined, but a simple analysis indicates that an additional inner region is required in which the solution must be described by the full Reynolds averaged equations.

The corresponding solutions for the position of the sonic line are given in Fig. 6. We note how increasing χ tends to dramatically increase the height of the sonic line behind the shock wave: there is a 15 fold increase in the thickness of the sonic layer across the corner for $\chi = 4.0$. The increase in upstream influence with increasing χ is also apparent in the solutions for the isobar patterns given in Fig. 7. These results show the coalescence of the isobars to form a shock wave that is driven forward as χ increases.

In Fig. 8 we compare the present and the Agrawal-Messiter composite solutions for the surface pressure with Roshko-Thomke's data for $M_\infty = 3.93$, $\theta_W = 14.9°$ on an expanded scale. The results clearly demonstrate how the present theory eliminates the "spike" at the corner and provides a continuation of a solution across the corner.

Figure 9 compares the present solution with the $M_\infty = 2.87$, $\theta_W = 8°$, compression corner data of Settles, Fitzpatrick and Bogdonoff [13]. The overall agreement between theory and experiment is seen to be quite good on the scale employed in the figure. This data set is of particular interest in the present study because of its high resolution in the near corner region. This is evident in Fig. 10 where we present the results on an expanded scale that more clearly shows the details of the pressure distribution near the corner. The results indicate that although the present theory provides a good prediction of the gradual part of the pressure distribution and the scale of upstream influence, it significantly underpredicts the actual extent of upstream influence in the experiment.

6. Conclusions

Results obtained to date indicate that the asymptotic theory generally provides an accurate prediction of the gradual part of the pressure distribution. The success of the rational theory provides strong support for the simple ad hoc inviscid model of Roshko-Thomke and Elfstrom for such flows. One of the principle contributions of the asymptotic theory, in this regard, was the demonstration that an empirical "slip" velocity is not required in a rational inviscid theory.

The present theory tends to quantitively underpredict the extent of upstream influence. This is likely due to turbulent induced unsteady shock motions which are not accounted for in Reynolds averaging.

The present asymptotic theory is incomplete in two respects. First, the inner solution also breaks down at the corner because of the appearance of a singularity at the origin. Although the nature of the singularity has not yet been determined, evidence from the numerical solutions suggests a very slow, unbounded growth in pressure at the origin. A scaling analysis suggests the need for an additional inner region, of dimensions of a wall layer thickness in both directions, in which the flow must be described by the full Reynolds averaged Navier-Stokes equations.

Of greater importance is the restriction of the present theory to attached flows. Unfortunately, to the authors knowledge, there has been almost no progress made in the development of a rational asymptotic theory for turbulent separated flow.

Acknowledgement

The authors wish to acknowledge the help of Antony Jameson in the development of the numerical scheme employed in this study.

References

1. Roshko, A. and Thomke, G.J., "Supersonic, Turbulent Boundary-layer Interaction with a Compression Corner at Very High Reynolds Number," Proc USAF ARL Symp on Viscous Interaction Phenomena in Supersonic and Hypersonic Flow, Univ of Dayton Press, Dayton, OH (1969) 109-138.

2. Elfstrom, G.M., "Turbulent Hypersonic Flow at a Wedge-compression Corner," J. Fluid Mech, 53 (1972) 113-127.

3. Agrawal, S. and Messiter, A.F., "Turbulent Boundary-layer Interaction with a Shock Wave in a Compression Corner," J. Fluid Mech, 143, (1984) 23-46.

4. Melnik, R.E. and Grossman, B., "Analysis of the Interaction of a Weak Normal Shock Wave with a Turbulent Boundary Layer," AIAA Paper 74-598 (1978).

5. Adamson, T.C. and Feo, A., "Interaction between a Shock Wave and a Turbulent Boundary Layer in Transonic Flow," SIAM J. Appl Maths, 29 (1975) 121-145.

6. Messiter, A.F., "Interaction Between a Normal Shock Wave and a Turbulent Boundary Layer at High Transonic Speeds. Part I: Pressure Distribution," Z. angew Math. Phys, 31 (1980) 204-226.

7. Melnik, R.E., "Turbulent Interactions on Airfoils at Transonic Speeds - Recent Developments," AGARD CP 291 Paper 10 (1981).

8. Adamson, T.C. and Messiter, A.F., "Analysis of Two-dimensional Interactions Between Shock Waves and Boundary Layers," Annual Rev of Fluid Mech, 12, Annual Reviews Inc, Palo Alto, CA (1980) 103-139.

9. Melnik, R.E., Chow, R.R., Mead, H.R. and Jameson, A., "An Improved Viscid/inviscid Interaction Procedure for Transonic Flow Over Airfoils," NASA CR-3805 (1985).

10. Melnik, R.E. and Grossman, B., "On Turbulent Viscid-inviscid Interaction at a Wedge Shaped Trailing Edge," Proc of Symp on Numerical and Physical Aspects of Aerodynamic Flows, I, Springer-Verlag, New York (1982) 211-235.

11. Sykes, R.L., "An Asymptotic Theory for Incompressible Turbulent Boundary-layer Flow Over a Small Hump," J. Fluid Mech, 101 (1980) 647-670.

12. Murman, E.M. and Cole, J.D., "Calculation of Plane Steady Transonic Flows," AIAA J., 9 (1971) 114-121.

13. Settles, G. S., Fitzpatrick, T. J., and Bogdonoff, S.M., "Detailed Study of Attached and Separated Compression Corner Flow Fields in High Reynolds Number Supersonic Flow," AIAA J., 17 (1979) 579-585.

14. Cusic, R., "An Asymptotic Theory of Turbulent Boundary Layer Interaction in a Compression Corner at Supersonic Speeds," PhD thesis, Univ of Colorado, Boulder (to be submitted).

15. Melnik, R.E., Cusic, R.L., and Siclari, M.J., "An Asymptotic Theory of Supersonic Turbulent Interactions in a Compression Corner," Grumman Corporate Research Center, to be published (November 1985).

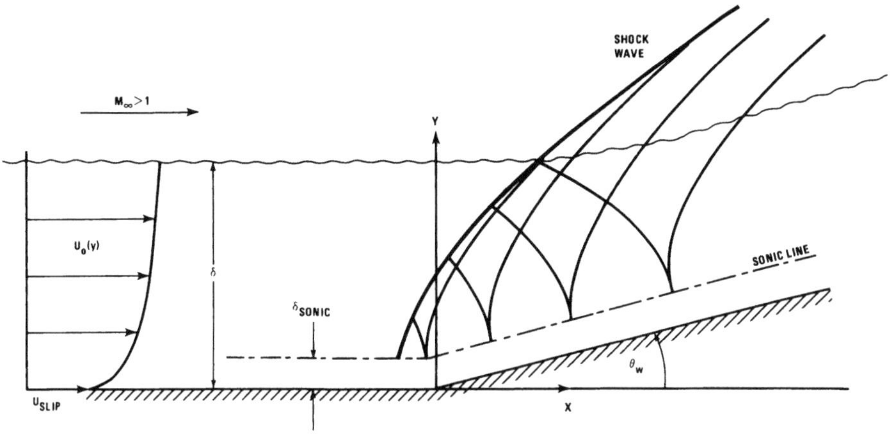

Fig. 1. Sketch of a turbulent boundary layer interaction in a compression corner at supersonic speeds

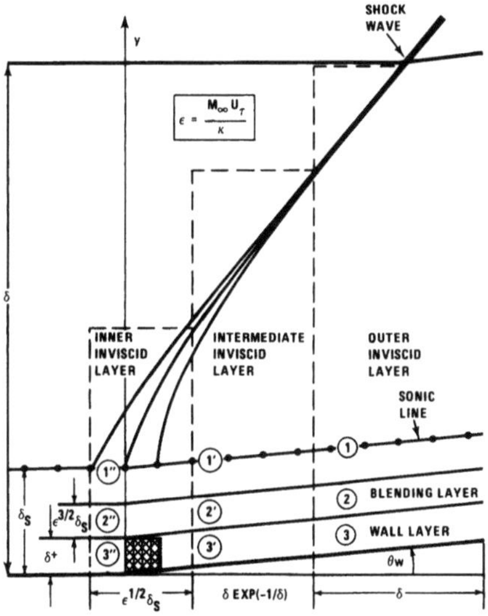

Fig. 2. Asymptotic flow field structure for high Reynolds number and small wedge angles

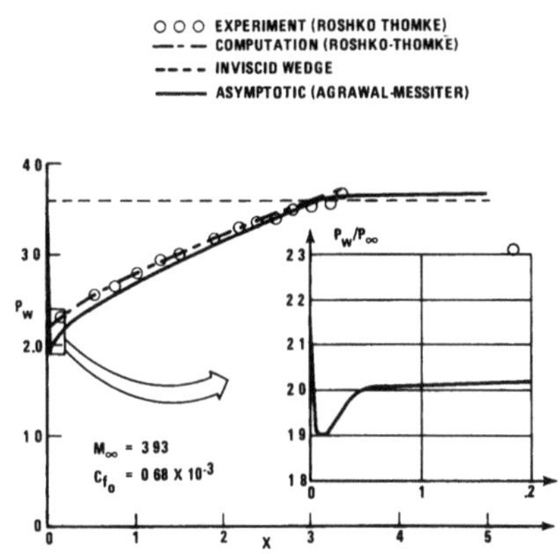

Fig. 3. Surface pressure distribution for the 14.9° compression corner

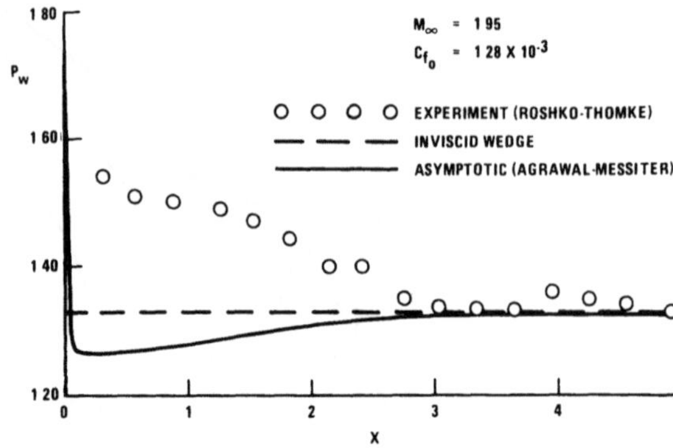

1161 0078

Fig. 4. Surface pressure distribution for the 5.16° compression corner

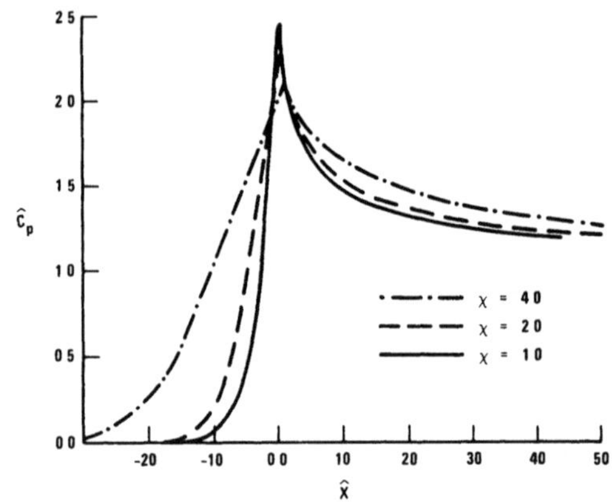

1161 0098

Fig. 5. Surface pressure, variation with interaction parameter, χ , from the present theory

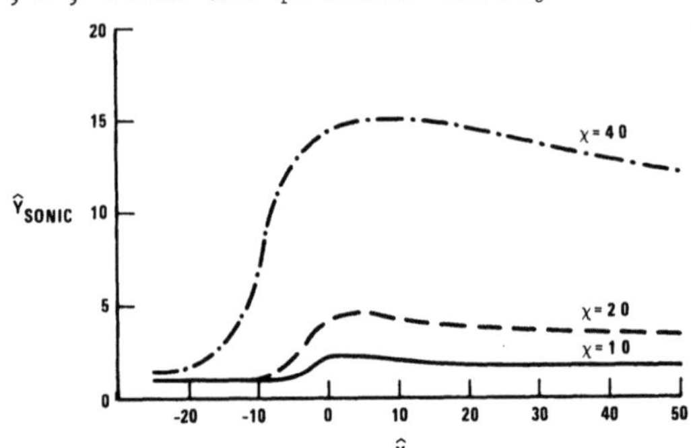

1161 0118

Fig. 6. Sonic line shape variation with interaction parameter, χ , from the present theory

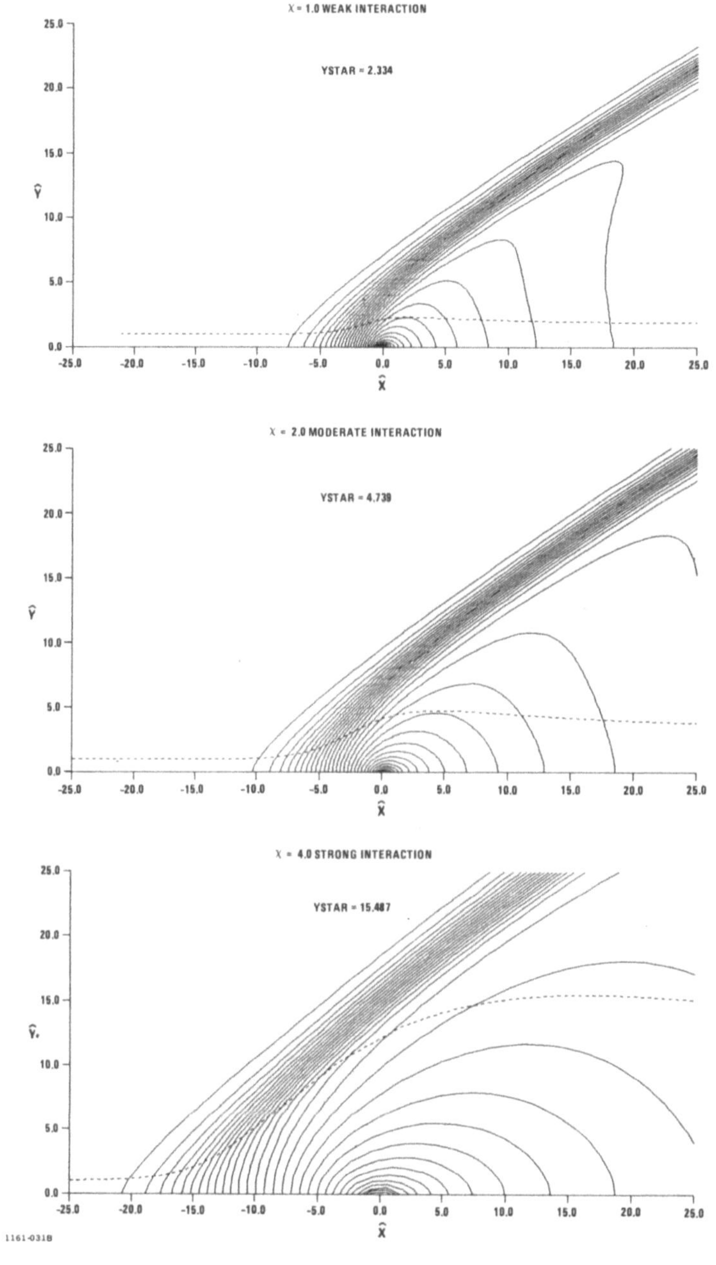

Fig. 7. Isobar patterns from the present theory

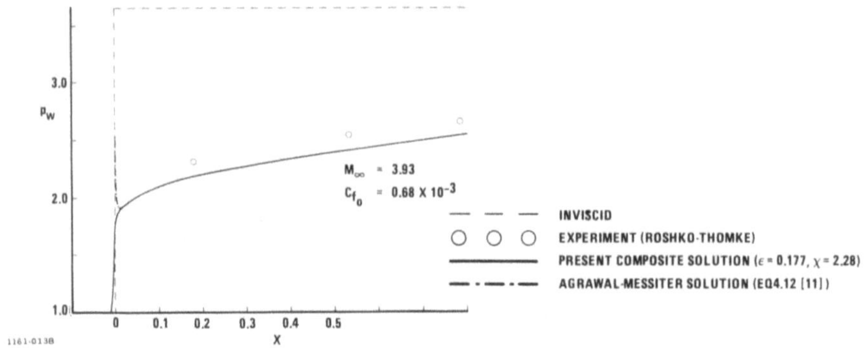

Fig. 8. Surface pressure distribution near the origin of the 14.9° compression corner

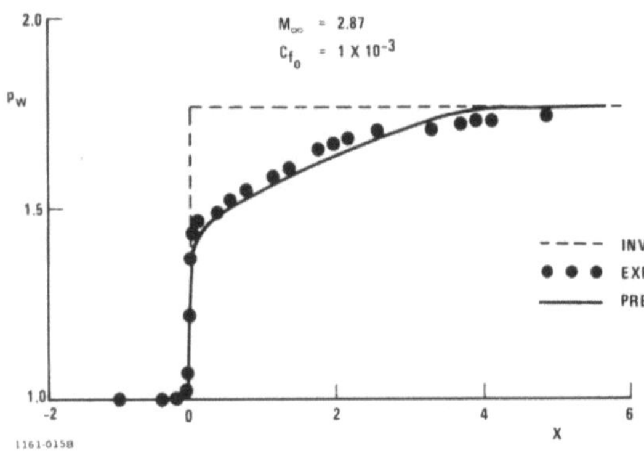

Fig. 9. Surface pressure distribution for the 8° compression corner

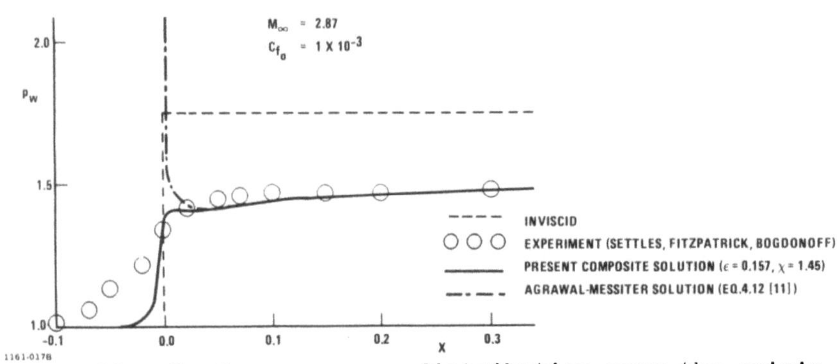

Fig. 10. Surface pressure distribution near the origin of the 8° compression corner

Second Order Computation of Shock-Wave/ Boundary-Layer Interactions in Curvilinear Coordinates

D. AYMER DE LA CHEVALERIE

Laboratoire d'Etudes Aérodynamiques U.A. 191
C.E.A.T. - 43 rue de l'Aérodrome
86000 POITIERS, France

Summary
Navier stokes equations are expanded in internal and external approximations in two orthogonal coordinate systems. Coordinates based upon one limiting external streamline give rather straightforward relations and can be used when the Mach number is not too large. Coordinates fitted on all the external streamlines, require much computational work but gives good agreement in hypersonic range.

1. General discussion

It is well known that parabolic equations require less storage and computing time than elliptic ones. As Davis notices [1], if Navier-Stokes equations are replaced by parabolic approximations, the main simplification has been done, and second order or turbulent terms add no further difficulties. The ellipticity which remains in the whole problem must only be treated by means of an efficient coupling algorithm which is independent of these terms.

Present study consists in second order approximation of boundary layer concept applied to the computation of shock wave boundary layer interactions with four basic ideas :
- the flow is divided in two parts : one governed by Euler equations and the other by boundary layer approximation. For having a well posed problem for the Euler equations, the enthalpy must be prescribed on the boundaries. So, we use as matching line a streamline of the external flow.
- The matching along the displacement surface leads to a singular supercritical behaviour of the solution above Mach 2 for turbulent flows. On an other hand, it is well known that the

Turbulent Shear Layer/Shock Wave Interactions
IUTAM Symposium Palaiseau 1985
Editor: J. Délery

singularity at separation can be removed by taking into account the interaction between the external flow and the boundary layer. In the same way, one can expect that supercritical singularity may be avoided by connecting together the first and the second order in boundary layer equations.

- when elliptical equations are approximated by parabolic ones, the solution is dependent of the coordinate system in which the parabolization has been done. Kaplun [2] proved the existence of an optimal system linked with external streamlines for asymptotic developments.
- the peculiarity of second order developments is the appearance of a pressure gradient in the direction normal to the flow which is noticed by all experimentalists. The bad results obtained in turbulent range with first order boundary layer concept and recent improvements given by Le Balleur [3] allowing both normal pressure gradient, and a greater overlapping domain emphasize those ideas.

This study consists in writing the Navier-Stokes equations in general curvilinear coordinates, and finding the best coordinate systems in which they must be developed. This is reported in the next section. Part 3, 4 and 5 are related to the external flow, boundary layer and the coupling between them. Comparisons between results obtained with this method and other computation of experiments are shown in part 6.

2. Effect of coordinates

Let us consider viscous steady two-dimensional compressible flow on solid boundaries. Navier-Stokes equations are non-dimensionalized and written on general orthogonal curvilinear coordinates (s,n) with $h_1(s,n)$ and $h_2(s,n)$ the scale factors. Wall coordinates are expressed in this new system of reference as (s_w, n_w).

Taking the limit $\varepsilon = Re^{-1/2} \rightarrow 0$, n being fixed, the second order approximation of Navier-Stokes equations are usual Euler equations for the functions U, V, R, H.

$$\frac{\partial}{\partial s} (h_2 RU) + \frac{\partial}{\partial n} (h_1 RV) = o(\varepsilon) \tag{1a}$$

$$RU\frac{\partial U}{\partial s} + \frac{RV}{h_2}\frac{\partial h_1 U}{\partial n} - \frac{RV^2}{h_2}\frac{\partial h_2}{\partial s} + \frac{\partial P}{\partial s} = o(\varepsilon) \tag{1b}$$

$$h_2\frac{RU}{h_1}\frac{\partial V}{\partial s} + \frac{RUV}{h_1}\frac{\partial h_2}{\partial s} - \frac{RU^2}{h_1}\frac{\partial h_1}{\partial n} + RV\frac{\partial V}{\partial n} + \frac{\partial P}{\partial n} = o(\varepsilon) \tag{1c}$$

$$RU\frac{\partial H}{\partial s} + h_1\frac{RV}{h_2}\frac{\partial H}{\partial n} - U\frac{\partial P}{\partial s} - h_1\frac{V}{h_2}\frac{\partial P}{\partial n} = o(\varepsilon) \tag{1d}$$

For laminar flows, the internal approximations written in variable $N=(n-n_w)/\varepsilon$ with N being fixed and $\varepsilon \to o$ gives usual second order boundary layer equations for the functions u, v, ϱ, h.

$$\frac{\partial}{\partial s}(h_2\varrho u) + \frac{\partial}{\partial N}(h_1\varrho v) = o(\varepsilon) \tag{2a}$$

$$\varrho u\frac{\partial u}{\partial s} + \frac{\varrho v}{h_2}\frac{\partial h_1 u}{\partial N} + \frac{\partial p}{\partial s} - \frac{1}{h_1 h_2}\frac{\partial}{\partial N}\left[\frac{h_1^2}{h_2}\left(\mu\frac{\partial u}{\partial N}\right) - \frac{\mu u h_1}{h_2}\frac{\partial h_1}{\partial N}\right] = o(\varepsilon) \tag{2b}$$

$$-\frac{\varrho u^2}{h_1}\frac{\partial h_1}{dN} + \frac{\partial p}{\partial N} = o(\varepsilon) \tag{2c}$$

$$\varrho u\frac{\partial h}{\partial s} + \frac{h_1}{h_2}\varrho v\frac{\partial h}{\partial n} - u\frac{\partial p}{\partial s} - \frac{h_1}{h_2}v\frac{\partial p}{\partial N} - \frac{1}{h_2}\frac{\partial}{\partial N}\left[\frac{h_1}{h_2}\frac{\mu}{Pr}\frac{\partial h}{\partial N}\right] - \frac{h_1\mu}{h_2^2}\left[\frac{u}{h_1}\frac{\partial h_1}{\partial N} - \frac{\partial u}{\partial N}\right]^2 = o(\varepsilon) \tag{2d}$$

We try now to find the most appropriate coordinate system. We will define it as the one which gives the largest matching domain between the internal and the external approximation. We use for this intermediate functions $u_i(s,N^*,\varepsilon)$, $h_i(s,N^*,\varepsilon)$, $v_i(s,N^*,\varepsilon)$... of an intermediate variable $N^* = (n-n_w)/\eta$ where η is a small parameter to be defined.

Let these function obey the non viscous part of the internal approximation (left hand side of system 2) with the following boundary conditions on the n=0 line :

$$u_i(s,0,\varepsilon) = U(s,0,\varepsilon), \quad h_i(s,0,\varepsilon) = H(s,0,\varepsilon), \quad v_i(s,0,\varepsilon) = V(s,0,\varepsilon)$$

One consider a first coordinate system based upon one limiting streamline so that :

$$V(s,N^*,\varepsilon) = \eta N^*\frac{\partial V}{\partial n}(s,o,\varepsilon) + o(\eta)$$

and if : $\quad 0<\eta\ll\varepsilon^{1/3}$,

then $u_i(s,N^*,\varepsilon) = U(s,N^*,\varepsilon)+o(\varepsilon)$, $hi(s,N^*,\varepsilon) = H(s,N^*,\varepsilon)+o(\varepsilon)$...

This can be easily proved if one see that the left hand side of equation 2 becomes identical with equations 1 up to the second order with identical boundary conditions.

In the same way, one can consider a second coordinate system based upon all streamlines such as $v_i(s,n,\varepsilon) = 0$, systems 1 and 2 are always identical for $0 \leq \eta \leq 1$.

Boundary layer equations are written in a defect form to have both sides of equation equal to zero for $N^* \to \infty$. For S-Momentum equation, the left hand side becomes :

$$\frac{1}{h_1 h_2} \frac{\partial}{\partial N} \left[\frac{h_1}{h_2} \left(\mu \frac{\partial h_1 u}{\partial N} - \mu_i \frac{\partial h_1 u_i}{\partial N} \right) - \frac{2h_1}{h_2} \frac{\partial h_1}{\partial N} (\mu u - \mu_i u_i) \right]$$

If $\varepsilon \ll \eta \ll 1$, the equations for intermediate and boundary layer equations are identical to second order with the same boundary conditions.

Then for all N^*, $u = u_i$, $h = h_i$, $v = v_i$.

Then, one can write that if $V(s,0,\varepsilon) = 0$ the overlapping domain is $D =]\varepsilon, \varepsilon^{1/3}[$ and if $V(s,n,\varepsilon) = 0, \forall n$, it becomes $D =]\varepsilon, 1]$. This can also be seen on fig. 1a and 1b where u, U and u_i are shown as a function of n for both coordinate systems.

3. External flow computation

3.1. Coordinates linked with one limiting streamline

The limiting streamline is caracterized by its coordinates in the cartesian system (x,y). The values of h_1 and h_2 are :

$$h_1 = 1 + Kn \quad , \qquad h_2 = 1$$

where $K = - \frac{\partial \theta}{\partial s}$ is the curvature of streamline.

Oblique shock wave theory, tangent wedge or Prandtl Meyer relations can be used to relate pressure and deviation :

$$p = f(\theta)$$

If incoming flow is isentropic and no focalisation occur in the computational domain, a very simple analytical solution of equation 2 is given by [5] :

$$u_i(s,n) = \frac{u_0(s)}{h_1(s,n)} \tag{3a}$$

$$h_i(s,n) = h_0(s)\left[1 + \frac{u_0^2(s)}{2h_0(s)}\left[1 - \frac{1}{h_1^2(s,n)}\right]\right] \tag{3b}$$

$$\frac{p_i(s,n)}{p_0(s)} = \left[\frac{\varrho_i(s,n)}{\varrho_0(s)}\right]^{\gamma} = \left[\frac{h_i(s,n)}{h_0(s)}\right]^{\gamma/\gamma-1} \tag{3c}$$

3.2. Coordinates linked with all streamlines

One has to solve Euler equations (2a,b,c,d) for $v_i=0$, with geometrical relations between h_1, h_2 and θ, owing to the orthogonality of coordinates

$$\frac{\partial h_1}{\partial h} = -h_2\,\frac{\partial \theta}{\partial s} \quad , \qquad \frac{\partial h_2}{\partial s} = h_1\,\frac{\partial \theta}{\partial n} \tag{4a,b}$$

In the second coordinate system, the characteristic lines are given by :

$$\frac{dn}{ds} = \pm\,\frac{h_1}{h_2}\,\mathrm{tg}\,\alpha \text{ where } \sin\alpha = \frac{1}{M} \text{ and M the local Mach number}$$

As the pressure-deviation relation is used here on the ψ=0 line, this implies non usual boundary conditions to make it compatible with complete Euler calculation.
Those conditions are :
on n=0 line : $h_1 = 1$, θ , $p = f(\theta)$ on s=0 line : $h_2 = 1$, S

Taking first h_1 and h_2 equal to 1 and computing eq.1a,b,c,d with v=0, by characteristic methods above the n=0 line, one obtain θ and then h_1 and h_2 from eq.4a,b. The process is repeated with the new values of h_1 and h_2 two or three times until convergence is achieved. Under the n=0 line, the values of θ are extended from above and h_1 and h_2 deduced.

Then h_1 and h_2 are fitted in polynomial expressions of n at each step s by a last square method to be used in boundary layer computation. One can do the same for u_i, h_i, ϱ_i, p_i but if one suppose the external flow isentropic, eq.3a,b,c can again be used.

4. Boundary layer

A new coordinate transformation is performed to fit the mesh with the wall. It is given by :

$$\eta = \frac{n-n_w}{n_e-n_w} \qquad \text{where } n_e \text{ is the edge of the boundary layer}$$

The computing domain has now a rectangular shape.

Boundary layer equations become :

$$* \ \frac{\partial}{\partial s}(h_2 \varrho u) + \frac{\partial v^*}{\partial \eta} - \frac{h_2 \varrho u}{n_e-n_w}\frac{dn_w}{ds} = 0$$

$$* \ \varrho u \frac{\partial u}{\partial s} + \frac{v^*}{h_2 h_1}\frac{\partial h_1 u}{\partial \eta} - \varrho_i u_i \frac{\partial u_i}{\partial s} + \int_\eta^1 \frac{\varrho_i u_i^2 - \varrho u^2}{h_1}\frac{dh_1}{d\eta}\, d\eta$$

$$= \frac{1}{Re(n_e-n_w)^2 h_1 h_2}\frac{\partial}{\partial \eta}\left[\frac{h_1}{h_2}\mu\frac{\partial h_1 u}{\partial \eta} - \frac{2h_1}{h_2}\frac{\partial h_1}{\partial \eta}(\mu u - \mu_i u_i)\right]$$

$$* \ \varrho u \frac{\partial h}{\partial s} + \frac{v^*}{h_2}\frac{\partial h}{\partial \eta} - \frac{u^2 v^*}{h_1 h_2}\frac{\partial h_1}{\partial \eta} + u \varrho_i u_i \frac{\partial u_i}{\partial s} - u\frac{\partial}{\partial s}\int_\eta^1\left(\frac{\varrho_i u_i^2 - \varrho u^2}{h_1}\right)\frac{\partial h_1}{\partial \eta}\,\partial\eta$$

$$= \frac{1}{Re(n_e-n_w)^2 h_2}\left[\frac{\partial}{\partial \eta}\left[\frac{h_1}{Prh_2}\left[\mu\frac{\partial h}{\partial \eta} - \mu_i\frac{\partial h_i}{\partial \eta}\right]\right] + \frac{h_1}{h_2}\left[\mu\left[\frac{u}{h_1}\frac{\partial h_1}{\partial \eta} - \frac{\partial u}{\partial \eta}\right]^2 - \mu_i\left[\frac{u_i}{h_1}\frac{\partial h_1}{\partial \eta} - \frac{\partial u_i}{\partial \eta}\right]^2\right]\right]$$

The boundary conditions are :

at the wall :

$$u = 0\ , \qquad h = h_w\ , \qquad v^* = \frac{\varrho h_2}{n_e-n_w}\left[u\frac{dn_w}{\partial s}(\eta-1) + \frac{h_1}{h_2}v\right] = 0$$

at the outer edge, inverse method requires a condition on v. We take here :

$$\lim_{N\to\infty}(h_1 \varrho v) = \lim_{N\to\infty}(h_1 \varrho_i v_i)$$

which gives by integrating the continuity equations :

$$\frac{d}{ds}\left[(n_e-n_w)\int_0^1 h_2(\varrho_i u_i - \varrho u)d\eta\right] = -\varrho_{iw}u_{iw}\left[h_{2w}\frac{\partial n_w}{\partial s} - h_{1w}\,\mathrm{tg}\,\theta_{iw}\right]$$

θ_{iw} is the deviation at the wall of the intermediate fluid and can be obtained by the pressure deviation relations $\theta_{iw} = f(p_w)$ u_i, h_i, p_i, ϱ_i are functions of $u_0(s)$, $h_0(s)$ by relation 3a,b,c.

Additional relations used are :

$$\lim_{N\to\infty} (u-u_i) = 0 \quad , \quad \lim_{N\to\infty} (h-h_i) = 0$$

with state equation

$$\frac{p}{\varrho} = \frac{\gamma-1}{\gamma} \; h$$

the numerical procedure is then identical with ref. [4].

5. Coupling procedure

The coupling procedure has been described in ref.[6]. We shall here brievely summarize it. It is obtained by solving the functional relative $F[P(s,0)] = \text{grad}\ (p-P)(s,0) = 0$

For this, we use a Newton method and write that :

$$\left[\frac{\partial F}{\partial P}\right]^k P^{k+1}(s,0) = 0 = \left[\frac{\partial F}{\partial P}\right]^k P^k(s,0) - F[P^k(s,0)]$$

One can show (ref.1) that the Jacobian matrix $\frac{\partial F}{\partial P}$ can be approached by :

$$\begin{bmatrix} F_1+1 & F_1-1 & & & & & \\ F_2 & F_2+1 & F_2-1 & 0 & & & \\ \vdots & \vdots & \vdots & \vdots & & & \\ F_j & \cdots & F_j & F_j+1 & F_j-1 & 0 & \cdots \end{bmatrix}$$

where $F_j = \frac{\partial (\text{grad } P)J-1/2}{\partial PJ}$ are obtained by two inverse boundary layer computations. With P prescribed at the end of the interaction, the above relations allow pressure to be corrected.

6. Results and conclusions

A supersonic case is first computed ($M_\infty = 3$, $R_e = 1.68.10^4$ for a 10° ramp) with the first coordinate system, and is compared with Navier Stokes results of CARTER [7]. Wall pressure along the ramp has been reported on fig.2a. The accordance between the two computations is good. On fig.2b, isobars and streamlines of the internal flow are reported in cartesian coordinates. They are also in good agreement with results of ref.[7].

When Mach number becomes higher, one can see (fig.3) that either the first order wall coupling computation or the second order method based on the limiting streamline show discrepancy with LEWIS measurements [8]. With streamlines fitted coordinates, the results are better and quite in agreement with Navier-Stokes computation of CARTER [7] (not reported here) except in the very downstream part of the interaction due to the imposed unperturbed external pressure.

On fig.4, first results for turbulent flow with the first coordinate system, and a frozen eddy viscosity model are presented. One can see the good agreement of the second order computation in the free interaction part and the important effect of curvature terms on the plateau-pressure. The rough turbulence model is probably the cause of the discrepancy of the downstream results and effort is now being done on this field.

In conclusion, it is shown that the choice of the coordinate system in which second order approximation of Navier-Stokes equations is developed is very important. In supersonic range the first coordinates system based upon the displacement surface concept seem sufficient, but for hypersonic Mach numbers one must use streamlines fitted coordinates.

Acknowledgment

The author gratefully acknowledge Professor T. Alziary de Roquefort of University of Poitiers for his help during this work.

References

1. Davis R.T.; Rubin S.G. : "Non Navier-Stokes viscous flow computation", Comp. and Fluids Vol.8, p. 101

2. Kaplun S. : "Fluid Mechanics and singular perturbations", Academic press, 1967

3. Le Balleur J.C. : "Viscid-inviscid coupling calculations for Two and Three- Dimensional Flows", VKI LS 1982-04, 1982

4. Ardonceau P.; Alziary de Roquefort T.; Aymer de la Chevalerie D. : "calcul de l'interaction onde de choc-couche limite avec décollement", AGARD CP 291, 1980

5. Davis R.T.; Werle M.J.; Wornom S.F. : "A Consistent Formulation of Compressible Boundary-Layer Theory with Second Order Curvature and Displacement Effects", AIAA Journal, Vol.8, No.9, 1970

6. Aymer de la Chevalerie D. : "Algorithme de couplage inverse pour le calcul d'une interaction onde de choc couche limite", C. R. Acad. Sc. Paris t. 297 (26 septembre 1983)

7. Carter : "Numerical Solutions of the Supersonic, Laminar Flow over a Two-Dimensional Compression Corner", Lecture Notes in Physics, Vol.19, Springer Verlag (NY) 1973

8. Lewis J.E.; Kubota T.; Lees L. : "Experimental Investigation of Supersonic Laminar, Two-Dimensional Boundary Layer Separation in a Compression Corner with and without Cooling" AIAA Journal, Vol.6 No.1 1968

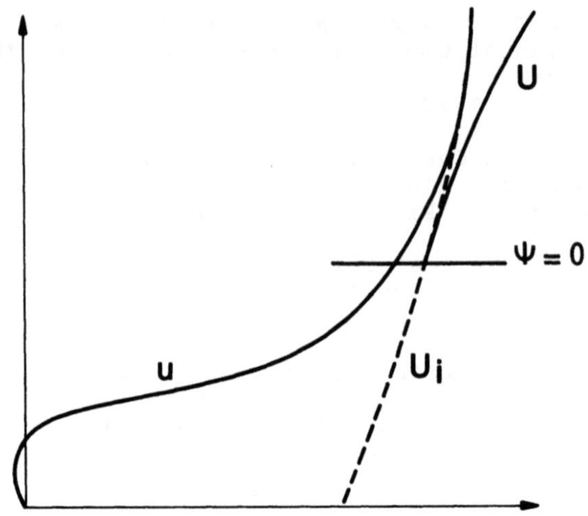

a) Coordinates linked with one streamline

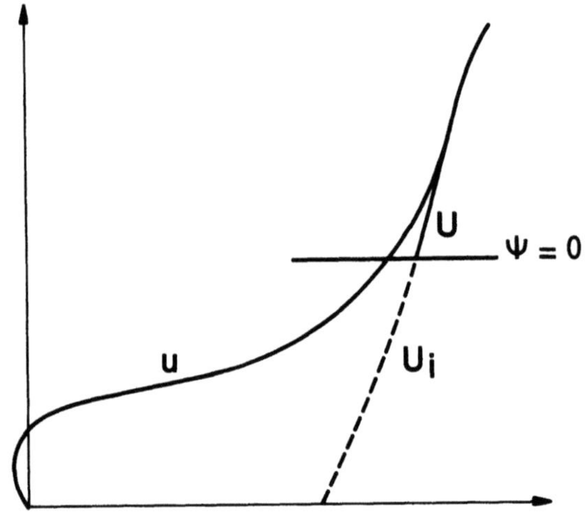

b) Coordinates linked with all streamlines

FIG. 1 - VISCOUS INVISCID MATCHING OF VELOCITY PROFILES

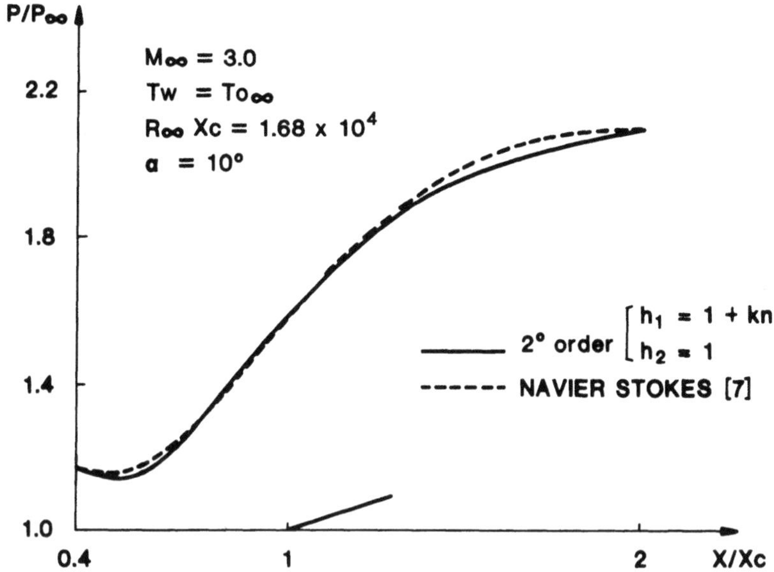

a) Wall pressure

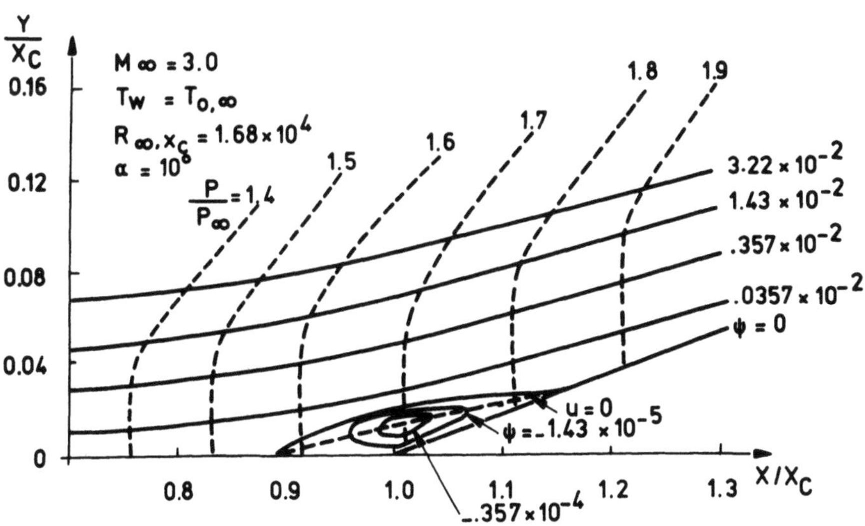

b) Isobars and streamlines of external flow

FIG.2 - LAMINAR SUPERSONIC FLOW

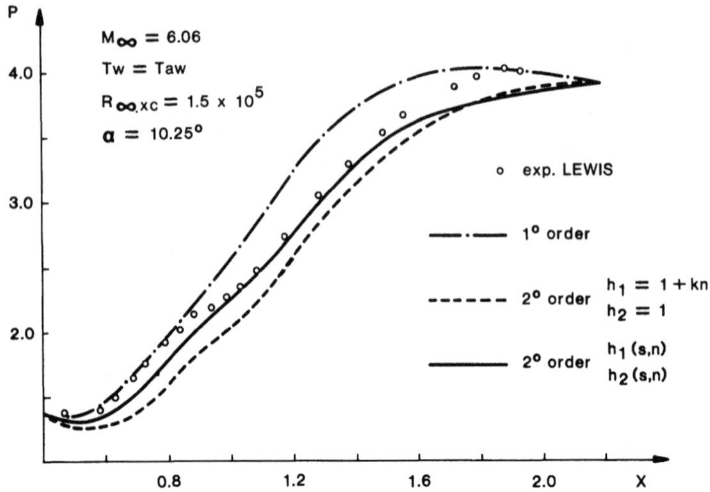

FIG.3 - LAMINAR HYPERSONIC FLOW

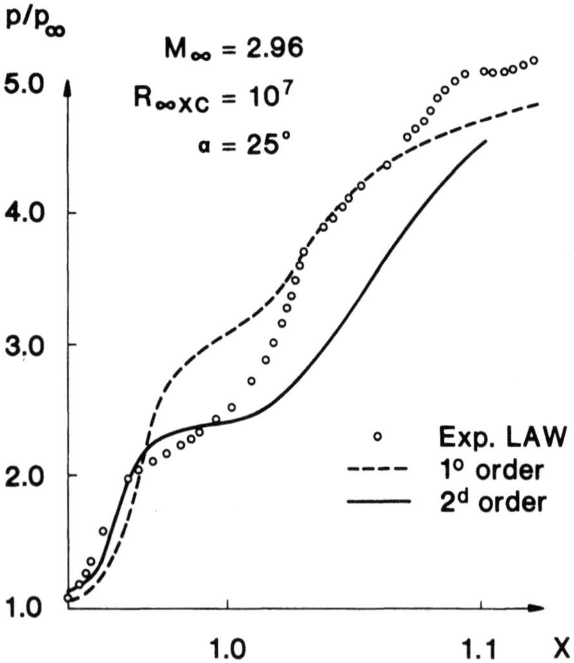

FIG.4 - TURBULENT SUPERSONIC FLOW

Navier-Stokes and Experimental Modeling of Blunt-Base Rocket Nozzle Flow

C. K. Forester

I. Strom

Boeing Military Airplane Co.

P. O. Box 3707 M/S 33-14

Seattle, Washington 98124

(206)241-4402

Summary

The influence of grid placement and grid density on the ability to compute an underexpanded jet in a supersonic afterbody flowfield is studied using a conservation-based body-fitted computational technique. The thin-shear-layer formulation of the compressible, Reynolds-averaged, Navier-Stokes equations together with mass and energy conservation equations are modeled using an artificial time-dependent, explicit numerical algorithm with the turbulence approximated by a two-layer algebraic model with wall functions for the solid boundaries and whose properties are tailored to the three physically distinct mixing zones. Solutions are obtained for supersonic flow over an axisymmetric conical afterbody with a blunt base, containing a centered propulsive jet where the freestream Mach number is 2.0 and the jet exit Mach number is 2.5. Exhaust exit-plane static pressures are considered in the range of one to nine times the freestream static pressure. The conical nozzle-exit half-angles are zero and twenty degrees. Comparisons are made between computed and experimental results for base pressure, separation length, afterbody pressure distribution, and flowfield structure. The numerical solutions are found to be sensitive to the computational grid structure and the mixing (turbulence) model. Error norms are applied to aid the detection of inappropriate grid choices. The best results are obtained with adaptive grids that track both free shear layers and a mixing model which is germain to the local flow features.

Introduction

Jet engine and rocket motor exhaust plumes dramatically influence the afterbody aerodynamic drag by complex interactions between the jet plume and the external flow. The aerodynamic drag is influenced by the afterbody geometry, the nozzle geometry and the exhaust nozzle pressure ratio. Where exhaust plumes are underexpanded, their influence on base pressure and flow over the afterbody can be large, inducing in some instances extensive regions of flow separation. The flowfields for these configurations are complex with regions of strong inviscid/inviscid,

Turbulent Shear Layer/Shock Wave Interactions
IUTAM Symposium Palaiseau 1985
Editor: J. Délery

viscous/inviscid, and viscous/viscous interaction. Free shear layers, wall boundary layers, shocks, pressure dominated viscous flow, potential flow, and inviscid rotational flow can all interact. The goal of the present computational method is to model these interactions to engineering accuracy using a field grid formulation of the governing differential equations. The central ingredients of this are grid placement, grid density, residual error constraints and the mixing zone models. This report summarizes the search for correct combinations of these four ingredients.

Nozzle Geometry and Flow Phenomena

Figure 1 shows the cross section of the type of axisymmetric rocket nozzle which has been studied extensively (references 1-5). The geometry of the nozzle consists of a cylindrical body with a conical boattail. The convergent/divergent nozzle channel is smooth with no abrupt slope changes. It has a conical exit. The base of the nozzle is blunt due to a truncation normal to the centerline. Figure 2 shows the principle flow features from experimental studies. Comparisons of computations[1-3] with schlieren photographs show generally good agreement between the computed and experimental results with respect to the shock wave structure. These contributors concluded that further work was required for the accurate prediction of the base pressure and the associated separation zone length. In general, these contributors predicted separation lengths and base pressures that were too small. Nevertheless, the trends are correctly predicted - the base pressure and separation length increase with the increase of the nozzle pressure ratio and the increase of the nozzle exit angle.

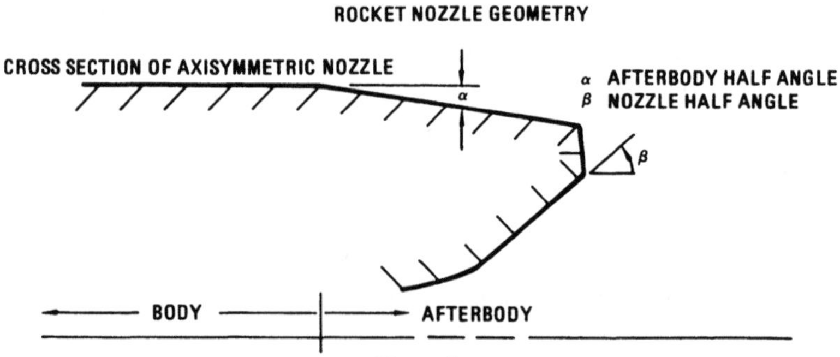

Figure 1.

Neglecting for the moment the effect of turbulent losses on the total pressure decay of the nozzle flow, the base pressure of a supersonic nozzle flow in a supersonic external stream is governed by the balance between the separated boundary layer of the afterbody and the resultant pressure of the Prandtl-Meyer centered expansion at the inner corner of the nozzle base. More Prandtl-Meyer turning leads to lower base pressure and visa versa for a given nozzle pressure ratio. Small nozzle exit half-angles promote more Prandtl-Meyer turning to match the outer flow. Thus smaller nozzle exit half-angles lead to lower base pressure for a given nozzle pressure ratio. Upon this trend is superimposed the total pressure loss mechanisms by mixing layers and wall boundary layers which modify the base pressure from the Euler equation model stated above.

The surface area dividing the low and high speed flow regions (see Figure 2) increases with increased distance from the nozzle base to the point of reattachment at which the two free shear layers merge into the final mixing zone between the nozzle plume and the external flow. The exposure of the base flow region to mixing is related to this surface area and the intensity of the mixing process. Favorable pressure gradients reduce mixing zone intensity while adverse pressure gradients excite mixing. Thus, the mixing between the high speed nozzle flow region and the low speed base flow region is very low because of the favorable pressure gradients that are associated with the Prandtl-Meyer expansion. Because of the adverse pressure gradients between the high speed external flow and the low speed mixing region near the base of the nozzle, the mixing can become very intense over a substantial surface area. Clearly the mixing model must include at least three types of mixing flow:

1. nearly laminar flow layer between the high speed nozzle flow and the low speed base flow region
2. the vigorous flow mixing region between the high speed external flow and the low speed base flow region
3. the development of the single turbulent free shear layer downstream of the reattachment point.

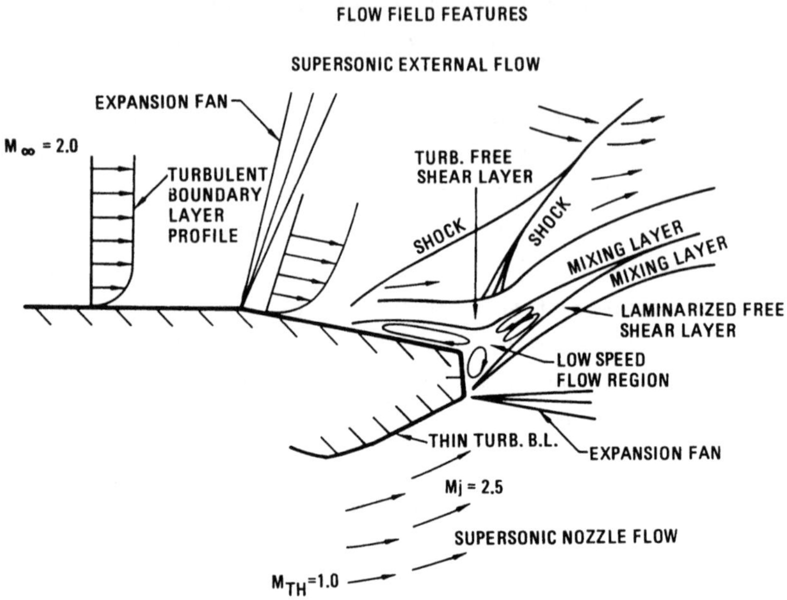

Figure 2.

The Numerical Study

In the present study the two-layer algebraic mixing model is applied to situations (2) and (3) except that the scale factor of each region is different. The basis for chosing the scale factors is provided by Birch.[6] Laminar flow is assumed for situation (1). Euler flow modeling is achieved by setting all momentum diffusion coefficients to zero. In the present study, Euler and Navier-stokes flow modeling is performed to assess the effects of the mixing upon the base pressure, reattachment length of the base flow and the detachment length of the external boundary layer.

In the present study truncation errors that are inherent in the numerical model are examined by the use of three processes:

1. varying the grid density
2. varying the grid distribution
3. computing the truncation error by a multigrid process.[7]

The first of these is implemented by the usual method of changing the number of grid points. The range of 600 to 6000 cells is used in the present study. The second of these is implemented by two approaches for adaptive grid distributions:

1. wake-adaptive grid
2. free shear layer adaptive grid

In both of these grid generation methods, the vertical grid coordinates are adjusted continuously during the solution process.

The base flow region grid is composed of parallelograms, see Figure 3, for the wake-adaptive grid design. For the second grid pattern, shown in Figure 4, the base flow region is composed of quadrilaterals. The grid in this base flow region is squeezed by forcing the edges of the grid for the base flow region to conform to fitting both free shear layers along the plume edges for continuous static pressure across the plume edges. Figure 5 shows an overview of the grid of the analysis domain including the nozzle flow, the base flow and the exterior flow regions.

WAKE-ADAPTIVE GRID,
BASE REGION

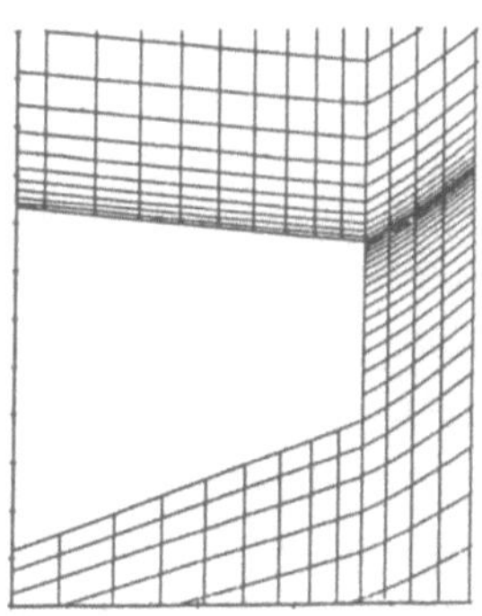

Figure 3.

SHEAR LAYER ADAPTIVE GRID,
BASE REGION

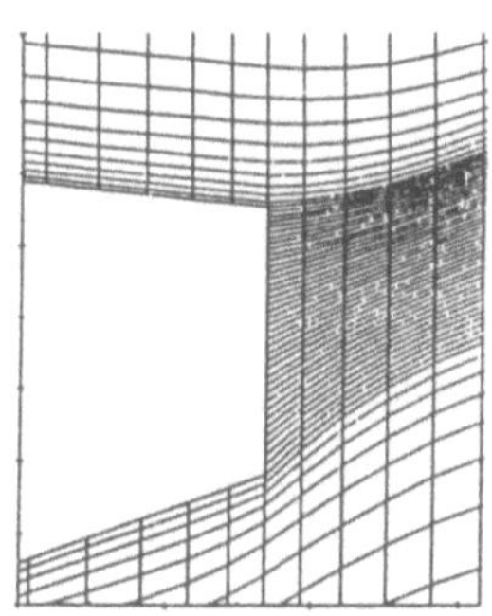

Figure 4.

ANALYSIS DOMAIN GRID

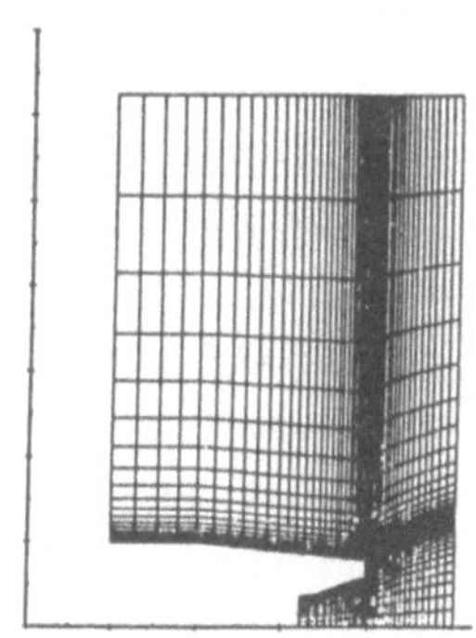

Figure 5.

The search for grid patterns with minimum truncation error effects has been aided by the use of error norms which are being developed for that purpose.[7,10] For example the artificial diffusion ratio[10] (smoothing flux divided by total flux) is a measure of how much smoothing is being applied for stable, convergent computations of the convection terms which include vorticity, linear momentum, mass density, and energy density. Clearly grids are sought in which local integrals of this ratio are minimum. Other error norms employed included the ratio of the local residual (see the next section) to local truncation error (evaluated by a multigrid process[7]), the ratio of the diffusion and convection of vorticity, and local integrals of these quantities as necessary. To our knowledge, the use of these error detection algorithms is new (see the acknowledgement). They show promise for assessing the numerical accuracy of the results of these and related computations.

The control of residual error is also necessary. When the discretized equations are solved, each equation has a residual error. These errors are reduced by iteration processes. Arbitrary cut-off points are used to stop the iteration process. The impact of residual error upon the numerical solution for the grid coordinates and for the values of the dependent variables is assessed by chosing a sequence of residual tolerances for complete solutions. An error table[7] is constructed from this data to define the residual error spectrum and control residual error contamination.

Computational Approach

The computational approach in the present study is an extension of the work of Forester[8] and Peery[9]. It uses a conservation-based body-fitted adaptive grid model of the thin-shear-layer formulation of the compressible, Reynolds-averaged, Navier-Stokes equations together with mass and energy conservation equations. Control of the residual errors is achieved by an artificial time relaxation approach. Control of the truncation errors is through the use of a sequence of solutions on varying grid densities and varying grid distributions. The formulation of the governing thin-shear-layer equations, the formulation of the finite volume explicit predictor-corrector finite difference algorithm, the formulation of the boundary conditions, the formulation of the two-layer algebraic turbulence model and the associated wall functions, and the formulation of the mesh generator and the adaptive mesh mover are detailed by Peery and Forester[9]. The current numerical process has been enhanced by the use of the local cell-related time step (constant CFL over the analysis domain) to accelerate the convergence. Also smoothing has been applied to the coefficients of the smoothing algorithm to obtain consistant convergence. Without this feature limit cycles of the smoothing coefficient would occasionally prevent convergence to steady state. Convergence to steady state is now reliable.

All the numerical results reported here-in employed the axisymmetric version of the computer code. The computer program is setup for a choice up to three coupled computational regions. In the present study computational region one is assigned to the nozzle flow field, computational region two is assigned to the base flow region and computational region three is assigned to the remaining flow field which includes the freestream. The grid is composed of the intersections of parallel vertical lines and warped pseudo-stream lines. The coordinates of the pseudo-stream lines are adjusted to satisfy the condition of constant static pressure between computational regions one and two and between computational regions two and three. The nozzle wall boundary layer is applied at the top edge of computational region one. The afterbody wall boundary layer grid is continued downstream of the nozzle corners. The boundary layer grid for the free shear layers is adaptively warped to keep the free shear layers in this grid. Migration of the free shear layers from this grid artificially exaggerates the mixing rate of the free shear layers in the coarse grids which lay near the boundary layer grids. Forcing the boundary layer grid to track the free shear layer dramatically improves the resolution compared to most nonadaptive grid choices. As is shown in the computational results, dramatically improved accuracy is achieved as a result.

Use of the nonadaptive grid choice may take many forms - the grid coordinates may be frozen during the solution process, the grid coordinates may be adjusted constantly to various constraints which do not lead to fitting the free shear layers to the boundary layer grids.

The last option was used in the so-called wake-adaptive grid computation for the present study. The constraint upon the free shear layer grids was: 1) a constant spacing was imposed between the free shear layer grids of the amount of the dimension of the base of the nozzle and 2) the wake was tracked but the free shear layers were not. This approach to generating the wake-adaptive grid was selected to provide grids that most nearly agree with Deiwert's[3] choice of grid configuration and spacing relationships.

Computational Studies

Solutions are obtained for supersonic flow over an axisymmetric conical afterbody containing a centered propulsive jet where the freestream Mach number is 2.0 and where the jet exit Mach number is 2.5. Exhaust exit-plane static pressures range from one to nine times the freestream static pressure. Both the external flow and the jet exhaust flow are ideal air at a common total temperature of 285.6 degrees Kelvin. The Reynolds number of the external flow is about one million based upon the afterbody diameter. The boundary layer thickness along the internal nozzle wall is assumed to be negligible because the experimental boundary layer thickness is small relative to the dimensions of the nozzle. The boundary layer thickness on the afterbody is defined by the experimental data for thickness and shape factor. Resolution of this boundary layer is achieved within the framework of wall function analysis. The nozzle throat Mach number is prescribed at a uniform Mach number of one. The freestream Mach number is uniform at 2.0. The computations correspond, in part, to a series of experiments by Agrell and White[5] and others of the AGARD[1,2] study for identical flow conditions and for nearly the same configuration. Agrell and White[5] used an afterbody half-angle of eight degrees. The AGARD[1] study used six degrees for the afterbody half-angle. This study used six, eight and 11.7 degrees for the afterbody half-angles. Three afterbody half-angles are chosen for the study of the sensitivity of the base pressure to the half angle of the afterbody. The nozzle exit half-angle are zero and twenty degrees for all the computations in this report.

Figure 6 shows the velocity vectors for nozzle pressure ratios of one through nine. Notice the dramatic change in flow field structure with respect to the nozzle pressure ratio. Figure 6d corresponds most nearly to the schematic drawing in Figure 2 with respect to the magnitude of the upstream influence of the base flow region. These figures clearly show the change in the size of the low Mach number nearly-constant-pressure region. Figure 7 shows the base flow region for the wake-adaptive grid computations with the pressure ratio set to nine. A comparison of Figures 6 and 7 illustrate the substantial change in the solution of the base flow region by simply changing the grid from fitting the wake to fitting the free shear layers with the boundary layer grid region. Flow separation does not exist in the solution with the fitted wake. The solution with the fitted free shear layers shows a substantial region of stagnated flow with the result that the shock structure is displaced upstream by a significant amount. This upstream shift agrees more nearly with the experimental data. To further illustrate the severity of the flow gradients in the base region, Figure 8 show the Mach number contours contours of the free shear layer adaptive grid solution. Sharp flow turning near the nozzle corners are accompanied by steep gradients normal to the stream lines. Figure 9 shows the artificial diffusion ratio (smoothing flux divided by total flux) in the neighborhood of the base corners with the wake-adaptive grid computations. This corresponds to a loss in total pressure and vorticity. This loss is reduced by the shear layer adaptive grid computations.

Figure 10 shows the results of the computations for wake-and free shear layer adaptive grid options. Dramatic differences between these results are indicated. The wake-adaptive grid results agree with Deiwert's results (fixed and gradient adaptive grid) which under-predict the base pressure. Also shown in Figure 10 are the Euler and Navier-Stokes results. Notice that the base pressure is changed by the mixing. The Navier-Stokes solutions most nearly agree with the experimental data for the base pressure prediction.

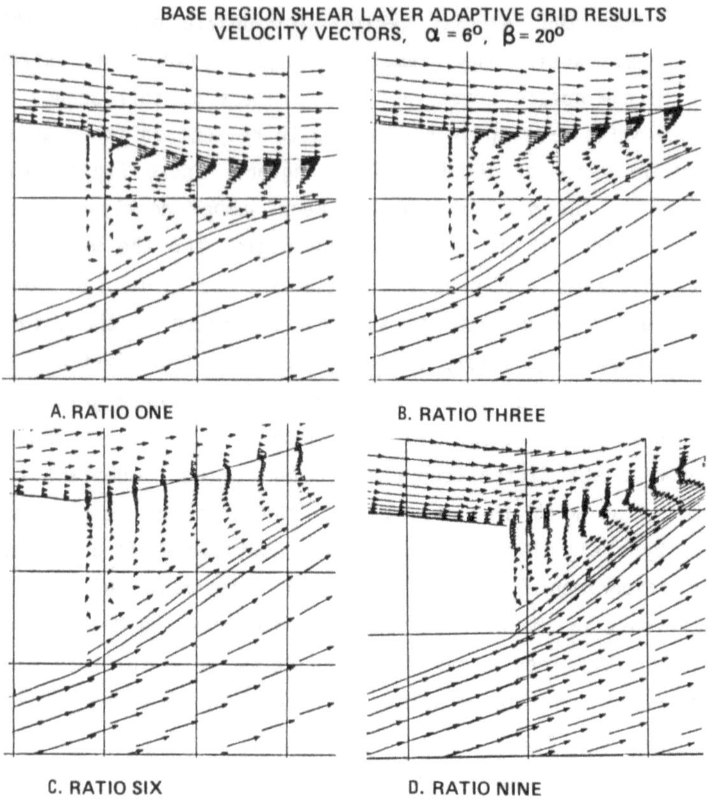

Figure 6.

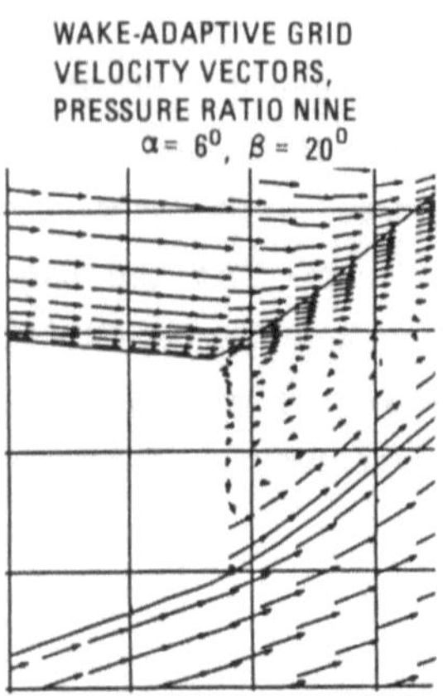

Figure 7.

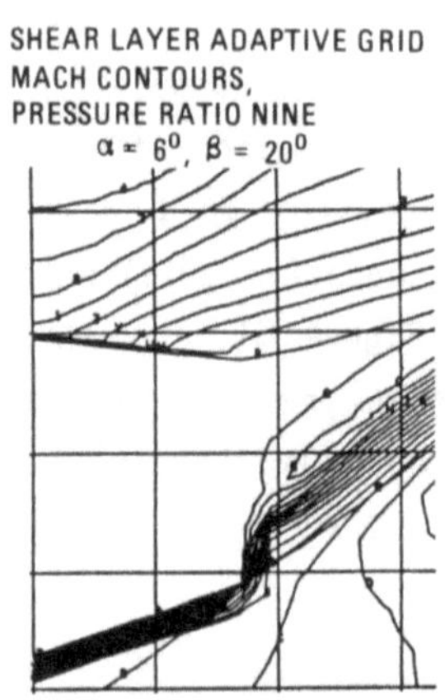

Figure 8.

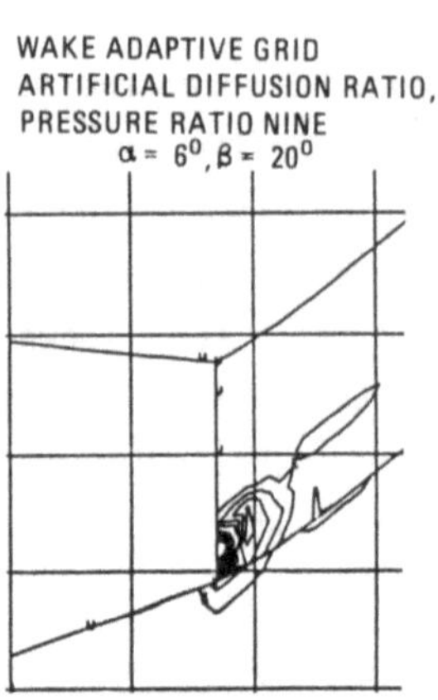

Figure 9.

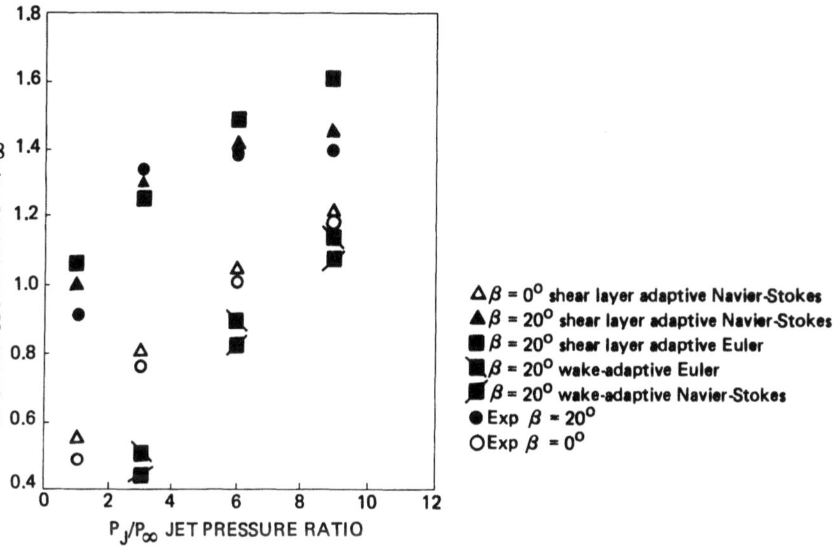

Figure 10. Base Versus Jet Pressure Ratio

The only anomally found in the present results has to do with the prediction of the position of the shock system and the location of separation point. Both features are too far downstream. For example for a pressure ratio of nine and a nozzle angle of twenty degrees, the separation point should be at about one half of the length of the boattail. In the present result, the separation point is near the seven eights point on the boattail for these nozzle conditions with free shear layer adaptive analysis. The wake adaptive Navier-Stokes analysis for these nozzle conditions shows no separation. Consequently, the shear layer adaptive grid provides improved analysis accuracy. The reason why the separation point is not accurately predicted when the base pressure is accurately predicted has not been determined at this time.

It is desirable to reduce the complexity of the analysis process that provides accurate results. To this end, error norms are being developed to monitor solution accuracy.[7,10]. Results on the utility of these error monitors will be given elsewhere.

Conclusions

The computations of axisymmetric rocket nozzle flow in the presence of supersonic exterior flow for nozzle exit plane to freestream static pressure ratios of one through nine have been presented. The accuracy of the computations are shown to be strongly related to grid selection, residual error tolerances, and the mixing process near the base of the nozzle. For the first time, error norms have been applied to aid the search for the grid configurations that minimize the local truncation errors. The use of these error norms has been found to be beneficial. In summary, an approach has been presented which provides accurate computations of rocket nozzle flows for engineering purposes for the base pressure. Further work is required to accurately predict the separation length.

Acknowledgement

Support for this work comes in part from AFOSR; Dr. Jim Wilson, Technical Monitor.

References

1. Wagner, N. and L. E. Putnam (Eds.), "Comparison of Experimental and Computational Modeling of Boattail/Nozzle Flow," AGARD WG-08, 1985.

2. Putnam, L. E. and Bissinger, N. C., "Results of AGARD Assessment of Prediction Capabilities for Nozzle Afterbody Flows," AIAA paper 85-1464.

3. Deiwert, G. S., "Supersonic Axisymmetric Flow Over Boattails," AIAA Journal, Vol 22, pp. 1358-1365, 1984.

4. Petrie, H. L. and Walker, B. J., "Comparison of Experiment and Computation for a Missile Base Region Flowfield with a Centered Propulsive Jet," AIAA paper 85-1618.

5. Agrell, J. and R. A. White, "An Experimental Investigation of Supersonic Axisymmetric Flow over Boattails Containing a Centered Propulsive Jet," The Aeronautical Research Institute of Sweden, Stockholm, Tech. Note AU-913, 1979.

6. Birch, S. F., private communication, 1985.

7. Forester, C. K., "Error Norms for the Adaptive Solution of the Navier-Stokes Equations," NASA Langley Research Center Report NASA CR-165 828, under cotract NASI-16408, May 1982.

8. Forester, C. K. "Numerical Simulation of the Interaction of Jet and Freestream Flows in Engine Exhaust Systems," AIAA paper 78-144, presented at AIAA 16th Aerospace Sciences Meeting, Huntsville, Alabama, Jan. 16-18, 1978

9. Peery, K. M. and C. K. Forester, "Numerical Simulation of Multistream Nozzle Flows," AIAA Paper 79-1549, 1979.

10. Forester, C. K., "Error Norm Guided Flow Analysis," Report #1, AFOSR Contract No. F49620-84-C-0037, April 1985.

Interactions in Three-Dimensional Steady Flows

A "Preview" of Three-Dimensional Shock-Wave/ Turbulent Boundary-Layer Interactions

Alexander J. Smits and Seymour M. Bogdonoff

Department of Mechanical and Aerospace Engineering
Princeton University
Princeton, New Jersey 08544

1. Introduction

The interaction of a shock wave with a turbulent boundary layer can drastically change aerodynamic loads, alter combustion processes, and produce intense local heating, buffeting, and loss of control. Shock-wave/boundary-layer interactions have therefore been the subject of extensive research. For the two-dimensional case, numerous mean flow data sets exist, over a wide range of Mach and Reynolds numbers, and for a number of flows detailed turbulence measurements have become available. Recently, the unsteady nature of many two-dimensional interactions has also received considerable attention.

By comparison, our knowledge of three-dimensional interactions is still very primitive. The existing work in three-dimensional shock wave/boundary layer interactions was recently reviewed by Settles & Dolling (1985). It is clear that we have only begun to understand the structure of these flows and to appreciate the influence of the interaction geometry and the effect of the initial flow conditions. In this paper, we aim to complement the work by Settles & Dolling with a critical assessment of current understanding and to formulate strategies for future research in this area.

2. Current Experimental and Computational Status

In three-dimensional interactions, there exists a large discrepancy between the information we need and the information actually available. The large number of geometric variables, the general complexity of the flow structure and the high degree of experimental difficulty means that this situation is likely to continue in the near future. Hence, given limited resources, we need to choose our future investigations carefully.

Turbulent Shear Layer/Shock Wave Interactions
IUTAM Symposium Palaiseau 1985
Editor: J. Délery

We are almost immediately restricted to studies of simple geometries. Two of the simplest are the swept compression corner and the sharp fin at an angle of attack (see Fig. 1). The interactions produced by these geometries have been studied under a wide variety of initial and boundary conditions (see, for example, Bachalo 1974, Peake 1975, Oskam 1976, Settles et al. 1979, and Kubota & Stollery 1982), and have been the subject of many analytical and computational studies (for recent examples see Korkegi 1976, Horstman 1984, Inger 1984, Knight 1984, and Stalker 1984). We therefore recommend that future studies should pay particular attention to these two geometries.

Experiments have identified many flow features, such as the upstream influence line, the primary and secondary "separation" and "reattachment" lines, the inception region, and regions of conical and cylindrical similarity (see Fig. 2). The empirical scaling laws developed by Settles & Teng (1984) describe these features as a function of Mach number, Reynolds number based on boundary layer thickness, angle of attack and angle of sweep. The generality of these scaling methods is the subject of current debate but it seems clear that empirical scaling laws can only provide a limited physical understanding of the flow behavior.

We lack a deeper understanding primarily because most experiments on sharp fin and swept compression corner interactions were restricted to surface flow visualization and measurements of the mean wall static pressure. Visualization of the flow field away from the wall is extremely useful (see, for example, Kubota & Stollery 1982, Settles & Teng 1982) but it is often very difficult to implement in practice and also difficult to interpret unambiguously. In addition, very few flow-field measurements are available. For example, mean flow surveys for sharp fins and swept wedges are available for only six different cases (Peake 1975, Oskam 1976, McClure & Dolling 1983, Settles et al. 1984). Turbulence data are available for only one case: the 10 degree fin interaction at Mach 2.9 by Tan (1985), who measured the component of the mass-flow intensity in the direction of the mean flow. Finally, the sole information on the unsteadiness of these flows comes from the recent experiments on the sharp fin interaction by Tan et al. (1985) and Tran et al. (1985), and both these studies were confined to a single Mach and Reynolds number.

To improve our understanding of the flow behavior, we urgently require information on the instantaneous structure of the flow field. This kind of data is difficult and expensive to obtain. Hence, we must further limit the number of cases considered, and the task of choosing the most informative experiment becomes even more difficult.

In this respect, recent computational work can provide a particularly valuable guide. The efforts by Horstman (NASA-Ames) and Knight (Rutgers) are especially interesting. Both use the full time-averaged Navier-Stokes equations, with different turbulence models and wall treatments. Their calculations have been most revealing. It seems that in some cases the agreement with experiment is generally better than that observed in calculations of two-dimensional flows. For example, Horstman & Hung's (1979) calculation of the 10 degree fin flow captured most of the details, even though a relatively coarse grid was used. Subsequent computations for the 24/40 degree swept wedge (Settles et al. 1984) have also shown good agreement with experimental data. Recently, Horstman (1984) calculated the interactions produced by swept wedges at Mach 3 for a wide range of angles of attack and sweep. His results, using a two-equation eddy-viscosity turbulence model, showed good agreement with the experimental static pressure distribution when the angle of attack and the angle of sweep were not too large. Knight (1984), using the Baldwin-Lomax turbulence model, computed a number of sharp-fin interactions at Mach 3, and he too found good agreement with experimental data.

Recently, Horstman calculated the 10 and 20 degree sharp fin interactions first computed by Knight, using a different turbulence model, wall treatment and computational grid. Both Horstman's and Knight's calculations agreed closely with the measured Pitot pressure, yaw angle, and surface presssure distributions, and they also agreed with each other, except within the inner portion of the boundary layer (Knight et al. 1986). This agreement occurred despite differences in the calculated eddy viscosity of up to a factor of twenty (see Fig. 3).

3. "Weak" and "Strong" Interactions

It appears that in some cases the roles of the turbulence and unsteadiness are not important in describing the behavior of the mean field. There exist two complementary explanations. First, the flow is essentially pressure driven, so that gradients of turbulent stresses are small compared to the forces caused by pressure differences. These ideas form the basis of successful triple deck analyses by Stalker (1984) and Inger

(1984). Second, the interaction is "rapid" in the sense that the distortion of the turbulence field occurs over a time which is much smaller than the characteristic time of the turbulent motions. Hence, the role of the turbulence within the interaction is essentially passive, in that it is distorted by the shearing of the mean field, without interacting with it, and nonlinear interactions can be neglected. These concepts have been used with some success by Dussauge & Gaviglio (1981) and Jayaram et al. (1985) to describe two-dimensional expansions and compressions, and this approach may also prove useful in three-dimensional flows.

It is therefore possible to classify interactions into two broad categories: "weak" and "strong". In weak interactions, the mean flow field is governed primarily by inviscid mechanisms. The turbulence plays an essentially passive role and unsteadiness is not an important mechanism. A necessary condition for an interaction to be "weak" exists if two computations, using significantly different turbulence models, agree. A sufficient condition exists if both calculations agree with experiment. The sharp fin interaction at Mach 3 for angles of attack less than or equal to 20 degrees is a good example.

In strong interactions, the mean flow can not be calculated using existing turbulence models. Generally, the turbulence is severely distorted, the flow is unsteady, and there exists strong coupling between the mean and turbulent fields. A good example is the interaction produced at Mach 3 by a swept wedge at high angles of attack and angles of sweep less than 20 degrees.

Many practical problems still exist before we can claim to understand weak interactions. Calculations cannot be assumed to be correct until they are validated by experimental work. At present, we cannot say where the calculations are reliable and where they are not. Many practical difficulties associated with grid resolution remain, especially near the wall and in the vicinity of shock waves. Nevertheless, it is very encouraging to find that computational methods have evolved to the point where many important flows can be calculated with some confidence. These computations can also be used to develop our physical understanding. By drawing streamlines in different planes, the flow topology can be mapped out (see, for example, Horstman 1984). Vorticity contours can be drawn, and other flow variables can be evaluated with a resolution impossible to obtain experimentally.

Despite these considerations, the study of strong interactions clearly requires the most urgent attention. We believe that a prime subject for future research (at Mach 3) is the 24 degree swept wedge for angles of attack less than 20 degrees. In the rest of this paper, we use this particular example to suggest the kind of studies required to improve our understanding of strong interactions. We begin with a general description of the flow field development, starting with the two-dimensional case of zero sweep.

4. The Strong Interaction Produced by a Swept Compression Corner

The two-dimensional case is separated and contains a significant region of subsonic flow, as well as experiencing strong unsteady effects (Settles et al. 1979, Dolling & Murphy 1982). The shadowgraph in Fig. 4 shows the instantaneous flowfield. Qualitatively, the turbulent mixing appears to be considerably enhanced across the shock. The incursions of freestream fluid are very deep, and the shock appears to fan out and break up into a number of subsidiary shocks. Shadowgraphs, however, represent a spatial average across the flow and do not give a good indication of the behavior at any given streamwise plane. Muck et al. (1985) showed that the shock front in a two-dimensional 20 degree compression corner was not actually split but wrinkled in the spanwise direction. In addition, Dolling and Murphy (1982) found that the shock sheet exhibits significant streamwise flapping, introducing large fluctuations in the instantaneous pressure field at points ahead of the corner, thereby causing the mean pressure to rise. This "upstream influence" is seen in the wall pressure distribution (Fig. 5). The turbulence measurements by Andreopoulos et al. (1985) seem to indicate that this unsteadiness is occasionally so violent that five boundary layer thicknesses downstream of the corner undisturbed fluid, originating from the upstream freestream, can be found close to the wall. In other words, the distortion of the shock wave can be sufficient to reduce its instantaneous strength over some region to zero.

The surface flow pattern (Fig. 6) displays a cellular character in the separated zone. There is evidence to suggest that the nature of the pattern depends on upstream conditions. However, these spanwise variations may indicate the presence of longitudinal roll cells which are known to develop in concavely curved shear layers. Similar surface flow patterns were observed near reattachment by Roshko & Thomke (1966) and Shamroth & MacDonald (1970), and they seem to be a general feature of supersonic, separated flows. The presence of three-dimensional features in nominally two-dimensional flows underlines the now widely accepted notion that there are no "truly" two-dimensional flows, especially when the flow is separated.

In the two-dimensional case, side-wall fences were used to simulate the infinite-span condition. As the corner is swept back, fluid begins to flow along the corner line, and because of the finite length of the model the infinite span condition immediately breaks down. The model becomes semi-infinite, and the inviscid shock shape is now conical. An inception region appears near the apex of the model, and this region is particularly noticeable when the upstream fence is removed (see Figs. 7 and 8). For low angles of sweep the far field is nearly cylindrical, despite the apparent conical nature of the inviscid shock wave. An asymptotic state is expected but it is likely to scale with the local boundary layer thickness, rather than the thickness measured near the apex. A truly cylindrical interaction, therefore, seems unlikely to occur at any spanwise position.

With increasing sweep, the size of the inception region grows. This growth is caused by the spreading influence of the pressure disturbances initiated at the apex. At this point, the boundary layer first encounters the swept shock wave, and the pressure rise across the shock sets up a lateral pressure gradient which abruptly turns the flow in the subsonic, separated region. The turning angle of the freestream flow, however, is approximately equal to the yaw angle, and it is always smaller than the flow near the wall. These differences in flow direction generate secondary motions. But, away from the wall, the flow is supersonic, and pressure disturbances can only propagate in the downstream direction along characteristics. Hence, the growth of the inception region must be a function of Mach number.

At small angles of sweep, there remains a substantial region of subsonic flow. As the sweep angle is increased, the secondary flow becomes stronger under the action of the increased lateral pressure gradient, and the region of subsonic flow must therefore decrease in size and ultimately shrink to negligible proportions. However, for sweep angles up to 10 degrees, the wall pressure distribution does not significantly change

(Fig. 5), suggesting that the region of subsonic flow is not changed significantly. Significant unsteadiness in two-dimensional flows has been associated with the presence of large subsonic regions, as found in separated flows. Thus, the interaction at large angles of attack and small angles of sweep should also display significant unsteadiness. As the angle of sweep increases, and the subsonic region shrinks in size, we may expect that the unsteadiness will be reduced.

It is not clear what connection exists between the surface features and the flow a small distance away from the wall. From continuity, they obviously indicate that above these lines of convergence and divergence there is transport of fluid either away from or towards the surface. The magnitude of the pitch angle is unknown. However, if we adopt the simplest possible interpretation of the flow field structure, the lines suggest the presence of a significant vortical motion aligned in the direction of the corner. This interpretation is supported by the flow visualization of Settles & Teng (1982), and the calculations by Horstman (1984).

5. Recommendations for the Study of Strong Interactions

The 24 degree swept compression-corner flowfield described in Section 4 illustrates many of the problems encountered in strong interactions. Turbulence measurements are urgently required, both for the purpose of studying the distortion and straining of the turbulence structures and to investigate the influence of the incoming turbulent length and time scales on the structure and steadiness of the flow field. In addition, we need to consider the problems associated with instrumentation and measurement techniques. How do we relate the instantaneous structure of the flow field to surface flow visualization and mean wall pressure distribution results? Is this process possible? What measurements are required to do so? Exensive flow visualization is obviously required, preferably in high speed sequences to show the flow development in time. Presently available flow visualization techniques are simply not adequate, and considerable effort should be devoted to the development of new techniques and the improvement of existing ones.

These recommendations also apply to the area of quantitative measurements. We need to characterize the instantaneous flow field at a number of different points in the flow. Multiple hot wire probes, combined with arrays of high frequency pressure transducers should be used. New methods of analyses are required to distinguish the flow structure. What is the nature of the flow-field unsteadiness? What is the instantaneous distortion of the separated zone? Does the recirculation zone simply move backwards and forwards, with little distortion, or is it increasing and decreasing its size in some oscillatory fashion? Can the unsteadiness be forced by an externally imposed perturbation so that conditional sampling techniques can be used? What is the effect of Mach number? The interactions produced by similar geometries at lower Mach numbers could behave quite differently because of the rapid increase in sonic layer thickness as the Mach number decreases. Another factor that may enter is the effect of boundary layer thickness, or the Reynolds number.

We can see that the detailed consideration of a single strong interaction can generate many questions, and stimulate a great deal of experimental work. We feel that such detailed studies are essential to gain a full understanding of these very important flows, and recommend that future research proceed along these lines.

Acknowledgements

This paper was prepared with support from the AFOSR under contracts monitored by Dr. James Wilson. The authors wish to thank Dr. David Tan, Dr. Jannis Andreopoulos, and Mr. Shuyi Wang for their helpful comments on an earlier draft.

References

1. Andreopoulos, J, Muck, K.C. and Dussauge, J.-P., 1985. "The Structure of an Unsteady, Separated Compression Corner Flow". In preparation.

2. Bachalo, W.D., 1974, "Three-Dimensional Boundary Layer Separation in Supersonic Flow", Univ. of Calif., Berkeley, Report FM74-10.

3. Dolling, D.S. & Murphy, M., 1982, "Wall Pressure in a Supersonic Separated Compression Ramp Flowfield", AIAA Paper 82-0986.

4. Horstman, C.C., 1984, "A Computational Study of Complex Three-Dimensional Compressible Turbulent Flow Fields", AIAA Paper 84-1556.

5. Horstman, C.C. & Hung, C.M., 1979, "Computation of Three-Dimensional Turbulent Separated Flows at Supersonic Speeds", AIAA Paper 79-2.

6. Inger, G.R., 1984, "Analytical Investigation of a Swept Shock-Turbulent Boundary Layer Interaction in Supersonic Flow", AIAA Paper 84-1555.

7. Jayaram, M., Dussauge, J.-P. & Smits, A.J., 1985, "Analysis of a Rapidly Distorted, Supersonic, Turbulent Boundary Layer", Fifth Symp. on Turbulent Shear Flows, Cornell Univ., Ithaca, NY.

8. Knight, D.D., 1984, "Numerical Simulation of 3D Shock Turbulent Boundary Layer Interaction Generated by a Sharp Fin", AIAA Paper 84-1559.

9. Knight, D.D., Horstman, C.C., Shapey, B. & Bogdonoff, S.M., 1986, "The Flowfield Structure of the 3-D Shock Wave-Boundary Layer Interaction Generated by a 20° Sharp Fin at Mach 3", to be presented at the 24th Aerospace Sciences Meeting, Reno, Nev., Jan. 1986.

10. Korkegi, R.H., 1976, "On the Structure of Three-Dimensional Shock-Induced Separated Flow Regions", AIAA J. 14:597-600.

11. Kubota, H. & Stollery, J.L., 1982, "An Experimental Study of the Interaction Between a Glancing Shock Wave and a Turbulent Boundary Layer", J. Fluid Mech., 116:431-58.

12. McClure, W.M. & Dolling, D.S., 1983, "Flowfield Scaling in Sharp Fin-Induced Shock-Wave Turbulent Boundary Layer Interaction", AIAA Paper 83-1754.

13. Muck, K.C., Dussauge, J.-P. & Bogdonoff, S.M., 1985, "Structure of the Wall Pressure Fluctuations in a Shock-Induced Separated Turbulent Flow", AIAA Paper 85-0179.

14. Oskam, B., 1976, "Three-Dimensional Flowfields Generated by the Interaction of a Swept Shock-Wave with a Turbulent Boundary Layer", Ph.D. Thesis, Dept. Mech. & Aero. Eng., Princeton Univ.

15. Peake, D.J., 1975, "The Three-Dimensional Interaction of a Swept Shock Wave with a Turbulent Boundary Layer and the Effects of Air Injection on Separation", Ph.D. Thesis, Carleton Univ.

16. Roshko, A. & Thomke, G.J., 1966, "Observations of Turbulent Reattachment behind an Axisymmetric Downstream-Facing Step in Supersonic Flow, AIAA J 4:975-80.

17. Settles, G. S. & Dolling, D.S., 1985, "Swept Shock Wave-Boundary Layer Interactions". To appear in AIAA Progr. in Astro. and Aero.

18. Settles, G.S. & Teng, H.Y., 1984, "Cylindrical and Conical Upstream Influence Regimes of Three-Dimensional Shock/Turbulent Boundary Layer Interactions", AIAA J. 22:194-200.

19. Settles, G.S., Fitzpatrick, T. J . & Bogdonoff, S.M., 1979a, "Detailed Study of Attached and Separated Compression Corner Flowfields in High Reynolds Number Supersonic Flow", AIAA J. 17:579-85.

20. Settles, G.S., Perkins, J.J. & Bogdonoff, S.M., 1979b, "Investigation of Three-Dimensional Shock/Boundary Layer Interactions of Swept Compression Corners", AIAA J. 18:779-85.

21. Shamroth, S.J. & MacDonald, H., 1970, "A New Solution of the Turbulent Near Wake Recompression Problem", AIAA Paper 70-0228.

22. Stalker, R.J., 1984, "A Characteristic Approach to Swept Shock-Wave/Boundary-Layer Interactions", AIAA J. 22:1626-32.

23. Tan, D.K.M., 1985, "Turbulence Measurements in a Swept Shock Wave/Turbulent Boundary Layer Interaction". In preparation.

24. Tan, D.K.M., Tran, T.T. & Bogdonoff, S.M., 1985, "Surface Pressure Fluctuations in a Three-Dimensional Shock Wave/Turbulent Boundary Layer Interaction", AIAA Paper 85-0125.

25. Tran, T.T., Tan, D.K.M. & Bogdonoff, S.M., 1985, "Surface Pressure Fluctuations in a Three-Dimensional Shock Wave/Turbulent Boundary Layer Interaction at Various Shock Strengths". AIAA Paper 85-1562.

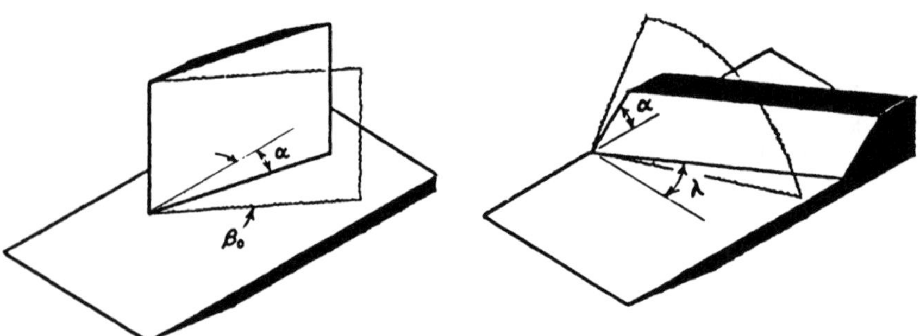

Fig. 1. Geometries for sharp fin and swept wedge interactions

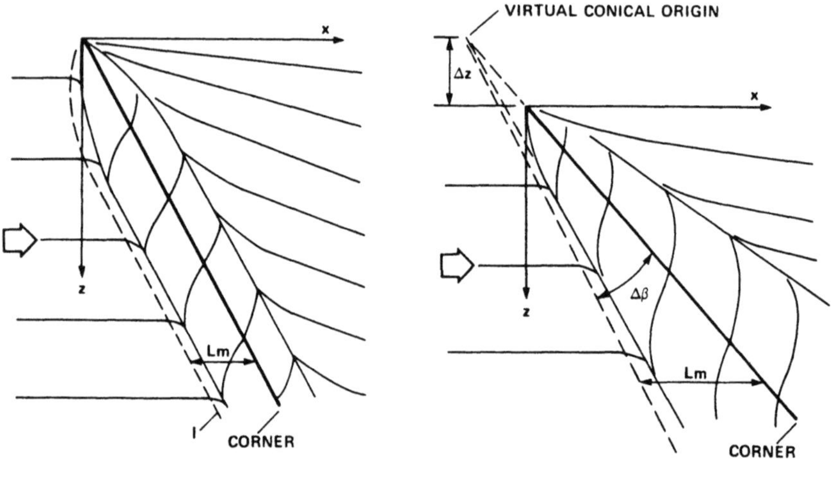

Fig. 2. Schematic of surface skin friction lines for cylindrical and conical flowfields (taken from Settles & Teng]984)

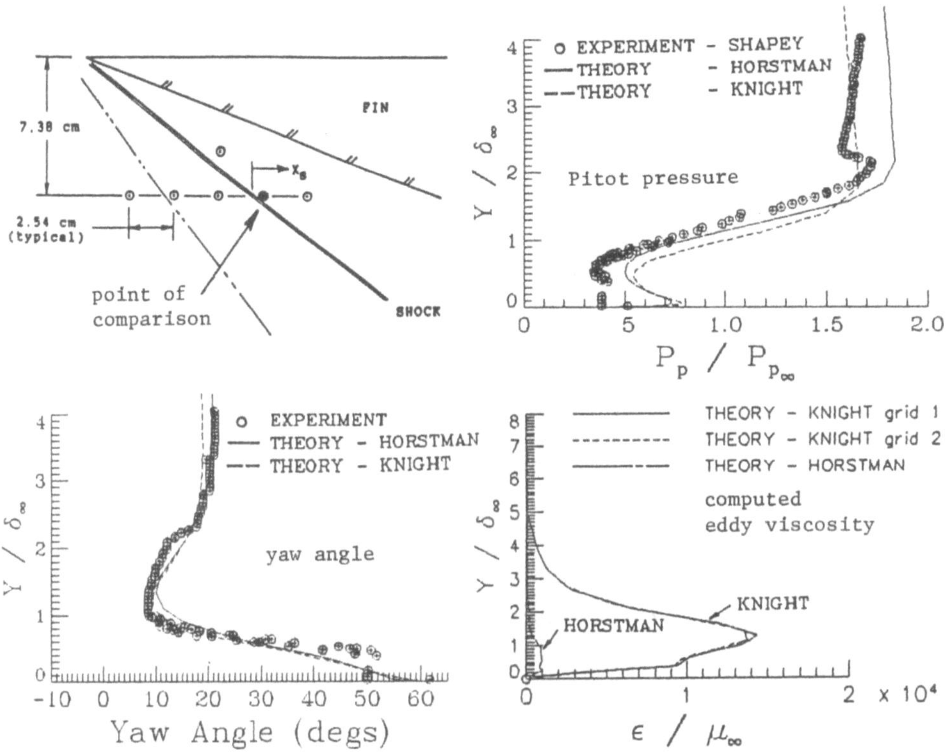

Fig. 3. Comparison of Computations by Horstman and Knight (from Knight et al. 1986)

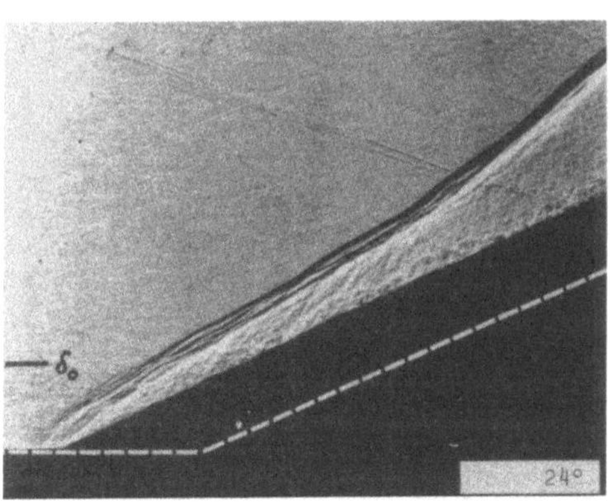

Fig. 4. Shadowgraph of 24 degree corner. Flow is from left to right (from Settles et al. 1979a).

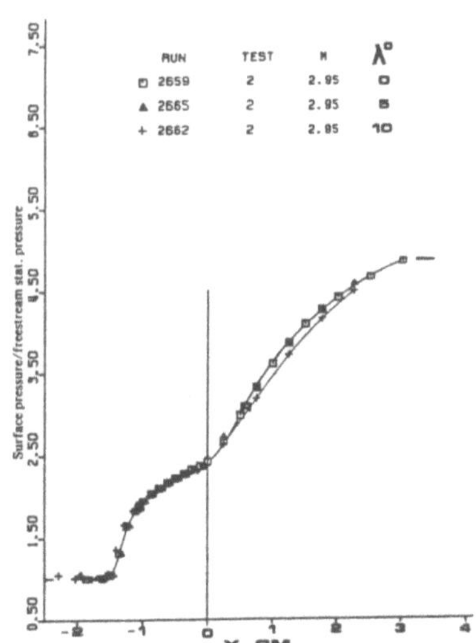

Fig. 5. Wall pressure (from Settles et al. 1979b)

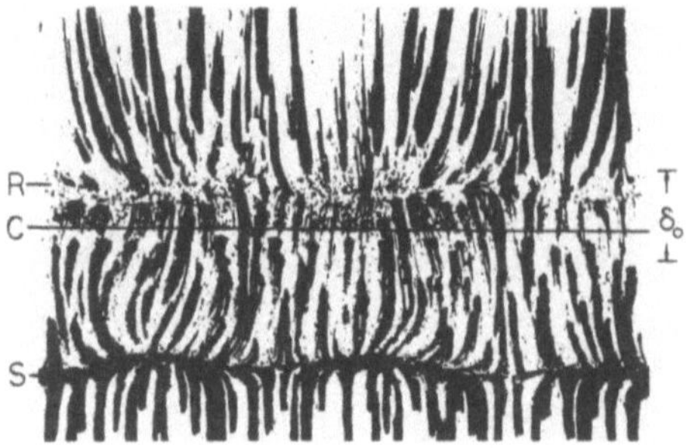

Fig. 6 Surface streaks for 24 degree corner flow, zero sweepback (from Settles et al. 1979a). Model on tunnel floor, δ_o = 25mm, width of tunnel 203mm.

Fig. 7. Surface flow visualization using the kerosene-lampblack technique for 24 degree corner flow, 10 degree sweepback (from Settles et al. 1979b). Model on plate, δ_o = 3.5mm.

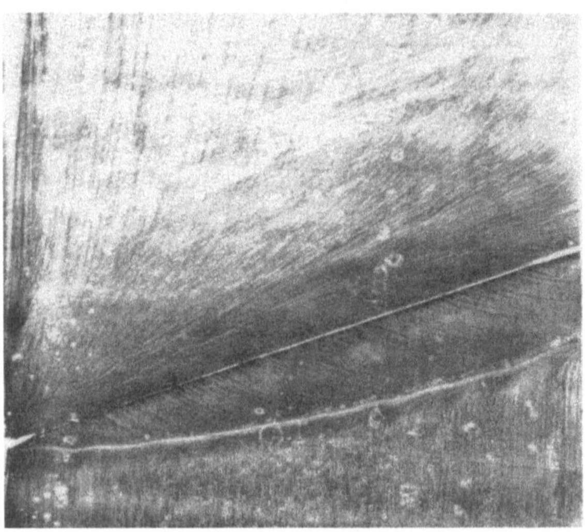

Fig. 8. Surface flow visualization using the kerosene-lampblack technique for 24 degree corner flow, 20 degree sweepback (from Settles et al. 1979b). Model on plate, δ_o = 3.5mm.

On the Inception Lengths of Swept Shock-Wave/ Turbulent Boundary-Layer Interactions

G. S. SETTLES

Mechanical Engineering Department
The Pennsylvania State University
University Park, Pennsylvania 16802 USA

Summary

Experimental data are shown for the inception lengths of swept compression corner- and fin-generated shock/boundary layer interactions at Mach 2.95. These results are found to correlate on the basis of three different flow regimes. The inception lengths of these flows are dominated by a singularity at the cylindrical/conical boundary for swept corners and by an elongation due to shock wave sweepback for fin interactions. Similarity rules for both Re_δ and shock generator geometry effects on inception lengths are demonstrated.

Introduction

Since the earliest experiments on swept shock/boundary layer interactions[1,2], researchers have observed these interactions to begin with an initial, fully three-dimensional region called an "inception zone". As illustrated in Fig. 1, this inception zone begins at the leading edge (apex) of a swept shock wave generator and extends for a distance in the sweep direction x given by ℓ_i, the "inception length". For distances $x > \ell_i$, recent experiments[3] have shown that either cylindrical or conical flow symmetry usually occurs. Such symmetry represents a simplification of complex 3-D interactions which has been useful in understanding their behavior[2,4]. However, the inception zone itself has not been studied in detail and remains essentially a mystery. The goal of this paper is to present experimental data on the inception lengths of 3-D interactions, to analyze their behavior, and to give a rational explanation of the flow phenomenology responsible for them.

Experimental Data

The data presented here are primarily from experiments[3-7] carried out by the author and colleagues at Princeton University during 1978-1983, using fin and swept compression corner shock generators. This data set consists of 58 different shock/boundary layer interactions at Mach 2.95.

Turbulent Shear Layer/Shock Wave Interactions
IUTAM Symposium Palaiseau 1985
Editor: J. Délery

The incoming boundary layer was turbulent with a typical thickness $\delta = 5$ mm, the wall was nearly adiabatic, and the Reynolds number Re was 6.3×10^7/m. The measurements pertinent to the present work included surface pressure distributions and surface-streak flow patterns obtained by the kerosene-lampblack technique[8].

Determination of Inception Lengths

Since the inception zones of these interactions merge aymptotically with outboard regions of either cylindrical or conical symmetry, their lengths are often not easy to discern with precision. A consistent criterion is called for. Here, ℓ_i values were extracted directly from the surface flow patterns by first fitting smooth curves to the interaction upstream influence lines by hand. This was done for all cases using segments of a single french curve (K&E 57168548) and a straight ruler. These curves were then digitized and curvefitted by least-squares regression, modeling the inception zone as an exponential decay* toward a linear asymptote:

$$\psi = \psi_0 \exp[-bx] \qquad [1]$$

where ψ and x (Fig. 1) are coordinates normal and tangential to the asymptote, respectively, ψ_0 is the ψ-intercept at the interaction apex, and b is the logarithmic decrement. ℓ_i was then chosen as the distance $\ln(\psi_0)/b$, where the upstream influence line lies 1 mm away from the asymptote in the ψ-direction. This method produced consistent ℓ_i values with a minimum of subjectivity in their determination.

Analysis of Inception Zone $Re\delta_0$ Similarity

It was shown by dimensional analysis in Ref. 6 that the length scales of swept interactions are properly described by the nondimensional parameter $(\ell/\delta_0)Re\delta_0^a$, where "a" was found to be approximately 1/3 for Mach 3 interactions. Fig. 7 of Ref. 6 clearly demonstrates that this nondimensionalization successfully correlates the inception lengths of various swept interaction experiments with respect to changes in both

---

* This is not to suggest that an exponential shape is unique to the physics of the inception zone; power-law or hyperbolic curves might have been used instead. Nonetheless, Eqn. 1 was successful in that acceptably small variances were found between the data and fitted curves everywhere except within 2-3 mm of the shock generator apex.

δ_0 and $Re\delta_0$. The resulting Reynolds number similarity rule for inception lengths may thus be written:

$$L_i \equiv (\ell_i / \delta_0) Re\delta_0^a = fn[\alpha, \lambda, M] \qquad [2]$$

Where L_i is defined as a nondimensional inception length parameter, α , λ are the shock generator angles of attack and sweepback, respectively, and M is the freestream Mach number.

In fact, the dimensional analysis[6] further demonstrates that δ_0 (or some length scale related to it) is the only length scale inherent in the boundary conditions of the swept interaction problem, so long as the shock generator itself is sufficiently large that it does not impose a dimension on the flow. For quasi-conical* flows, dimensional consistency thus requires the interaction to respond with a single length scale. It is proposed here that this length scale is, in fact, the observed inception length. It follows from this assertion that such swept interactions are otherwise nondimensional in nature aside from ℓ_i , and that the physical explanation for the appearance of an inception zone is that it is the response of the flow to an input boundary layer length scale. Support for this assertion lies in the success of Eqn. 2 in establishing the $Re\delta_0$ similarity of a wide variety of swept interactions tested thus far[2-7,9-11]. Further support lies in the limiting case of near-zero initial boundary layer thickness, where Eqn. 2 indicates that the inception zone should tend to vanish. This was, in fact, observed to occur in the fin interactions experiments of Zubin and Ostapenko[12].

<u>Analysis of Inception Zone/Shock Generator Similarity</u>

The influence of shock generator geometry, as expressed functionally in the right-hand side of Eqn. 2, is a more complex issue. The generators considered here ranged from near-2-D swept corners to highly swept fins. Ranges of the defining angles $5 \le \alpha \le 70$ deg and $0 \le \lambda \le 85$ deg resulted in L_i variations of more than an order of magnitude. It thus became clear that L_i , already nondimensionalized in terms of $Re\delta_0$ changes, remains a strong function of shock generator geometry. Further,

* Quasi-cylindrical interactions may also occur for certain semi-infinite swept compression corner shock generators[3]. In these cases an additional upstream influence length scale appears, which may also be described in a manner similar to Eqn. 2.

attempts to understand the trends of L_i in simple terms of α, λ or of shock wave strength fail because they do not take into account changes in interaction flow regimes.

As a compression corner is steepened and swept back progressively, it eventually becomes geometrically identical to a fin. Early in this process the swept corner interaction crosses a flow regime boundary from cylindrical to conical symmetry[3], at which the inception length naturally tends toward infinity. This regime boundary was earlier found to be a shock detachment phenomenon[3]. As such, it is a nonlinear function of α,λ. Since L_i is singular on this boundary, an expression for the boundary in terms of α,λ is needed in order to correlate values of L_i. Such a detachment expression is known from oblique shock theory (eg Eqns. 138 and 168 of Ref. 13), but is algebraically too clumsy to be useful for present purposes. Instead, through the coordinate transformation:

$$M_N = M\cos\lambda\ , \qquad A_N = \tan^{-1}(\tan\alpha/\cos\lambda) \qquad [3]$$

where M_N and A_N are the Mach number and shock generator angle in the normal direction, respectively, a simplified detachment expression may be obtained from a 2nd-order polynomial fit to the exact equations:

$$\zeta \equiv (\alpha_N + 38.53)/(M_N - 0.149\,M_N^2) \qquad [4]$$

ζ amounts to a "detachment similarity parameter[4]" which expresses the proximity of a given swept interaction to the cylindrical/conical flow regime boundary at $\zeta_{crit} = 43.6$. Eqn. 4, in which α_N is in degrees, varies less than 1% from the exact oblique shock detachment equations over the range $1.1 \le M_N \le 3.0$. Note that λ in Eqn. 3 represents the sweepback angle of both the compression corner and the shock wave for attached shocks. Once shock detachment has occured, however, a distinction must be made. For present purposes, λ_{shock} is used in computing M_N for these detached cases in order for Eqn. 4 to best represent interactions far beyond shock detachment.

One can now take account of the singular inception length behavior on the cylindrical/conical flow regime boundary by examining the L_i data in terms of ζ, as shown in Fig. 2. Low values of ζ represent mildly swept corner interactions which are always cylindrical outside the inception zone. As ζ approaches the boundary value ζ_{crit} = 43.6 from below, L_i grows rapidly without bound. In this first flow regime, Fig. 2 shows that 11 experiments with 10 ≤ α ≤ 24 degrees and 5 ≤ λ ≤ 50 degrees are correlated within the data accuracy.

Quasi-conical swept corner interaction symmetry is observed for 43.6 < ζ < 85, where L_i first decreases rapidly as the singularity on the flow regime boundary is left behind, then levels off. Thirteen different interactions with 10 ≤ α ≤ 70 degrees and 50 ≤ λ ≤ 70 degrees are correlated in this second flow regime in Fig. 2, which may be thought of as a transition regime between quasi-cylindrical swept corner and quasi-conical fin interactions.

Another flow regime boundary occurs at 80 < ζ < 90. Here, flow visualization experiments[4] reveal that the reattachment line of the separated flow migrates off the shock generator and onto the test surface. This new boundary occurs as the swept corner becomes more vertical and fin-like, reaching the point where the pressure on its surface is too high to support reattachment there. The boundary is marked by a sharp drop in L_i for λ = 70 degree swept corners in Fig. 2.

As ζ continues to increase, the swept corner shock generator is progressively distorted into the shape of a fin. The dimensionless inception length rises once again for ζ > 90 in quasi-conical flow. Data for 4 swept corners and 28 different fin interactions with 5 ≤ α ≤ 20 degrees and various leading edge sweepback angles are shown in this third flow regime of Fig. 2. Significantly, a strong connection between fin- and swept corner-induced 3-D interactions in terms of their inception lengths is shown here for the first time. The fin data are also seen to approach the limit of vanishing shock strength at about ζ = 150 in the coordinates of Fig. 2.

The rising trend of L_i in flow regime 3 of Fig. 2 can be examined in simpler terms than ζ. Stalker[14] first suggested to this author a simple argument for the variation of inception lengths with shock strength: The rate of fluid loss from a swept separation zone might be expected to scale with the tangential Mach number component M_T. Similarly, The rate of fluid input to that zone might be proportional to M_N. The ratio of these two rates, $M_T/M_N = \tan\lambda_{shock}$, might then express the spanwise growth rate of a swept interaction, and thus the inception length required for it to reach an equilibrium (cylindrical or conical) growth. Stalker showed limited evidence from his swept-step experiments[1] in support of this argument.

A more recent and rigorous triple-deck treatment of the problem by Inger[15], as yet unpublished, indicates that the inner viscous layer of a shock/boundary layer interaction will eventually respond to the (cylindrical or conical) boundary condition imposed upon it by the outer inviscid flow. The spanwise distance required for this to occur, according to Inger's analysis, should depend upon $\tan\lambda_{shock}$.

Given these indications, the present L_i data have been examined in terms of $\tan\lambda_{shock}$. The results show considerable scatter for swept corners due to the singularity in L_i at ζ_{crit}. However, as shown in Fig. 3, the fin data from the third flow regime of Fig. 2 are in good agreement with

$$L_i = 1133 \tan\lambda_{shock} \qquad [5]$$

which was arrived at from a least squares regression which passed arbitrarily close to the origin. This simple result appears to confirm the analyses by Stalker[14] and Inger[15] insofar as highly swept interactions are concerned. It further clarifies the inception length behavior shown in Fig. 2, which may now be characterized as the result of two disparate phenomena: the cylindrical/conical L_i singularity at low sweep angles and the spanwise "stretching" of L_i at high sweep.

It is interesting to note that, in the latter region described by Eqn 5, the similarity parameters $\tan\lambda_{shock}$ and ζ are closely related. Since α_N = 90 degrees for fin shock generators in this region, Eqns. 3 and 4 show that ζ amounts to a trigonometric function of λ_{shock} there. In fact, ζ and $\tan\lambda_{shock}$ are essentially linearly related for $\lambda_{shock} > 50$ degrees, corresponding to fin angles of attack up to 30 degrees at Mach 2.95. Thus Figs. 2 and 3 present equivalent representations of the L_i behavior in the third flow regime.

This makes it possible to write a single expression conveying the behavior of L_i across the full range of $M = 2.95$ results presented here:

$$L_i = \zeta^{2.5} / \mid \zeta_{crit} - \zeta \mid + 0.004\, \zeta^3 - 500 \qquad [6]$$

This equation, though entirely empirical, does illustrate the singularity at ζ_{crit} and provides, as well, a reasonable match with all the present L_i data, as shown in Fig. 4. Note that, for the third flow regime, Eqn. 5 is an equivalent but more physically meaningful criterion.

Comparison with Numerical Predictions

Horstman[16] has carried out numerical solutions of the Reynolds-averaged Navier-Stokes equations for several of the interactions described here. He used a k-ϵ model for turbulence closure, a freestream Mach number of 2.95, and an input boundary layer matching that of the present experimental data. His results included simulated surface flow patterns bearing a striking resemblance to those obtained by experiments. Six of these patterns, for unswept fins with $10 \leq \alpha \leq 20$ degrees, were used to extract ℓ_i data by the same procedure as described earlier. The L_i results are shown in Fig. 4.

Fig. 4 reveals that the low-α numerical L_i predictions are comparable to the experimental data. However, results for $\alpha = 18$ and 20 degrees fall above the data band and reveal a trend of rising L_i with increasing shock strength, contrary to Eqn. 5 and the data. Although this trend is obvious from inspection of the computed surface flow patterns, it is preliminary to judge the predictions on the basis of only a few points. Horstman has also computed other fin cases and about 30 swept corner cases[17], but these are not presently available to the author in a form suitable for ℓ_i extraction. A more thorough treatment of numerical inception length results should thus be available in the near future.

Conclusions

1) Experimental data revealing the behavior of the inception lengths of swept compression corner and fin interactions at Mach 2.95 are shown.
2) Results from mildly swept corners to highly swept fin interactions correlate on the basis of three different flow regimes.
3) The inception lengths for these flows are shown to be dominated by a singularity at the cylindrical/conical boundary for swept corners and by an elongation due to shock wave sweepback for fin interactions.
4) Similarity rules for both Re_δ and shock generator geometry effects on inception lengths are demonstrated.

5) Limited results of numerical predictions compare reasonably with the inception length data, but further study of this issue is warranted.

Acknowledgements

Discussions with Prof. G. R. Inger, Dr. C. C. Horstman, Mr. F. K. Lu, Mr. R. L. Kimmel, and Prof. R. J. Stalker are gratefully acknowledged. The experiments on which this work is based were carried out at Princeton University and were supported by various research grants from AFOSR (monitored by Dr. J. D. Wilson) and from NASA-Ames Research Center (monitored by Dr. C. C. Horstman).

References

1. Stalker, R. J., "Sweepback Effects in Turbulent Boundary-Layer Shock-Wave Interaction," Journal of Aeron. Sciences, 27 (1960) 348-356.

2. Settles, G. S., and Dolling, D. S., "Swept Shock Wave/Turbulent Boundary Layer Interactions," survey paper to appear in AIAA Progress Series in Astronautics and Aeronautics Volume on "Tactical Missile Aerodynamics", 1986.

3. Settles, G. S., and Teng, H.-Y., "Cylindrical and Conical Flow Regimes of Three-Dimensional Shock/Boundary-Layer Interactions," AIAA Journal, 22 (1984) 194-200.

4. Settles, G. S., and Kimmel, R. L., "Similarity Conditions for Conical Shock Wave/Turbulent Boundary Layer Interactions", AIAA Paper 84-1557, June 1984, to be published in AIAA Journal.

5. Settles, G. S., and Lu, F. K., "Conical Similarity of Shock/Boundary-Layer Interactions Generated by Swept and Unswept Fins," AIAA Journal, 23 (1985) 1021-1027.

6. Settles, G. S., and Bogdonoff, S. M., "Scaling of Two- and Three-Dimensional Shock/Turbulent Boundary Layer Interactions at Compression Corners," AIAA Journal, 20 (1982) 782-789.

7. Goodwin, S. P., "An Exploratory Investigation of Sharp Fin-Induced Shock Wave/Turbulent Boundary-Layer Interactions at High Shock Strengths," MSE Thesis 1687-T, M.A.E. Dept., Princeton Univ., 1984.

8. Settles, G. S., and Teng, H.-Y., "Flow Visualization Methods for Separated Three-Dimensional Shock Wave/Turbulent Boundary Layer Interactions," AIAA Journal 21 (1983) 390-397.

9. McClure, W. B., and Dolling, D. S., "Flowfield Scaling in Sharp Fin-Induced Shock Wave Turbulent Boundary Layer Interaction", AIAA Paper 83-1754, July 1983.

10. Settles, G. S., Horstman, C. C., and McKenzie, T. M., "Flowfield Scaling of a Swept Compression Corner Interaction – A Comparison of Experiment and Computation," AIAA Paper 84-0096, January 1984.

11. Degrez, G., and Ginoux, J. J., "Three-Dimensional Skewed Shock Wave/Laminar Boundary Layer Interaction at Mach 2.25," AIAA Paper 83-1755, July 1983.

12. Zubin, M. A., and Ostapenko, N. A., "Structure of Flow in the Separation Region Resulting from Interaction of a Normal Shock Wave with a Boundary Layer in a Corner," Izvestiya Akademii Nauk SSSR, Mekhanika Zhidkosti i Gaza, 3 (1979) 51-58.

13. NACA Report 1135, "Equations, Tables, and Charts for Compressible Flow," (1953).

14. R. J. Stalker, private communication, July 7, 1980.

15. G. R. Inger, private communication, August 21, 1985.

16. Horstman, C. C., "Computation of Sharp Fin-Induced Shock Wave/Turbulent Boundary Layer Interactions," to be pub. In AIAA Journal.

17. Horstman, C. C., "A Computational Study of Complex Three-Dimensional Compressible Turbulent Flow Fields," AIAA Paper 84-1556, June 1984.

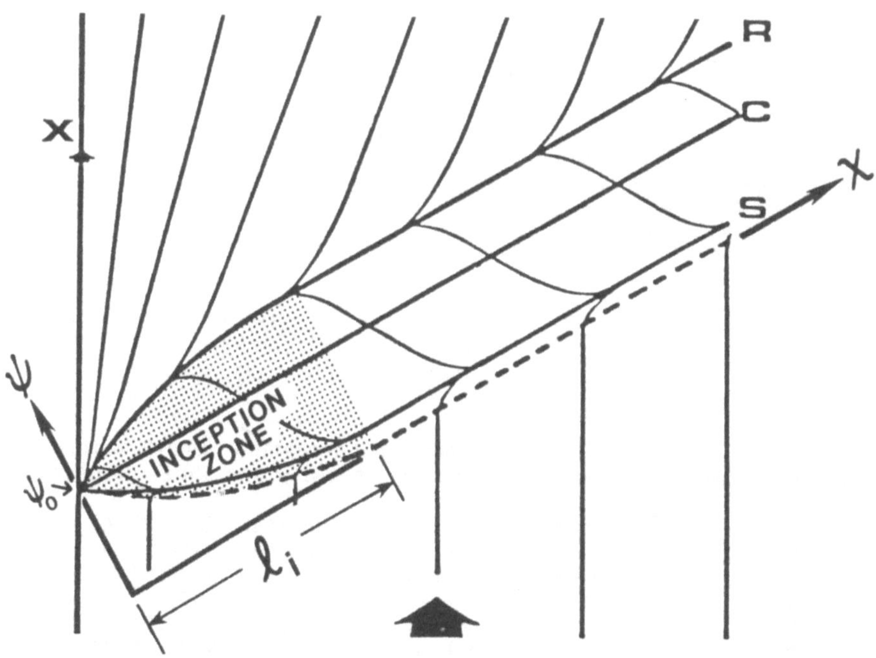

Fig. 1. Diagram of a typical (cylindrical) swept interaction surface flow pattern showing inception zone and inception length ℓ_i . (C = compression corner, R = reattachment line, S = separation line.)

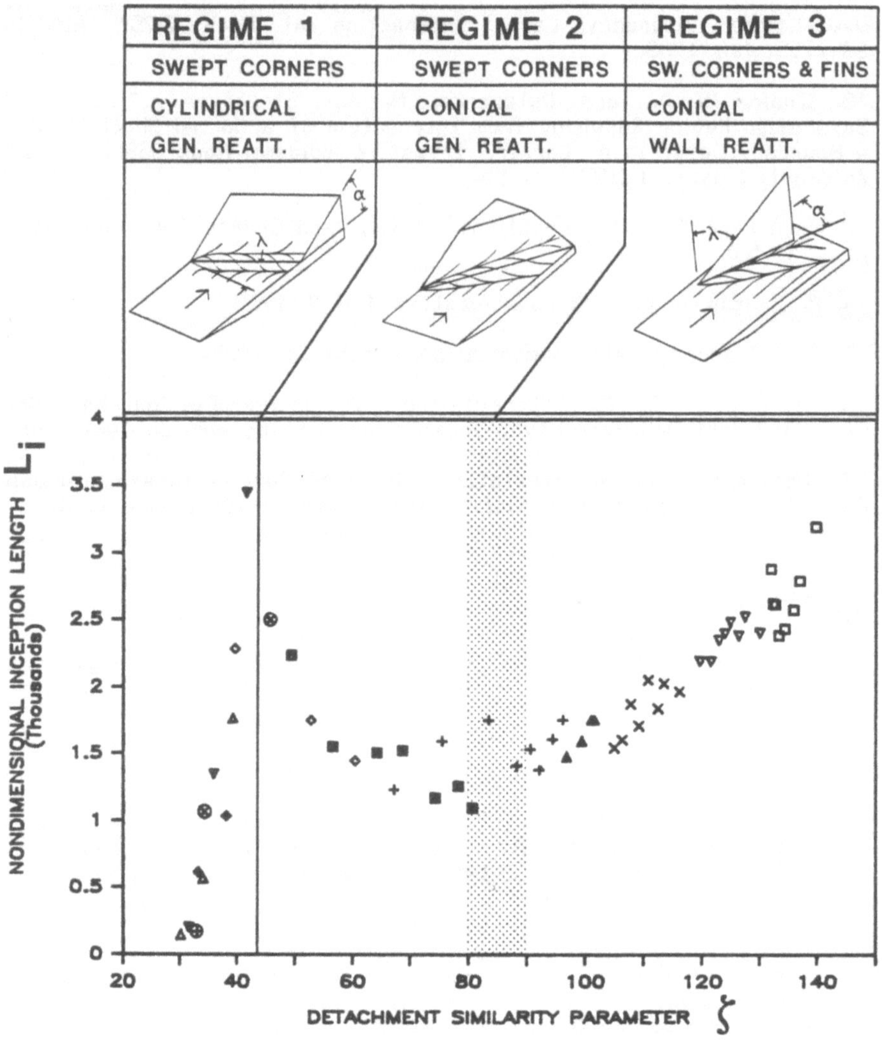

Fig. 2. Similarity plot of nondimensional inception length data, L_i, vs. detachment similarity parameter ζ, showing the three flow regimes.

- ⊕ – λ = 5 deg swept compression corners; Refs. 3, 4, 6.
- ◆ – λ = 10 deg swept compression corners; Refs. 3, 4, 6.
- △ – λ = 20 deg swept compression corners; Refs. 3, 4, 6.
- ▼ – λ = 30 deg swept compression corners; Refs. 3, 4, 6.
- ⊗ – λ = 40 deg swept compression corners; Refs. 3, 4, 6, 10.
- ◇ – λ = 50 deg swept compression corners; Refs. 3, 4, 6.
- ■ – λ = 60 deg swept compression corners; Ref. 4.
- + – λ = 70 deg swept compression corners; Ref. 4.
- □ – α = 5 deg fins, $0 \leq \lambda \leq 65$ deg; Ref. 5.
- ▽ – α = 9 deg fins, $0 \leq \lambda \leq 65$ deg; Ref. 5.
- × – α = 15 deg fins, $0 \leq \lambda \leq 65$ deg; Ref. 5.
- ▲ – α = 17.25, 17.5, 18.5, & 20 deg fins; Ref. 7.

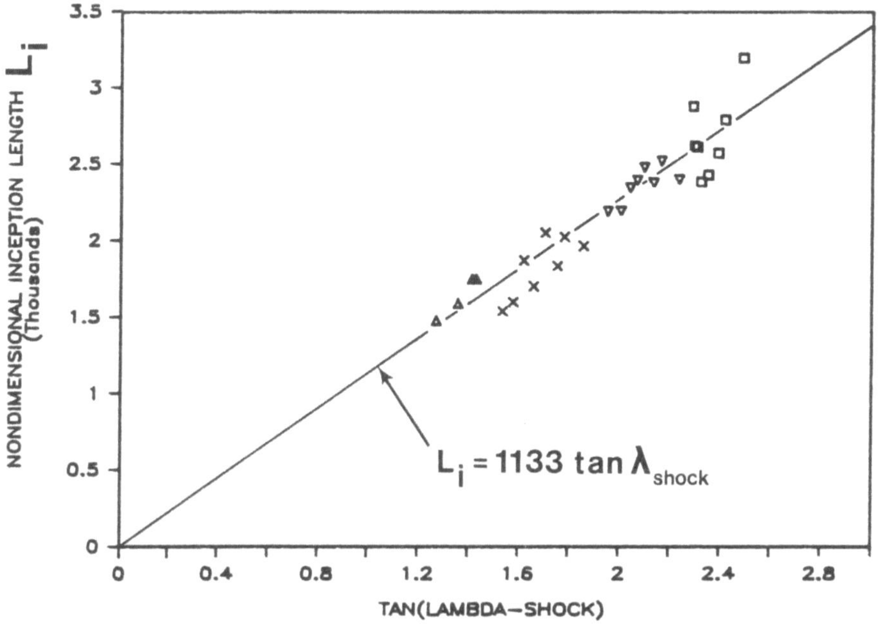

Fig. 3. Fin L_i data plotted vs. tangent of shock wave sweepback angle.

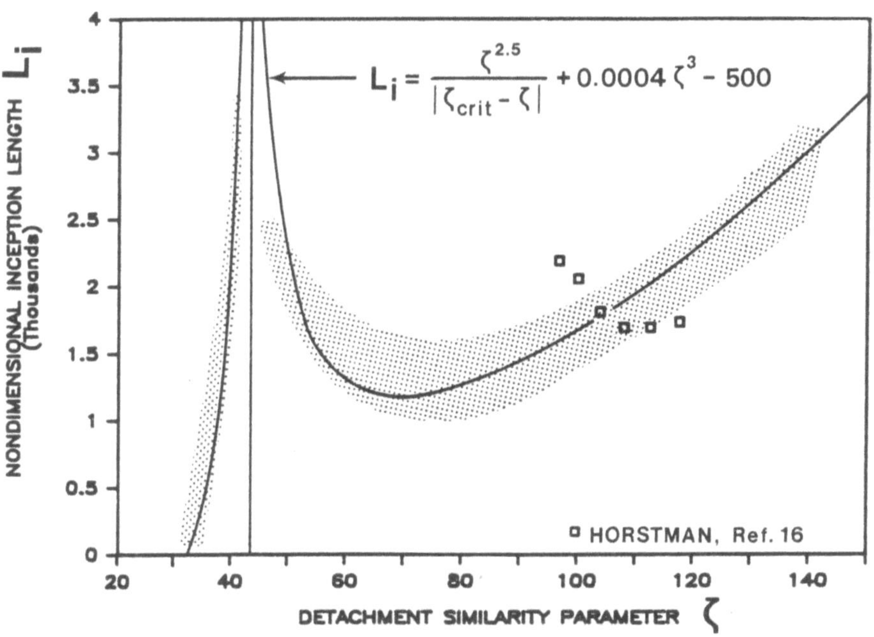

Fig. 4. L_i similarity plot comparing Eqn. 6, the band of the data of Fig. 2, and results of numerical predictions by Horstman.

Experimental Study of the Boundary-Layer Separation Conditions through a Shock-Wave on Airfoil and Swept Wing

A. MIGNOSI, J.B. DOR, A. SERAUDIE

Département d'Aérothermodynamique
ONERA / CERT
31055 - TOULOUSE CEDEX (FRANCE)

Summary

The experimental studies presented in this paper were obtained in T2 wind tunnel on two models equipped with LC 100 D profiles :
- an airfoil with a large chord (C = 400 mm) placed near the lower wall of the test section,
- a swept wing (C = 100 mm ; ϕ = 30° and ϕ = 45°).
Pressure measurements, boundary layer probings and wall visualizations around the separation conditions of the turbulent boundary layer interaction with the shock wave have been done.
The turbulent boundary layer separation occurs when the upstream local Mach number, normal to the shock, reaches approximately 1.3.

1 - Introduction

Transonic tests on airfoils (NACA 12, RA 16, LC 100 D) and on a swept wing (LC 100 D) at various swept angles, have been done at the ONERA/CERT wind tunnel T2 at Toulouse, in order to obtain good measurements on a turbulent boundary layer interacting with a shock wave around the separation conditions. These data were used by ONERA's Aerodynamic Division to improve theoretical methods.

- The transonic wind tunnel T2 is induction driven /1/, has a 0.38 x 0.4 m² test section and can be pressurized up to 5 bars. Downstream of the test section a sonic throat fixes the test Mach number.
- For all of these tests, wall pressure measurements on the model, oil wall streamlines visualizations and boundary layer explorations are performed.

In this paper, after a short presentation of the instrumentation and acquisition system, we will illustrate the most interesting results.

Turbulent Shear Layer/Shock Wave Interactions
IUTAM Symposium Palaiseau 1985
Editor: J. Délery

A first part concerns two dimensional flows on the airfoil LC 100 D. This model was chosen for its Mach number distribution of "Peaky" type. A negative speed gradient upstream of the shock wave gives a high value of the incompressible shape factor allowing to complete existing experimental data versus Hi near the separation conditions. Thanks to the dimension of the model (chord = 400 mm), the boundary layer thickness in the interaction region is rather important ($\delta \simeq$ 1 to 10 mm) allowing good explorations. The maximum Reynolds number R_e reaches 18 millions and all cases are obtained with a free transition upstream of the interaction.

For three-dimensional flows a second part presents results on a swept wing with a constant chord of 100 mm equipped with a LC 100 D profile. The swept angles ϕ = 30° and 45° are essentially considered to obtain data points around three-dimensional boundary layer separation. The transition is fixed near the leading edge.

2 - Experimental technique

For all of these tests, different measurements are performed. The pressures are measured for each test with a "Scanivalve system" driven by a mini computer as well on the model as on the test section walls.

The streamlines at the wall are obtained by painting the model with a thin film before the run. For that we use a mixture of oil and a black powder (a sort of soot).

During the run we can observe the evolution of the black oil and take photographies. We also have remarked that at the end of the run the flow can change the result particularly when there is an amount of mixture due to a separation.

The boundary layer explorations are performed with pressure probes equipped with Kulite mini transducers (see figure 1). These short response time probes can explore a boundary layer in 30 s. with about 100 measured points.

The probe is mounted with an hinge on a sting which moves in a vertical plane. To avoid vibrations a point of support is realized on the model.

Two types of mini probes are used : a stagnation one equipped with a pitot tube (thickness 0.12 mm) and a directional one (thickness 0.18 mm) which uses two mini transducers.

The directional probe comprises a central stagnation tube connected to a transducer P_t and two lateral pressure tubes whose difference $\Delta p = P_2 - P_1$ is measured.

This probe is calibrated and its front part is manually positionned at an angle near the flow direction (see figure 10). With successive probings at different angles β the value and the angle of the speed can be obtained (see §. 3). In all cases we suppose the static pressure constant normal to the wall and the stagnation temperature constant and equal to that of the upstream flow Tt.

3 - Two-dimensional tests with the LC 100 D (c = 400 mm) airfoil

3.1. Description of the experimental device

After a first study on RA 16 (c = 180 mm) and NACA 12 (c = 210 mm) airfoils in T2 wind tunnel, usually located in the center of the test section, a LC 100 D of large chord (c = 400 mm) was tested. The use of such a big model compared with the test section size (h = 380 mm, W = 400 mm) required a particular test device shown in figure 2.

The airfoil is mounted near the lower test section wall in order to increase the upperside flow region.

The upper test section wall is realized with a fixed metal sheet whose shape has been approximately calculated as a free stream flow line.

The lower wall is plane, but a part located under the airfoil has been perforated to reduce interferences (permeability $\simeq$ 22 %, hole diameter = 6 mm). The plenum is used without external suction.

Two stagnation pressures are performed P_t = 1.5 b and P_t = 3.2 b, giving Reynolds numbers of R_c = 8.5 and 17 millions at M_∞ = 0.8. Thanks to these high Reynolds numbers, free transition was used for all tests.

Different Mach number distributions on the airfoil are obtained with different angles of attack and throat positioning.

The wall pressure measurements on the airfoil are performed with a Scanivalve system driven by a mini computer, the separation is indicated by oil wall streamlines visualizations and checked with boundary layer explorations.

3.2. Experimental results : Mach number distribution

Two typical Mach number distributions are presented on figures 3 and 4.

In the first case the pitch angle and the throat opening varied so as to maintain a shock wave location near x/c = 30 %. In this case all flow visualizations indicated that when the separation begins, it is limited to a short bubble and the boundary layer reattached rapidly on all the rear part of the airfoil.
In the second case the pitch angle is fixed and the shock moves downstream with the throat opening. Oil flow patterns indicated a separation always extending from the shock wave to the trailing edge.
A first determination of the separation onset can be proposed from the oil flow visualizations and the Mach number distribution If we plot the local Mach number at a fixed location downstream of the shock wave as a function of the upstream Mach number M_1 (see figures 3 and 4), a change in the shape of these curves appears, corresponding to the separation onset indicated by the oil flow patterns.
For the shock wave interaction at x/c = 30 % the separation onset is badly determined, due to the regular evolution of the curves, nevertheless a value between 1.26 and 1.29 can be estimated. Comparatively oil flow visualizations indicate a value between 1.27 and 1.30, observations are also very difficult to do, due to the small extension of the separation bubble.
For the other case (shock wave at x/c = 65 %) this criterion is quite good and gives a value of 1.29 for P_t = 1.5 b. and 1.305 for P_t = 3.2 b. in good agreement with the wall oil flow visualization.

3.3. Experimental results. Boundary layer probing

After a first phase four principal configurations are analyzed, and for each of them we have chosen three or four cases : unseparated, separated and very near of the limit.
The four principal configurations correspond to two pressures, 1.5 and 3.2 bars and two shock locations :
- x/c = 30 % α = 1.75, 2.25 and 2.75°.
- x/c $\simeq$ 60, 65 % α = 2°

Examples of the boundary layer profiles $u/u_e(y)$ are given in figure 5. We have qualified the boundary layer at different locations : 20 - 27.5 - 32.5 - 37.5 - 40 - 56.25 - 75 - 87.5 %, but it was impossible to measure the separated region with a pitot probe. The boundary layer probings confirm the separation mecanism : short bubble when the shock is at x/c = 30 % and separation to the trailing edge when the shock is at x/c = 65 %.
Illustration of the case x/c = 65 % is given in figure 6 presenting the evolution of the integral parameters. $\delta 1$ (displacement thickness), $\delta 2$ (momentum thickness) and Hi (incompressible shape factor $Hi = \frac{\delta 1i}{\delta 2i}$) versus x. Each curve corresponds to an upstream Mach number M1.
Upstream from the interaction we have observed a quasi constant Mach number distribution which gives the same boundary layer for all cases : shape factor Hi = 1.35 with a displacement thickness $\delta_1 \simeq 0.7$ mm.
Downstream from the interaction the displacement thickness increases rapidly and is strongly dependent on M_1. In the same time δ_2 seems independent of M_1.
The results on the incompressible shape factor are probably essential. In the case $M_o = 1.28$ and 1.293 without separation Hi decreases slowly after the interaction, but at $M_o = 1.317$ it increases towards the trailing edge. Such an evolution is characteristic of adverse pressure gradient. In the case with a shock wave at x/c = 30 % the shape factor after the increase due to the interaction always decreases rapidly towards a value near 1.4.
In figure 7 we have plotted the incompressible shape factor Hi versus M_1 for two locations x/c = 75 and 87,5 % : we can see that the more the point is near the trailing edge the more the shape factor varies. The separation seems to appear for a shape factor around 3 which gives an upstream Mach number of 1.29 at 1.6 bars and 1.305 at $P_t = 3.2$ bars.
A semi empirical correlation /2/ could be found between all these different results by plotting the separated and unseparated cases as function of M_1 and Hi_1 (see figure 8).
It seems that we can observe two regions ; one for small values of M_1 without separation and the other with separation. A sort

of decreasing curve versus Hi divides the two parts and gives a good criterion for the separation onset.

4 - Three-dimensional tests on the swept wing LC100D c = 100 mm

4.1. Test device

The model (fig. 9) is a swept LC100D airfoil of constant chord c = 100 mm, equipped with three pressure hole lines, which are parallel to the flow for the sweep angle $\phi = 55°$: on each line, 32 upper surface holes and 9 lower surface holes, of diameter $\phi = 0.4$ mm.

The wing is mounted in the center of the test section, between solid plane walls. The upper wall is made of plexiglas and allows to photograph the oil visualization of the flow on the upper side, during the run.

For each sweep angle (30°, 45° or 55°), the model is equipped with a wing tip which is parallel to the upstream flow and a spanwise translation can be performed to keep constant the width taken by the wing in the test section. The wing tip is fixed to the lower wall by a steel wire, to reduce deformation and vibrations.

The wing heel is fixed on a turning table mounted on a lateral test section door : the angle of attack, measured in the vertical symmetry plane of the test section, is adjusted by rotating the wing around an axis parallel to its leading edge.

4.2. General results and separation condition

For $\phi = 55°$, we could only obtain interaction with a laminar boundary layer, the shock being located just behind the high speed peak, near the leading edge. It appeared very difficult to obtain good transition tripping upstream of this shock.

The interesting three-dimensional interaction cases between a turbulent boundary layer and a shock wave are obtained on the wing upper side in a region located near the wing tip, around the third pressure hole line, with the sweep angles $\phi = 30°$ and $\phi = 45°$. During these tests, transition is fixed at x/c = 5 % by a carborundum line (height : 0.12 mm). The total pressure is $P_t = 1.7$ bar and the Mach numbers of the tests are going from 0.75 to 0.95.

We must underline the great importance of the wall streamline visualizations for the general description of the flow and the study of the separation emergence in the shock foot when the infinite Mach number and essentially the angle of attack increase. Three examples of typical visualizations (unseparated - separation limit - separated) are given for $\phi = 45°$ on figure 11. The corresponding Mach number distributions, measured on the third pressure hole line, are given on figure 12.

The upperside flow includes :

- A high speed region with a peak followed by an isentropic compression and ended by a shock wave. The flow is quite two-dimensional and because of the leading edge acceleration, its direction trends to be normal to the wing.
- A shock wave whose general direction is parallel to the leading edge. In the shock foot, the streamlines strongly deviate towards the wing tip. Chordwise explorations of great amplitude in the flow field with the boundary layer probe allowed to check that the shock wave is rather normal to the wall, as shown on figure 14.
- Downstream of the shock, we can find a separated region. After reattachment, the flow comes back to the general direction of the test section flow.

When there is no separation, the wall streamlines cross the shock wave with a strong deviation, but without any interruption (figure 10(1)). When there is a boundary layer separation, the upstream wall streamlines gather on the shock line. Downstream of the shock, a reattachment line separates the wall streamlines moving up towards the shock (separated region) and the lines going downstream (figure 10 (2)). The separation limit corresponds to a gathering of the upstream lines along the shock, without separated region downstream of the shock (figure 11 - $M = .95$, $\alpha = 2°$). The separated region also appears on the wing Mach number distribution, downstream of the shock (figure 12 - $\alpha = 3°$).

During exploration with wall visualizations, of the test domain (M, α) for $\phi = 30°$ and $\phi = 45°$, it appeared clearly that the separation conditions are reached when the shock wave compression becomes important enough. This is essentially obtained when, for a given Mach number, the angle of attack is increased (figures 11 and 12). So, for $\phi = 45°$ and $M = 0.95$, the separation condi-

tion corresponds to $\alpha \geqq 2°$.
The normal component of the Mach number upstream of the shock seemed to be a good parameter to fix this value. It is obtained by the wall pressure measurements on the model (M) and by angle measurements on the visualizations : upstream two-dimensional flow and shock wave directions in the interaction region. The normal Mach numbers for all the separation limit test cases for $\phi = 30°$ and $45°$ are appreciably in the interval $1.32 \leqq M_N \leqq 1.34$.

4.3. Boundary layer exploration results

For each sweep angle $\phi = 30°$ and $45°$, boundary layer explorations are performed for two test cases corresponding to the separation limit and a quite separated boundary layer. The probing points are distributed along the chord, near the third pressure hole line, one point upstream and several points downstream of the shock. In spite of the small thickness of the explorated boundary layers on this swept wing (0.5 mm to 2 mm), we could obtain consistent results. However, the mechanical problems (vibrations, etc) to obtain a well defined movement become appreciable. The results given on the figures concern the sweep angle $\phi = 45°$.
The figure 13 gives a typical example of a directional boundary layer exploration, downstream of the shock, at x/c = 20 %, for the test case M = 0.95, $\alpha = 2°$, corresponding to the separation limit. The figure 14 recapitulates the exploration results for this same case.
Three explorations are done at the same point, each one with a given angle of the directional probe. These chosen angles correspond to the flow angle β at different heights in the boundary layer (wall, middle and outside regions). So the three curves measured across the boundary layer, of the pressure difference between the lateral holes, give the flow angle at each point by a linear zero method.
The second part of figure 13 gives the speed variation across the boundary layer, measured by the central pitot probe. It appears that the influence, on this measurement, of the small probe angles used, is negligible. Very close to the wall, when the probe thickness makes the exploration impossible, the measured points are completed by a logarithmic law.

On figure 14, the variation along the chord of the boundary layer flow angles shows that :

- upstream of the shock, the boundary layer is two-dimensional ;
- just downstream of the shock, in the region where separation will develop, the angle variation is very important, the wall streamlines deviating towards the wing tip ;
- far downstream of the shock (x/c = 40 %), the boundary layer trends to become two-dimensional ;
- the flow angles in the outside part of the boundary layer vary quite regularly along the chord, from $\beta \cong + 15°$ upstream of the shock to the direction of the general flow ($\beta = 0°$) at x/c = 40 %.

The speed curves in the boundary layer (figure 14) show characteristic variations of thickness and shape, along the chord.
The figure 15 gives the variation versus x/c of the integral boundary layer parameters (displacement thickness $\delta 1$ and incompressible shape factor Hi) computed from the exploration measurements, for the separation limit case ($\alpha = 2°$, shock located at x/c = 12 %) and the separated case ($M = 0.95$, $\alpha = 3°$, shock located at x/c = 20 %). $\delta 1$ and Hi increase rapidly when the boundary layer crosses the shock foot, with a peak downstream of the shock which is more important in the separated case.

5 - Conclusion

These experimental studies on airfoil and swept wing in a transonic wind tunnel allowed to qualify and to obtain consistent measurements on the turbulent boundary layer interacting with a shock wave, near the separation conditions, in two- and three-dimensional flows. These results are particularly interesting because of the high values of the shape factor Hi $\cong$ 1.4 obtained on the LC100D profile.
The oil visualization technique was systematically used and brought a lot of information.
A boundary layer exploration technique by thin pitot probes, with the possibility of directional analysis for three-dimensional flows, has been developed and used with success, particularly on the LC100D swept wing of 100 mm chord, for which the boundary layer thickness is of the order of 1 mm.

For the present time and the future, another transonic phenomenon related to the shock wave, which is now largely studied in the world, the buffeting, is also approached at the wind tunnel T2.

References

1. QUEMARD, C.; MIGNOSI, A. : Definition of a high flow quality injector driven tunnel : the pressurized transonic wind tunnel T2 at ONERA/CERT. 18th Israël Annual Conference 1976

2. GOBERT, J.L.; SERAUDIE, A.; MIGNOSI, A. : Etude de l'interaction onde de choc-couche limite sur profil LC100D de 400 mm de corde dans la soufflerie T2. ONERA R.T. N° 38/7078 AYD-Janvier 1980

3. MIGNOSI, A.; GOBERT, J.L.; SERAUDIE, A. : Etude préliminaire de l'interaction onde de choc-couche limite sur une aile à flèche variable. ONERA R.T. N° 43/7078 AYD-Octobre 1980

4. DOR, J.B.; SERAUDIE, A. : Etude de l'interaction onde de choc-couche limite sur une aile à 45° de flèche. P.V. d'essai OA N° 53/7078 AND - Avril 1982

5. DOR, J.B.; PLAZANET, M. : Approfondissement de l'étude expérimentale sur aile en flèche de profil LC100D. ONERA R.T. N° 58/7078 AND - Décembre 1982

6. DELERY, J. : Interaction onde de choc-couche limite turbulente en transsonique (Recherches en 78-79). ONERA R.S.F. N° 36/7078 AYD-Juillet 1980

7. DELERY, J. : Interaction onde de choc-couche cisaillée turbulente (Recherches en 79-80). ONERA R.S.F. N° 41/7078 AY-Novembre 1980

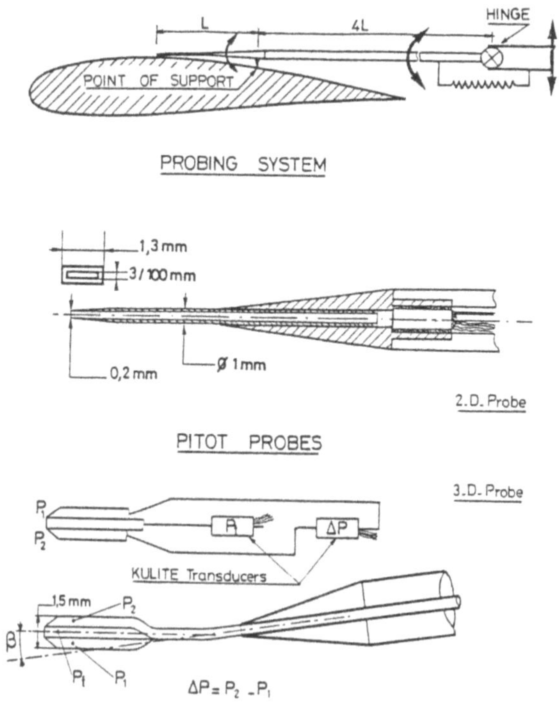

Fig. 1 - Boundary layer exploration system, simple and directional total pressure probes

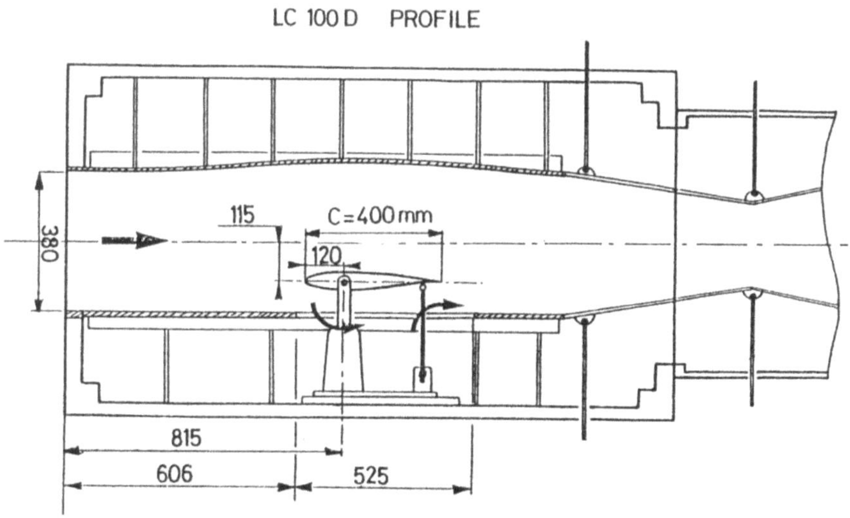

Fig. 2 - Large chord airfoil test device

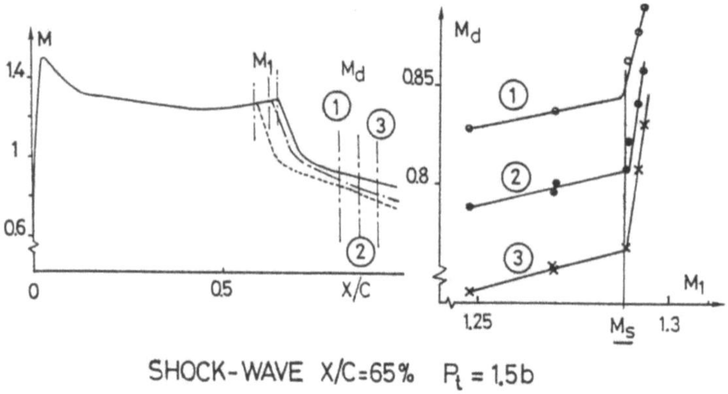

Fig. 3 - Separation emergence analysis on the Mach number distributions along the airfoil chord, for the shock location x/c = 65 %

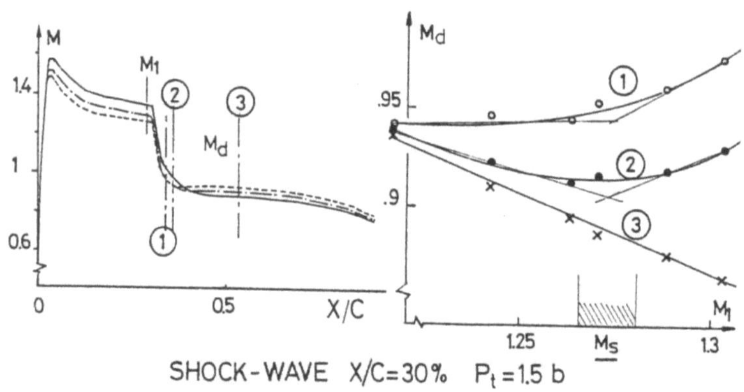

Fig. 4 - Separation emergence analysis on the Mach number distributions along the airfoil chord, for the shock location x/c = 30 %

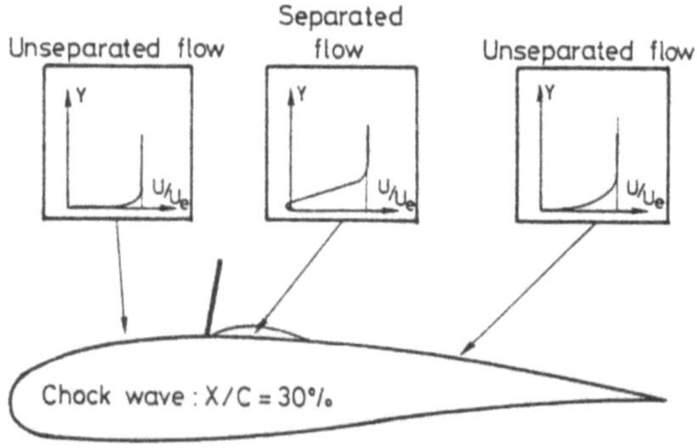

Fig. 5 - Schematic description of the boundary layer separation for the two shock locations x/c = 30 % and 65 % and shapes of the measured boundary layer speed curves

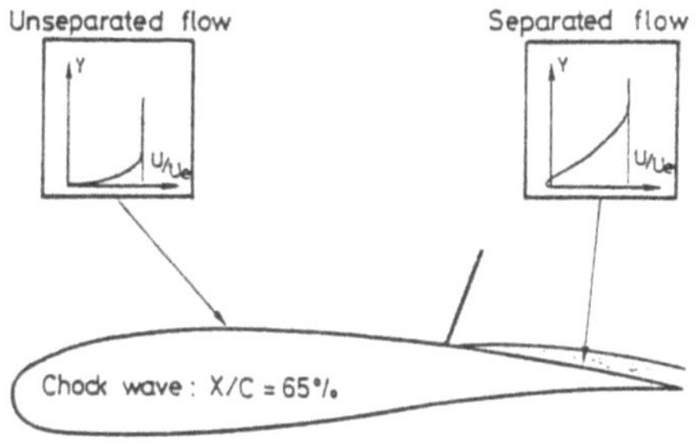

Fig. 5 - Schematic description of the boundary layer separation for the two shock locations x/c = 30 % and 65 % and shapes of the measured boundary layer speed curves

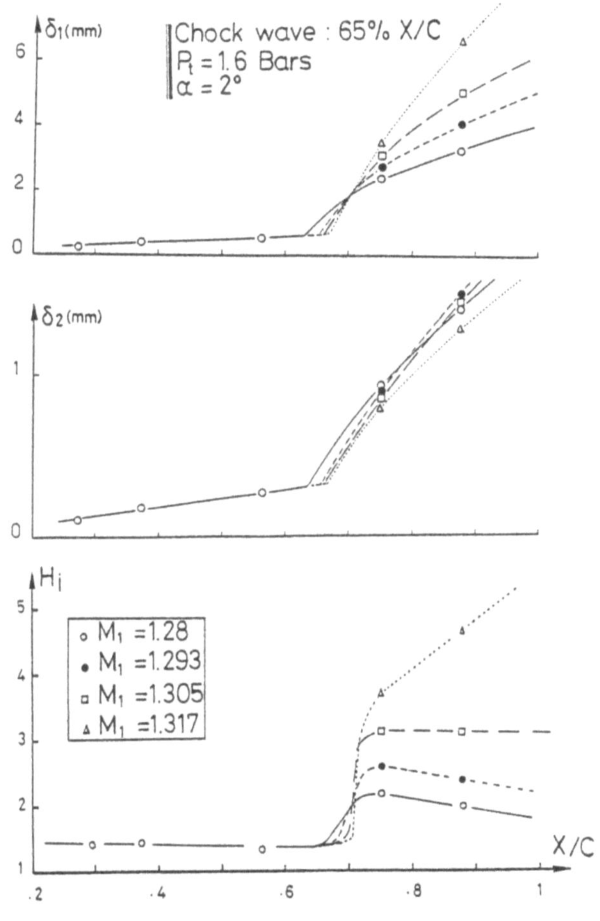

Fig. 6 - Variation along the chord of the measured integral parameters of the boundary layer, for the shock location x/c = 65 % and for different cases around separation

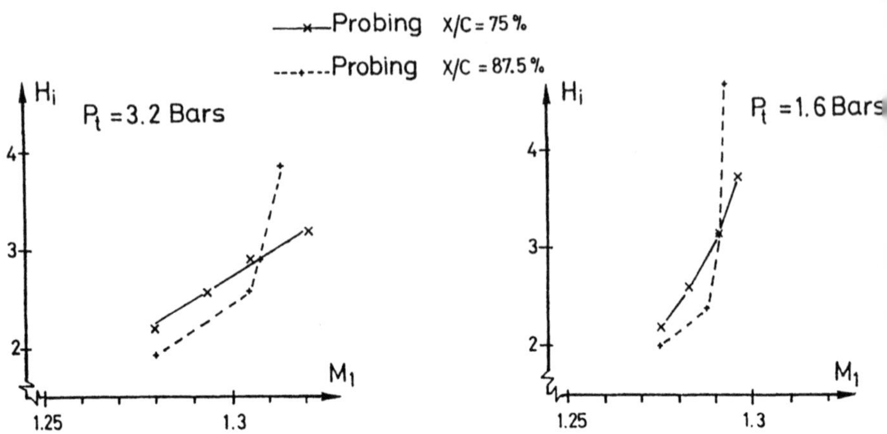

Fig. 7 - Variation of the incompressible shape factor Hi, measured at fixed position downstream of the shock (x/c = 65 %) with the local Mach number M_1 upstream of the shock

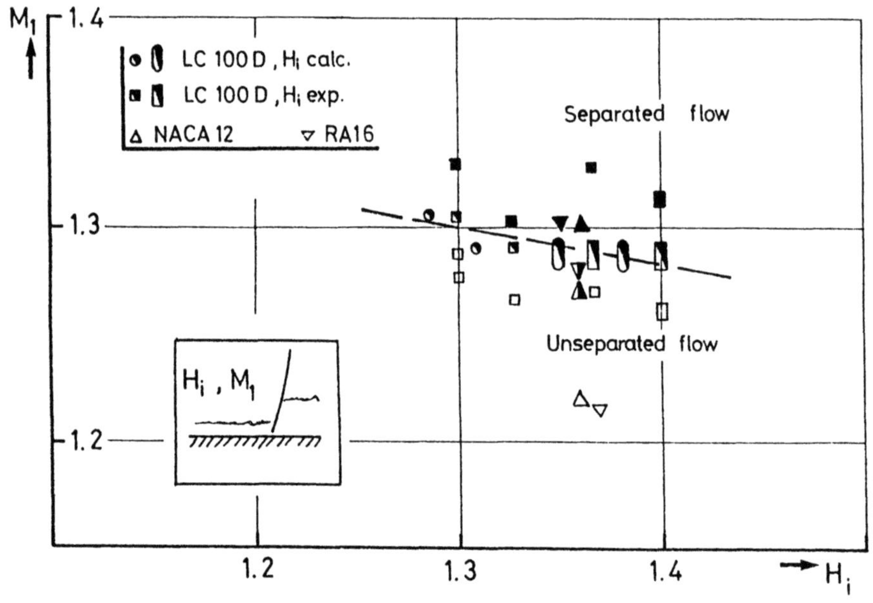

Fig. 8 - Separation condition in the plane (M_1, Hi) : local Mach number and boundary layer shape factor upstream of the shock

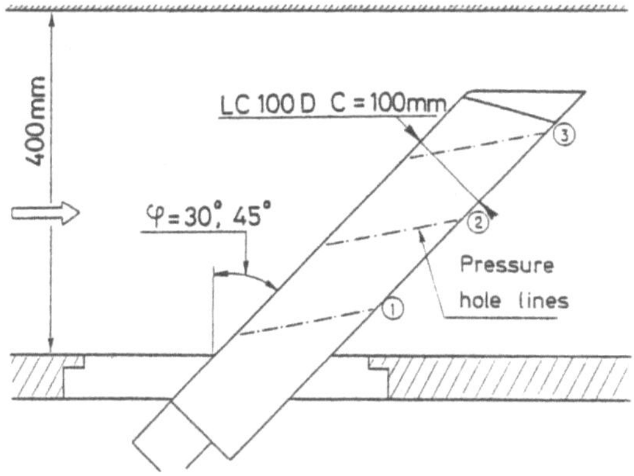

Fig. 9 - Swept wing test device

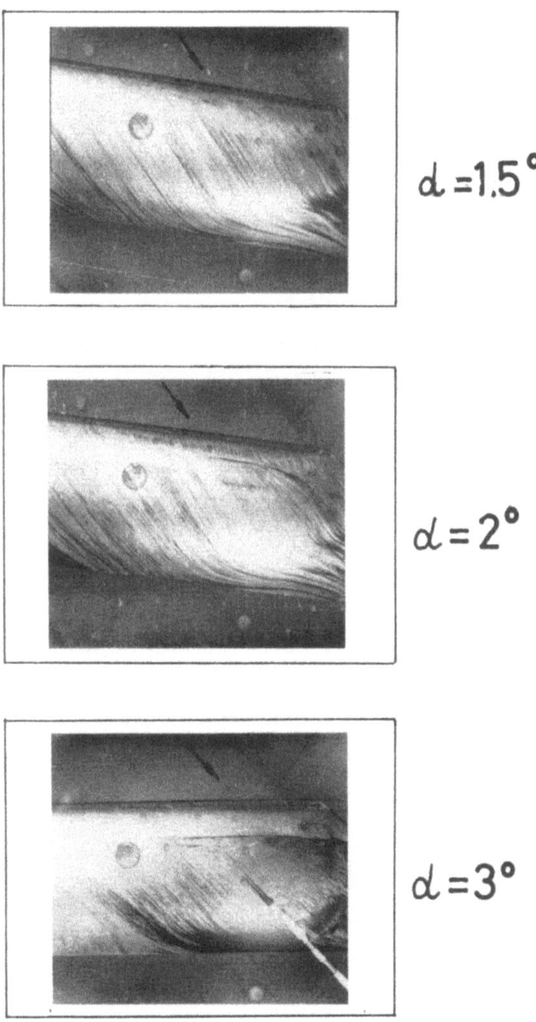

Fig. 11 - Wall streamline visualizations for different angles of attack (ϕ = 45° and M = 0.95)

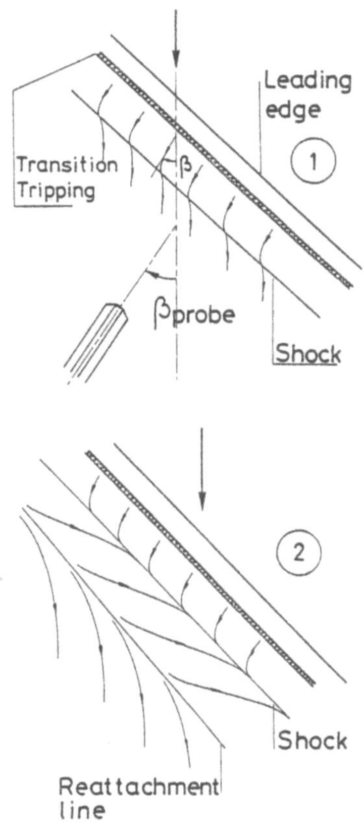

Fig. 10 - Wall streamline configurations : unseparated (1) and separated (2) boundary layer

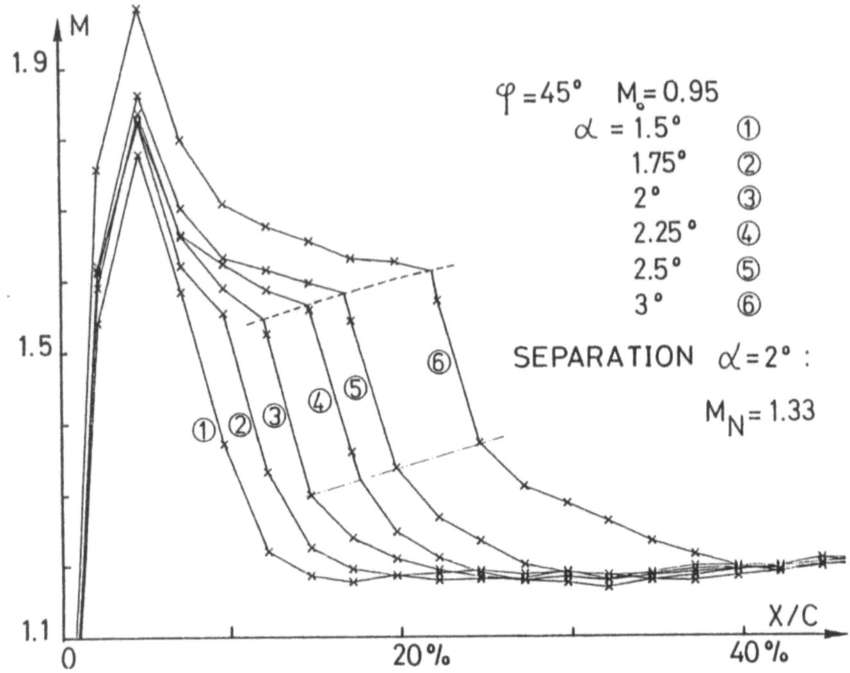

Fig. 12 - Mach number distributions along the model chord for different angles of attack (ϕ = 45° and M = 0.95)

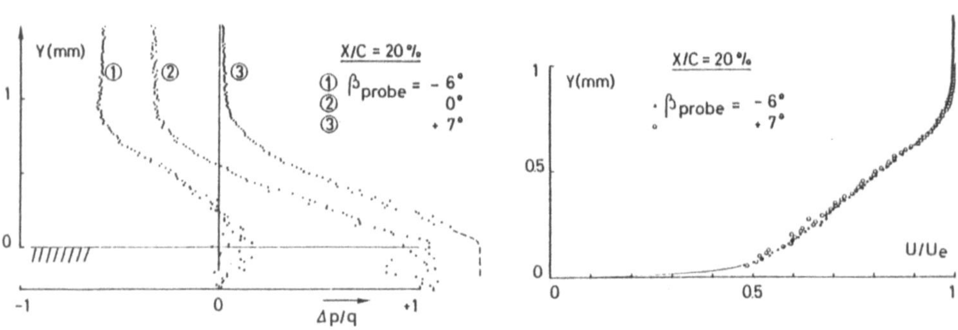

Fig. 13 - Example of a directional boundary layer exploration at x/c = 20 % for the separation limit case : ϕ = 45°, M = 0.95, α = 2°

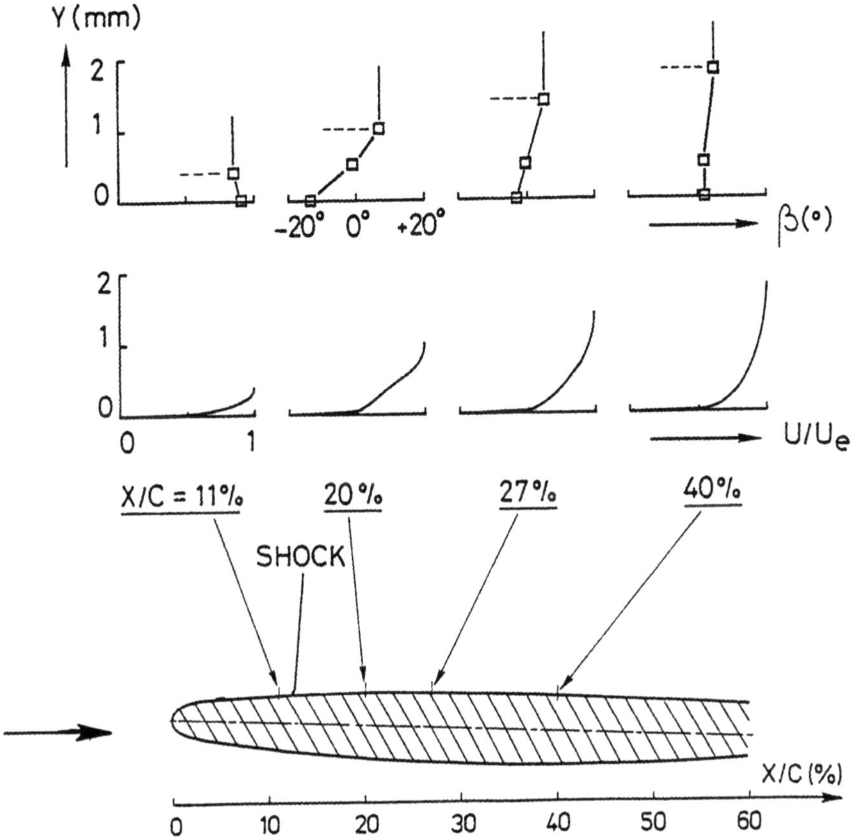

Fig. 14 - Boundary layer exploration results for the separation limit case $\phi = 45°$, $M = 0.95$, $\alpha = 2°$: flow angle and speed curves

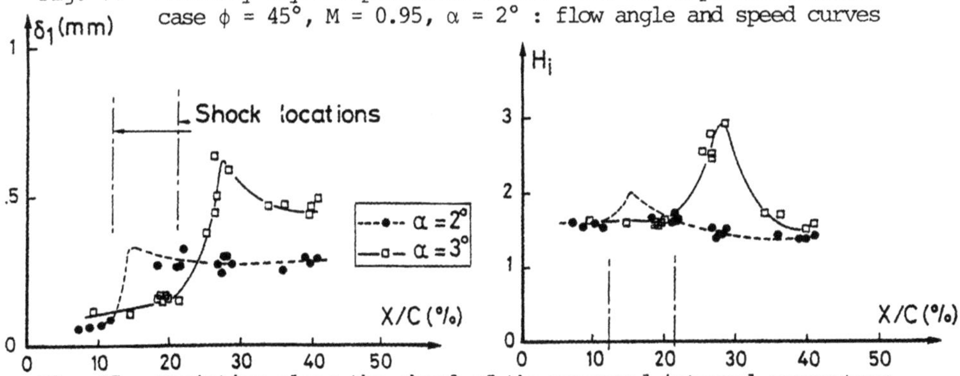

Fig. 15 - Variation along the chord of the measured integral parameters of the boundary layer $\delta 1$ and Hi : separation limit ($\alpha = 2°$) and separated boundary layer ($\alpha = 3°$) case for $\phi = 45°$, $M = 0.95$

A Model of the Flow over Swept Wings with Shock Induced Separation

J. L. FULKER and P. R. ASHILL

Royal Aircraft Establishment,
Bedford, UK

Summary

A new flow model is proposed for describing quasi two-dimensional flows with shock-induced separation on wings of modest sweep. Correlations of the main features of the flow model over a wide range of Reynolds number have been deduced from wind-tunnel tests on wings and aerofoils. These correlations have successfully been used as a basis for simulating 'full-scale' flows over wings in the wind tunnel.

1 Introduction

Uncontrolled separation of the flow from a smooth surface limits the manoeuvrability and operating altitude of aircraft[1]. In transonic flows, separations may originate from the intersection of the shock with the wing surface or close to the trailing edge. In either case, the region of separation grows with increasing incidence until the flow is wholly separated between the shock wave and the trailing edge; when this takes place 'flow breakdown' is said to occur.

In the absence, for the present, of entirely satisfactory methods of calculating flows with shock-induced separation, reliance has to be placed on wind-tunnel experiments to predict flow-breakdown conditions. However, transonic flows with separation may be especially scale sensitive, and hence associated experimental research has been concerned with improving our understanding of the fluid mechanics of such flows. In 1968, Pearcey, Osborne and Haines[2] made an important contribution in their categorisation of various types of separation and discussion of scale effects. They defined two broad groups of flow; in 'Model A', flow breakdown follows the progressive growth of a shock-induced separation while, in 'Model B', it is precipitated by a

Turbulent Shear Layer/Shock Wave Interactions
IUTAM Symposium Palaiseau 1985
Editor: J. Délery

separation moving upstream from the trailing edge. Although a valuable aid to interpretation, this rationalisation does not allow a quantitative assessment of scale effects to be made.

The starting point of the present work was the investigation of an hypothesis of Green[3] concerning scale effects on flows with shock-induced separation bubbles. As shown in section 2, this hypothesis is confirmed by the present study. The investigation also showed that neither of the two flow models of Ref 2 is adequate for wings of modern design, for which the adverse pressure gradients on the upper surface are mild except at the shock, in the associated separation bubble and in a region close to the trailing edge. In this Paper, a new flow model is proposed for such wings, and it is shown to provide a basis for a quantitative analysis of the development of separation leading to flow breakdown and of its sensitivity to scale effects.

The flow model has three main elements which are described in section 2. Following an extensive analysis of experimental data, correlations of high resolution, based on physically justified assumptions, have been derived for each of the elements. Details of the correlations are presented in section 2. As well as providing further understanding of the complex flow phenomena found on swept wings as flow breakdown is approached, the correlations are used to identify transition-trip positions on a wind-tunnel model which yield flows corresponding closely to those at full scale. The method used to determine these simulations and examples of its use are presented in section 3.

2 Description of Flow Model and Derivation of Correlations

The flow model is best described by reference to Fig 1. Consider the flow in the region of the wing outboard of the point on the wing surface where the two shock waves intersect to form a single shock wave (the 'outer' region). In this region, the flow is quasi two-dimensional except possibly in the near vicinity of the tip. At moderate incidence, condition 1, the shock is too weak for a separation bubble to occur. As incidence is increased to condition 2, the shock becomes strong enough to cause a bubble. The reattachment point, 'B', is upstream of the point 'R' which may be taken to define the start of the relatively rapid rise in pressure with streamwise distance near the trailing edge. As the shock gains in strength with increasing incidence, the bubble grows until, at condition 3, 'B' reaches a point at or close to 'R' where the bubble 'bursts' giving rise to the flow breakdown.

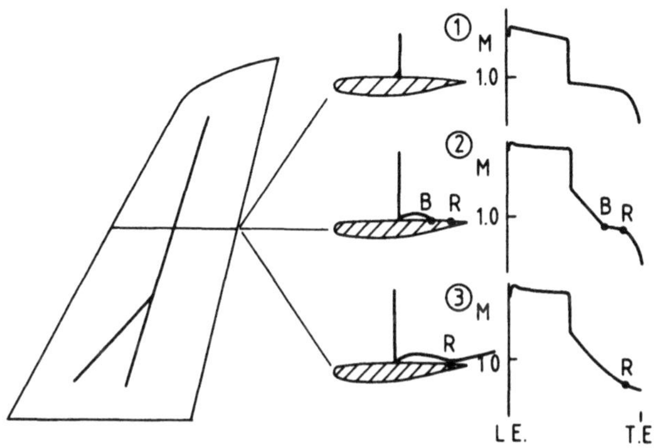

Fig.1. Flow model

The main features of the flow model are:

a. streamwise length of bubble or 'bubble length',

b. pressure rise through the shock,

and c. furthest downstream position of the bubble reattachment point before the bubble bursts.

Each of these features is considered for flows conforming to the flow model. The analysis is based on studies of static-pressure distributions on two wings of moderate sweepback (less than 30°) and of aspect ratio 8, as well as on a number of aerofoils. The measurements of pressure were made in tests at high-subsonic, freestream speeds, mainly in the 8ft × 8ft Wind Tunnel, RAE Bedford, for a wide range of mean chord Reynolds number, $R_{\bar{c}}$ (up to 13×10^6 for one of the wings and 20×10^6 for the aerofoils). In all cases, transition was fixed upstream of the shock, generally at positions close to the leading edge. However, for one of the wings, the transition trip on the upper surface was placed at various locations further downstream to study the effect of varying boundary-layer thickness at the shock without changing Reynolds number. Details of the models and tests are given in Ref 4.

2.1 Bubble Length

It is supposed that bubble length, ℓ_B , is independent of the streamwise static-pressure gradient downstream of the bubble. In addition, it is assumed that surface curvature does not affect the flow within the bubble. For the types of upper-surface curvature distribution typical of modern wing sections, both these assumptions are valid[4] provided that the shock is upstream of the rapid rise in pressure just ahead of the trailing edge. Thus bubble length may be supposed to depend on the flow conditions at the point 'a' just ahead of the shock (Fig.2).

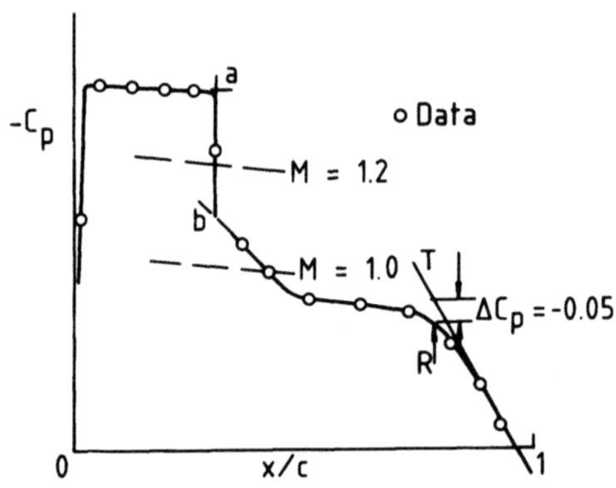

Fig.2. Definitions of points 'a', 'b' and 'R'

For two-dimensional flows with mildly-adverse pressure gradients ahead of the shock typical of those of modern wings, the boundary-layer state at this point is essentially defined by momentum thickness, θ_a , and the momentum-thickness Reynolds number

$$R_{\theta a} = u_{ea} \theta_a / \nu_a$$

Here ν is kinematic viscosity, suffixes a and e refer to the point 'a' and the outer edge of the boundary layer (external flow) and u is velocity in the direction of the external flow. In addition, the inviscid flow is defined by the Mach number of the external flow at the point 'a', M_a . Physical reasoning then leads to the following expression for the bubble length scale.

$$\ell_B/\theta_a = f\,(M_a\,,\ R_{\theta a}) \qquad (1)$$

This expression is implicit in Green's study[3] of shock-induced separation to be considered later.

Equation (1) is extended to swept-wing flows of a quasi two-dimensional character by interpreting M as the Mach number of the external flow component normal to the locus of points 'a' on the wing surface. Furthermore, for the interpretation of the boundary layer, use is made of the analogy between the flow along an external-flow streamline and a corresponding two-dimensional flow[5]. Thus θ_a is taken to be the momentum thickness of the flow in the direction of the external-flow streamline.

The parameters on the right-hand side of equation (1) and the momentum thickness θ_a are derived from measured static-pressure distributions using the relationships for adiabatic, isentropic flow and on the assumption that the flow ahead of the shock is of the infinite yawed-wing type. An infinite yawed-wing version[5] of the Lag Entrainment method of Green et al[6] is used to calculate θ_a and $R_{\theta a}$.

Bubble length is inferred from pressure distributions in the way described in Ref 4. Correlations of data for bubble length in the form suggested by equation (1) are shown in Fig 3 for various values of M_a . For clarity, only data for $M_a = 1.4$ are presented, the curves being least squares fits through the data. An analysis of sources of error indicates that the observed scatter about the mean line is consistent with uncertainties in the determination of shock position and sweep and of pressure ahead of the shock. The data for the aerofoils are consistent with those for the swept wing, indicating that sweep effects are well represented in the range of shock sweeps considered (0 - 25°). Moreover, the aerofoil data extend the correlations to $R_{\theta a} = 14000$, typical of cruise conditions on the outer wing of a subsonic transport aircraft with a mean chord Reynolds number $R_{\bar{c}} = 30 \times 10^6$.

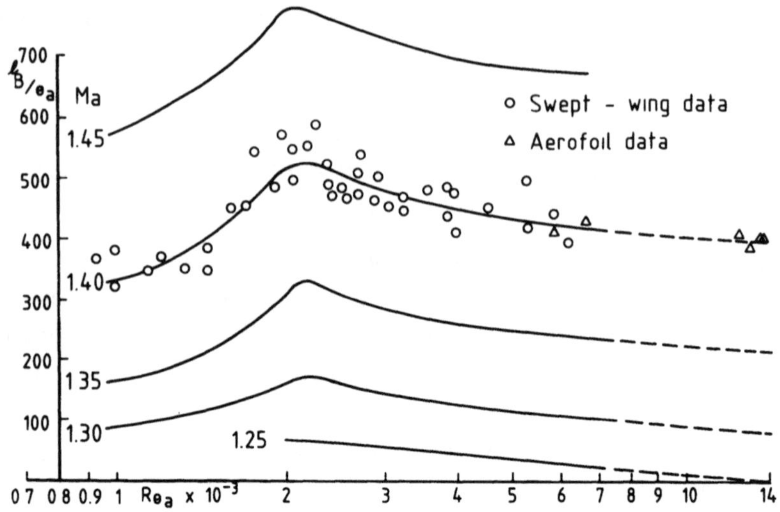

Fig.3. Correlations of bubble-length scale against Reynolds number

The correlations are in qualitative agreement with the variation in length scale with Reynolds number for shock-induced separation in two-dimensional flows proposed by Green[3]; he argued that the existence of a maximum length-scale for a given value of M_a is related to the change in character with Reynolds number of the velocity profiles of turbulent boundary layers on flat plates. Green's idea is based on the work of Coles[7] who showed that these velocity profiles conform to similarity rules for incompressible flow in such a way that they become more resistant to separation as Reynolds number increases. Below a value of $R_{\theta a}$ of 2000-5000, the similarity rules no longer apply, and the profile becomes increasingly 'full' as Reynolds number decreases. Green extended this approach to include compressible flows[8], and his analysis indicates that the turning value of $R_{\theta a}$ is largely independent of M_a over the range of interest, in agreement with the present correlations.

2.2 Shock Pressure Rise to Cause Separation

Arguments similar to those given in section 2.1 may be used[4] to justify a correlation of the form

$$M_a/M_b = g(R_{\theta a})$$

for the ratio of the Mach numbers at the points 'a' and 'b' (Fig 2).

The correlation obtained is shown in Fig 4; this is seen to be highly satisfactory, with the aerofoil data again consistent with those of the swept wings. A minimum value of M_a/M_b is observed at $R_{\theta a} = 5000$, although the variation over the range of $R_{\theta a}$ studied is small. A minimum was postulated by Green[3] for two-dimensional flows (at about the same value of $R_{\theta a}$) for the reasons given in section 2.1.

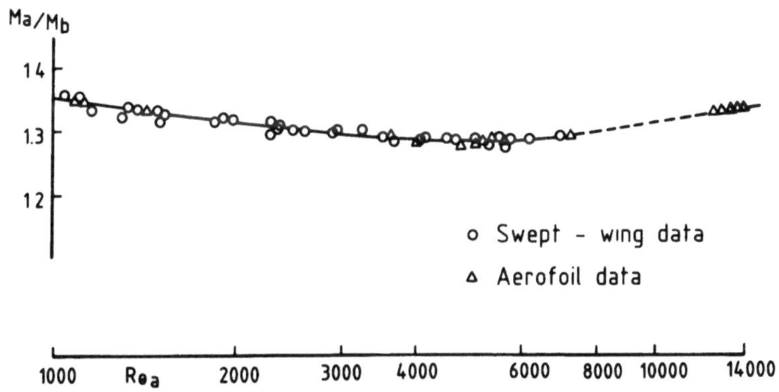

Fig.4. Variation of Mach-number ratio with Reynolds number

2.3 Flow Breakdown

As noted before, flow breakdown occurs when the bubble reattachment point 'B' reaches a point close to the start of the relatively rapid rise in pressure gradient 'R'. The point 'R' may be defined precisely by reference to Fig 2. A tangent 'T' is drawn to the pressure distribution downstream of the bubble at the point of maximum adverse pressure gradient. Provided that this gradient exceeds the pressure gradient in the region of the bubble, 'R' is defined as the point where 'T' diverges from the pressure distribution by $\Delta C_p = -0.05$. The point 'R' is difficult to define with certainty when the point 'B' is closer to 'R' than about 5% chord. However, for shorter bubbles, x_R does not vary significantly with lift. The position of 'R' is found not to be scale dependent, and this is to be expected since x_R is largely determined by the shape of the wing in this region.

In Ref 4 it is shown that the value of x_B at flow breakdown may be correlated with x_R and conditions of the flow at the point 'a' by an expression of the form

$$x_B^* = x_R + \theta_a^* \, h(M_a^*) \qquad (2)$$

where suffix * refers to flow breakdown and

$$h(M_a^*) = \begin{cases} 0, & M_a^* < 1.4 \\ 2000\,(M_a^* - 1.4), & 1.4 < M_a < 1.5 \end{cases} \qquad (3)$$

Since x_R is independent of scale, equations (2) and (3) imply that x_B^* is also scale independent if $M_a^* < 1.4$.

3 Simulation of 'Full Scale' Flows

The correlations have been used as a basis for simulating 'full scale' flows in the wind tunnel. Pressure distributions from the tests on wing/body 1 are used to illustrate the method; for this configuration, the flow on the lower surface is scale insensitive, and so only upper-surface conditions need be considered.

In order to ensure that the main features of the flow are simulated, it is necessary to represent in the wind tunnel.

a. shock position and pressure distribution upstream of the shock,

and b. the three features of the present flow model.

Considering the first requirement, Fig 5a shows pressure distributions on the outer wing for a case (C_L = 0.66, M_∞ = 0.78) in which shock-induced separation bubbles are observed and for which scale effects on shock position and strength are small but significant. At the somewhat higher lift coefficient 0.7 (Fig 5b), flow breakdown occurs at the two lowest Reynolds numbers but not at the highest, and thus scale effect on shock position is much more pronounced than in the previous case. However, as is shown later, bubble length decreases as transition position is moved downstream; hence aft-located trips may be used to inhibit flow breakdown and to represent the viscous/inviscid interactions downstream of the shock at 'full-scale'. Thus these trips may be used to simulate 'full scale' shock position and pressure distribution upstream of the shock, and evidence is presented in Ref 4 showing that this is achieved within ranges of trip positions at two Reynolds numbers ($R_{\bar{c}} = 5.1 \times 10^6$ and 13×10^6) for the case shown in Fig 5b.

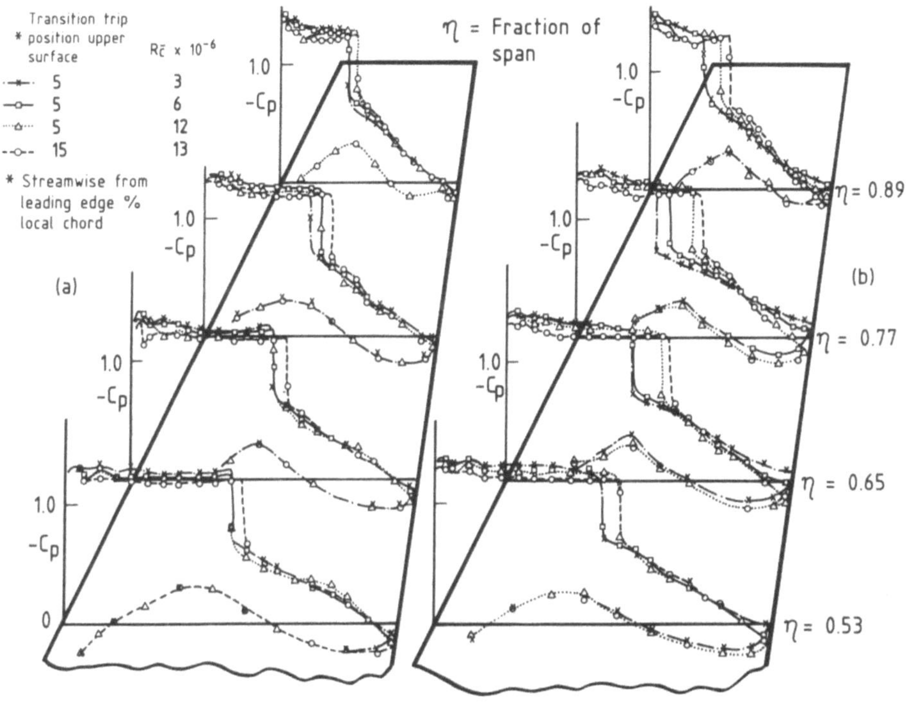

Fig 5 Scale effects on outer-wing pressure distributions, (a) C_L = 0.66, (b) C_L = 0.70

With the first requirement satisfied, it is possible to infer 'full scale' features of the present flow model using the correlations of section 2 and the boundary-layer calculation method on which they are based. The same procedure may then be used to determine the trip positions needed to simulate these same features in the wind tunnel.

In general, it is not possible to match all the conditions simultaneously. However, as noted above, there appears to be some freedom of choice of transition position to simulate shock position and shock upstream pressure distribution. Failure to simulate shock Mach-number ratio can be tolerated since M_a/M_b

varies slowly with $R_{\theta a}$. The position of bubble reattachment at flow breakdown, x_B^* , is either independent of scale ($M_a^* < 1.4$) or is found not to differ significantly between tunnel and 'full scale' when 'full scale' bubble length is simulated in the tunnel. Thus some latitude is allowable in transition position provided bubble length is matched to that at 'full scale'.

The technique is illustrated for the case of Fig 5b ($C_L = 0.70$, $M_\infty = 0.78$). 'Full scale' is taken to correspond to $R_{\bar{c}} = 30 \times 10^6$, with transition at 1% chord on the upper surface ($x_T/c = 0.01$) and with the wing aerodynamically smooth.

Calculated bubble length is plotted against trip position and mean chord Reynolds number in carpet form for the outer-wing station $\eta = 0.77$ (77% span) in Fig 6. This figure confirms the point made earlier about bubble length decreasing as transition is fixed further aft. At $R_{\bar{c}} = 5.1 \times 10^6$ a satisfactory match between tunnel and 'full scale' is obtained with a trip at 31% chord. This corresponds closely to a position that has been tested (30% chord). For $R_{\bar{c}} = 13 \times 10^6$ the required trip position is close to 15% chord which has also been tested. In both cases, the required trip positions are within the range of trip positions for which 'full scale' shock position and pressure distribution upstream of the shock are simulated[4].

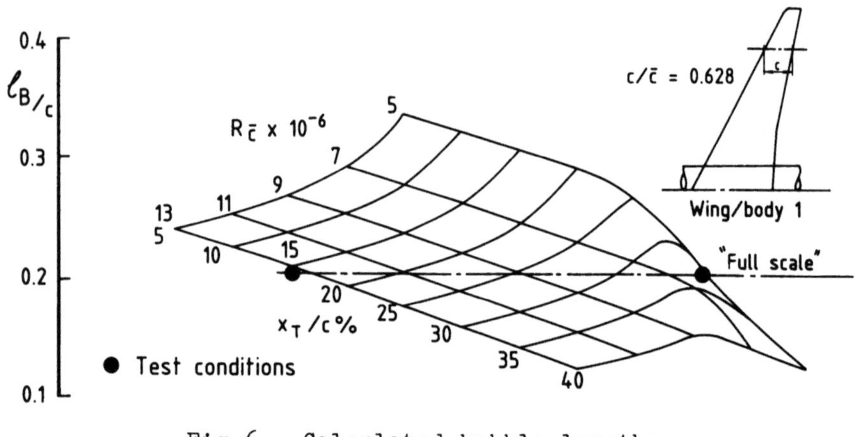

Fig 6 Calculated bubble lengths, $\eta = 0.77$, $M_\infty = 0.78$, $C_L = 0.70$

At these trip positions there is a slight mismatch between tunnel and 'full scale' values of M_a/M_b, the 'full scale' value being 1.33, and the tunnel values being 1.31 at $R_{\bar{c}} = 5.1 \times 10^6$ and 1.28 at $R_{\bar{c}} = 13 \times 10^6$.

For this flow, M_a is everywhere less than 1.4. Hence the position of the reattachment point of the bubble at flow breakdown is independent of scale.

Similar trip positions are indicated for other stations on the outer wing.

An indirect way of checking the effectiveness of the simulation is to compare the pressure distributions with the appropriate trips at the two Reynolds numbers. Since the two tests were made at different tunnel stagnation pressures, the effects of static aeroelastic distortion need to be taken into account. However, the main effect is to change the local incidence of the wing; hence aeroelastic effects may reasonably be ignored since the comparisons are made at substantially the same local lift coefficients on the outer wing. Fig 7 shows that the agreement between the two distributions is good, especially when viewed against the large scale effects exhibited in Fig 5b. Other examples providing support for the technique are given in Ref 4.

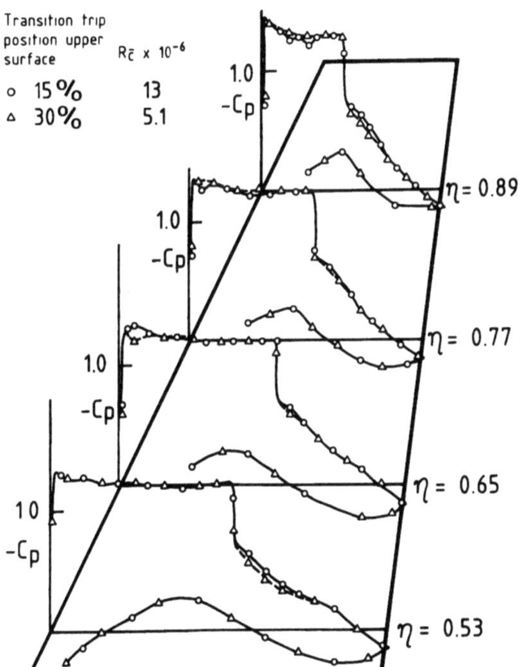

Fig.7. Two independent simulations of 'full scale' pressure distributions, outer wing $M_\infty = 0.78$, $C_L = 0.7$

4 Conclusions

A new flow model has been proposed for wings with shock-induced separation based on the notion that the streamwise length of the bubble is limited by the adverse pressure gradient near the trailing edge. The main features of the flow model have been studied for quasi two-dimensional flows on wings of low sweep ($< 30°$) over a wide range of Reynolds numbers and transition-trip positions, and have been correlated with flow quantities just upstream of the shock. The main conclusions from these correlations are as follows:

(i) The variation of bubble length with Reynolds number is in qualitative agreement with that postulated on physical grounds for two-dimensional flows by Green[3]. For a given shock upstream Mach number, bubble-length scale displays a maximum at a value of $R_{\theta a}$ of approximately 2000.

(ii) Shock Mach number ratio, M_a/M_b, varies slowly with Reynolds number.

(iii) A correlation has been obtained which allows the streamwise position of the reattachment point of the bubble at flow breakdown to be determined. For values of shock upstream Mach number, M_a, less than 1.4 this point coincides with the point 'R' denoting the start of the region of adverse pressure gradient close to the trailing edge.

The flow model has been used as a basis for simulating, in the wind tunnel, flows with shock-induced separation at high Reynolds number. In order to test the method, the simulations have been tried for flows about a swept-wing model at two different Reynolds numbers. Good agreement is obtained between the two simulations of 'full scale' pressure distributions, suggesting that the method has a sound physical basis.

References

1. Fluid Dynamics of Aircraft Stalling. AGARD CP-102 1972.

2. Pearcey, H.H.; Osborne, J.; Haines, A.B.; The interaction between local effects at the shock and rear separation - a source of significant scale effects in wind-tunnel tests on aerofoils and wings. Paper 11. AGARD CP-35 1968.

3. Green, J.E.; A discussion of viscous-inviscid interactions at transonic speeds. Paper 2. AGARD CP-35 1971.

4. Fulker, J.L.; Ashill, P.R.; A study of factors influencing shock-induced separation. RAE Technical Report to be published.

5. Smith, P.D.; A calculation method for the turbulent boundary layer on an infinite yawed wing in compressible, adiabatic flow. RAE Technical Report 72193 1972.

6. Green, J.E.; Weeks, D.J.; Brooman, J.W.F.; Prediction of turbulent boundary layers and wakes in compressible flow by a lag-entrainment method. ARC R&M 3791 1973.

7. Coles, D.E.; The turbulent boundary layer in a compressible fluid. USAF Project Rand Report R-403-PR 1962.

8. Green, J.E.; A note on the turbulent boundary layer at low Reynolds number in compressible flow at constant pressure. RAE unpublished.

Separation ahead of Blunt Fins in Supersonic Turbulent Boundary-Layers

N. SAIDA

Department of Mechanical Engineering
Collegé of Science and Engineering
Aoyama Gakuin University, Tokyo 157,
Japan

Summary

An experimental study of the three dimensional shock wave/turbulent boundary layer interaction induced by a blunt fin was made mainly at Mach 2.48. The effects of leading edge shape and angle of attack were studied. Surface pressure measurements, and flow visualizations were made. The results show that, on the test surface near the fin, the structure of the vortex in the reverse flow region depends strongly on the leading edge shape, but it does not affect the flow near the separation point. Furthermore the effects of the angle of attack appear downstream of the plateau region especially on the windward side.

Introduction

The problems of interaction between shock wave and boundary layer induced by a blunt fin in a supersonic turbulent boundary layer are important in such areas as external aerodynamics and turbomachinery. Therefore, a considerable effort has been made by many authors to investigate the matter [1]-[14]. Particularly for a circular cylinder or a hemicylindrical leading edge fin the effects of the geometrical parameters, such as fin height, leading edge diameter, sweep angle and the effects of the incoming flow, such as free stream Mach number, Reynolds number and boundary layer thickness are studied. From these experiments it was found that the dominant parameters controlling the centerline flow field are leading edge diameter and sweep angle. However as far as is known, no systematic studies have been made of the effects of leading edge shape. Furthermore the effects of angle of attack for a hemicylindrical leading edge fin are not well understood and detailed experimental data are scanty.

The purpose of the present experiment was to study the effects

Turbulent Shear Layer/Shock Wave Interactions
IUTAM Symposium Palaiseau 1985
Editor: J. Délery

of leading edge shape on the separated flow field and also the details of the flow near the junction of the flat plate and the hemicylindrical leading edge fin at angle of attack.

Experimental Apparatus and Procedure

The experiments were conducted in a 8x10 cm supersonic blowdown wind tunnel. The model geometry and the coordinate system used in the present experiments are shown in figure 1. The fin models are composed of double wedge or hemicylinder, with a fin thickness of d = 10mm and height h = 35mm. The wedge semi-vertex angle θ_W was 20, 30, 40, 60 and 90deg respectively.

In this study, almost all tests were conducted at free stream Mach number of 2.48±0.02. The undisturbed boundary layer growing on the test surface was fully turbulent with 99.5% thickness δ of 0.73cm. The Reynolds number per meter in the test section was 4×10^7. An additional tests at M_∞ = 1.98 were also made at α = 0 deg for hemicylinder and wedge models having a semi-vertex angle of 20, 30 and 40deg.

For measurements of the surface pressure, oil flow pattern and shock wave shape the model was fixed on the side wall of the test section. But for visualization of the center plane flow field the model was fixed on the floor of the wind tunnel. The tests were made in two separate series. In the first series, the fin was fixed at zero angle of attack. On the surface of one side wall, pressure holes of 0.6mm diameter were bored at intervals of 10mm in seven rows. By sliding the model in x and y directions detailed pressure measurements were obtained. Also on the surface of the fin, pressure holes of 0.3mm diameter were bored at intervals of 2∿4mm along the leading edge and for wedges

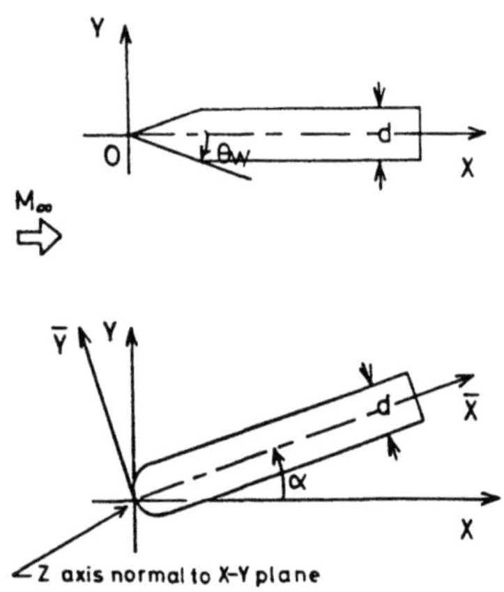

Fig.1. Sketch of model and coordinate system

along a line parallel to the leading edge. In all a variety of six fins were tested. In the second test series a hemicylindrical leading edge fin was used. The fin was fixed at an angle of attack of 0, 4, 8 and 12deg. The fin was instrumented with four vertical rows of pressure holes at x/d = 1.0, 2.0, 3.0 and 4.0. All surface pressure measurements were made with the conventional mercury manometer. To visualize the flow, an oil film technique was employed to determine the separation and the re-attachment lines of the surface. The oil was a silicone oil mixed with titanium dioxide and oleic acid. Schlieren photographs were taken to visualize shock waves and shear layers.

Results and Discussion

1. Flow in the plane of symmetry

Schlieren photographs were taken for each case. It reveals that as the wedge angle increases, the shear layer and the separated shock become distinct and at θ_w large enough to form a detached shock, the flow field becomes qualitatively similar to the hemicylinder case. A similar trend can also be seen at $M_\infty = 1.98$. Further, the separated shock angle β is insensitive to the leading edge shape and depends only on Mach number. The measured values are $\beta = 32°$ at $M_\infty = 2.48$ and $\beta = 40°$ at $M_\infty = 1.98$.

The effect of leading edge shape on the upstream pressure distribution along the centerline is presented in figure 2. The ordinate is the pressure p normalized by the free stream static pressure p_∞ and the abscissa the distance from the fin's leading edge x, normalized by d. For the θ_W = 20deg wedge the pressure increases monotonically. For θ_W above 40deg the pressure distribution tends to be similar in shape to that of the hemicylinder. However, for the flat faced fin after reaching the plateau region a shelf of constant pressure can be seen near the trough. It can be seen from figure 2 that the first peak pressure is almost the same for all cases. If one defines an upstream influence length Lu as a distance from the leading edge to the intersection of the maximum pressure rise slope with the x axis, Lu/d increases linearly with the increase of θ_w. This is shown in figure 3 and can be approximated as follows:

$$Lu/d = -0.291+0.062\theta_w \quad (\theta_w ;\ deg)$$

A similar trend can be seen at smaller Mach number.

The maximum pressure which is located on the centerline just ahead of a blunt fin moves sideways in the case of a sharp fin and is located near the junction of wedge and fin. In these cases it is difficult to obtain the exact value, but from the obtained data it is clear that the maximum pressure ratio P_{max}/p_∞ increases with the increase of Lu/d. This means that with the increase of Lu/d, the distance travelled by a stagnation streamline passing over the separated region and re-attaching to the flat plate increases and the momentum transferred from the outer shear layer increases. Therefore Pmax increases with the increase of Lu/d. In other words the maximum pressure depends strongly on the shape of the leading edge.

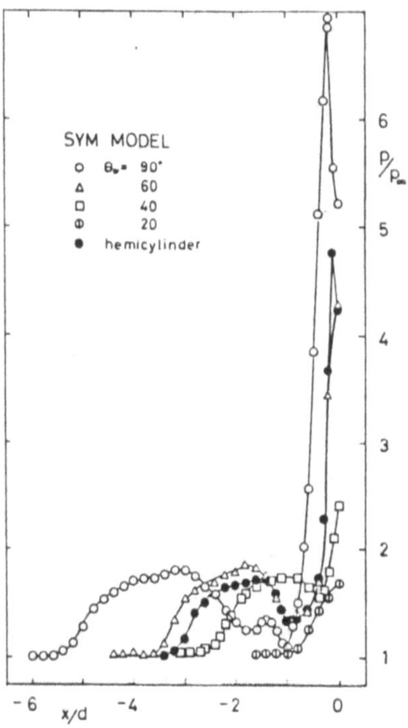

Fig.2. Centerline pressure distributions

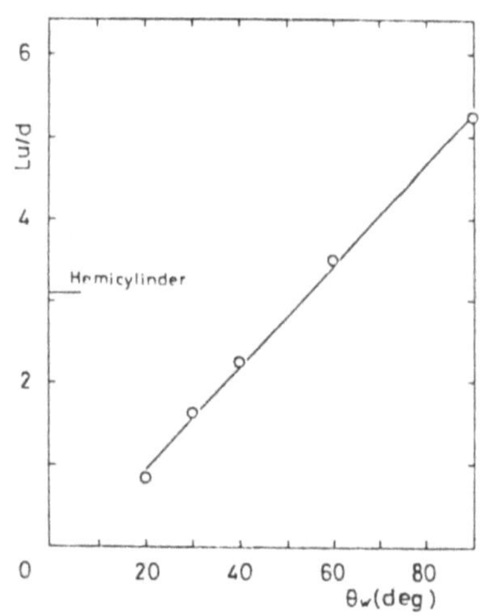

Fig.3. Variation of upstream influence distance with wedge angle

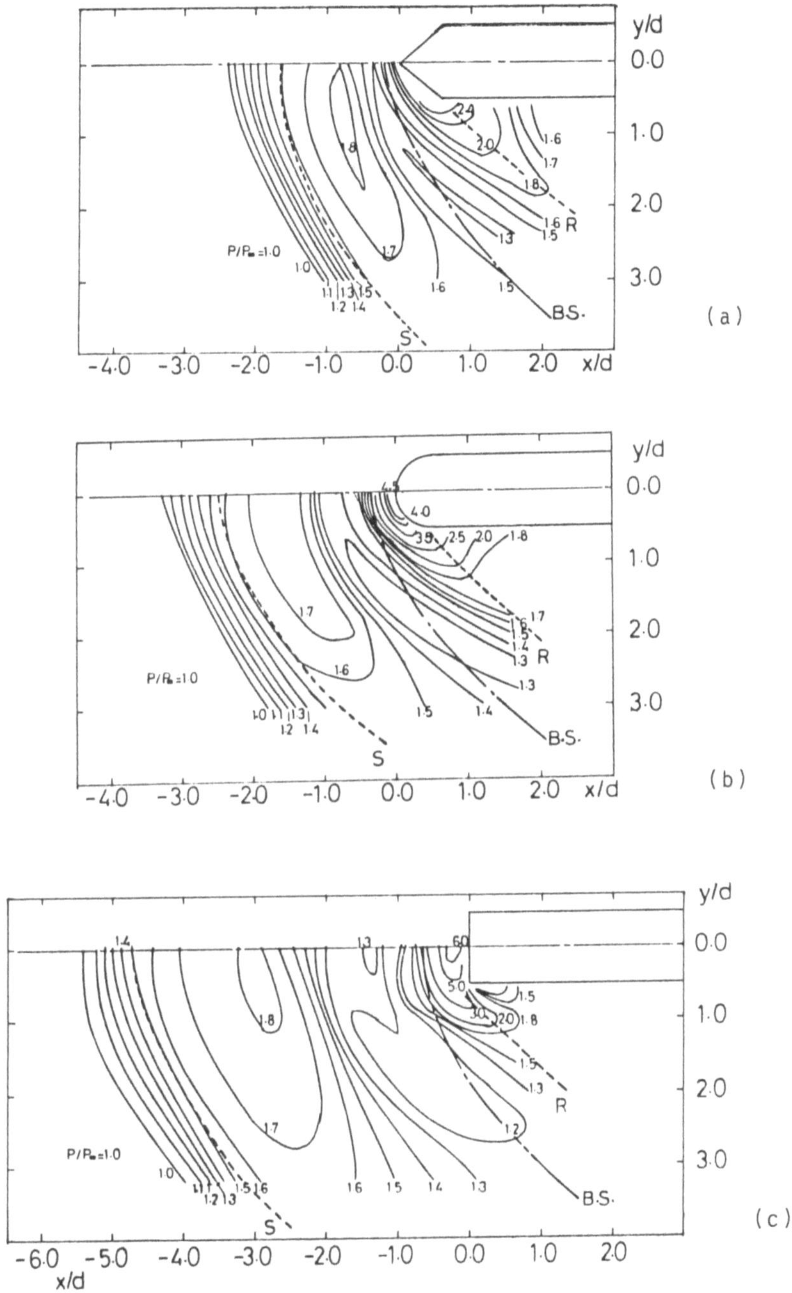

Fig.4. Pressure contours on flat plate surface
(a) 40° wedge; (b) hemicylinder; (c) flat faced fin

2. Flow over the flat plate

Figures 4(a),(b) and (c) show the isobars of the flat plate for 40deg wedge, hemicylinder and flat faced fin. Shown in these figures are the separation line S and the re-attachment line R obtained from the oil film flow photographs. The bow shock shape is also included. For all cases the separation line coincides with that of the constant pressure line corresponding to $p/p_\infty = 1.5\sim1.6$ and the distance between the upstream influence line and the separation line is nearly constant. Excluding the flat faced fin case the re-attachment line corresponds to a line connecting the second peak pressure points. Near the centerline the bow shock is located ahead of a rapid pressure rise point. As can be seen the separated region increases with the increase of bluntness.

In figure 5 the flat plate pressure distributions for different fins are compared by taking the origin at the upstream influence point x_u of each row. The pressure curves take a similar shape until the plateau region irrespective of the leading edge shape. But downstream of the plateau region, especially near the fin a large discrepancy appears. With increasing spanwise distance the second peak points move downstream and the peak pressures decay. From these results it is clear that the structure of the vortex in reverse flow region depends on the leading edge shape, but flow near the separation point is independent of the

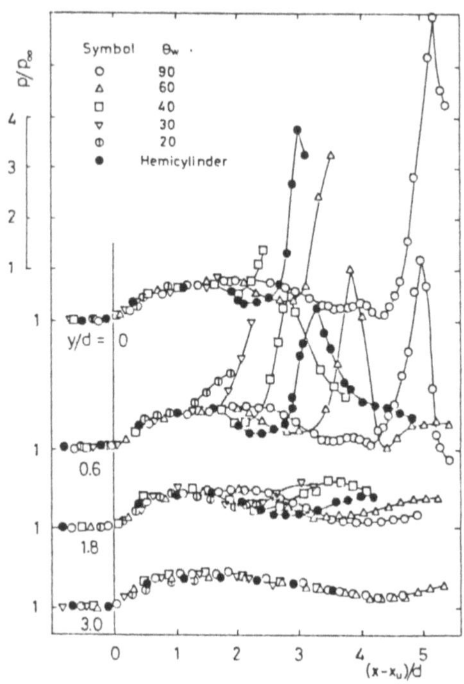

Fig.5. Effect of leading edge shape on streamwise pressure distribution

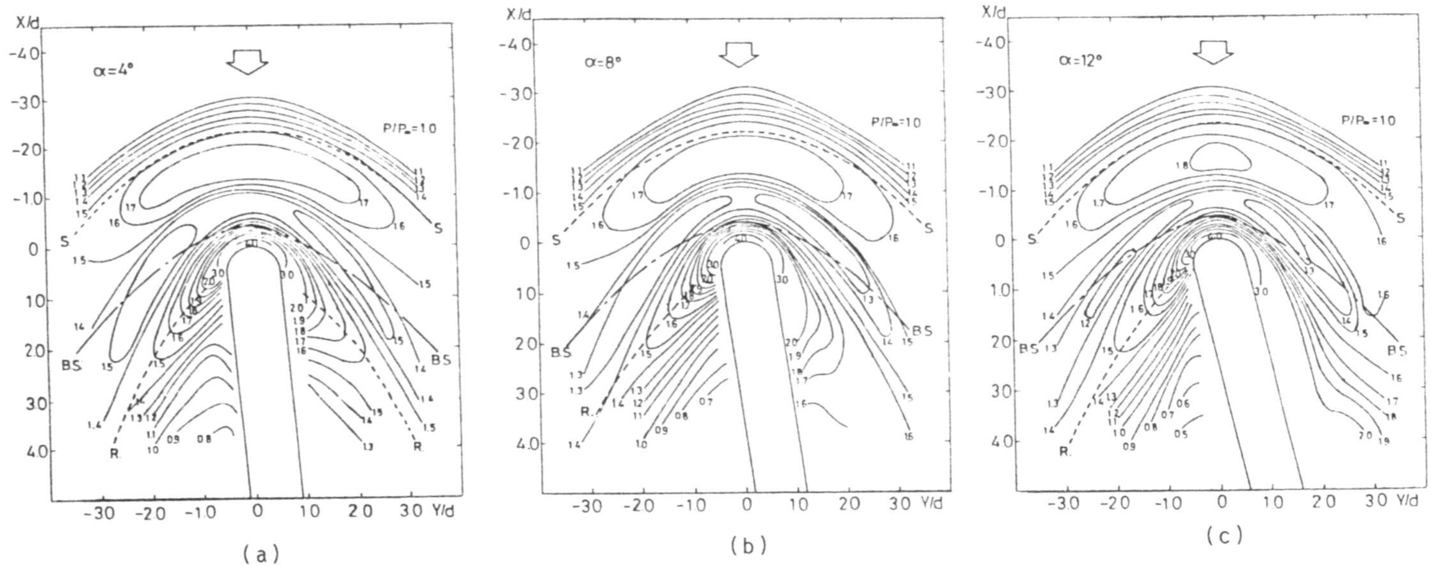

Fig.6. Pressure contours on flat plate surface induced by a hemicylindrical leading edge fin at angle of attack (a) α = 4deg; (b) α = 8deg; (c) α = 12deg

geometry causing the separation. That is as was mentioned by Chapman et.al.[15] free interaction regime exists.

Consider next the effects of angle of attack on the flat plate surface. Figures 6(a),(b) and (c) show the isobars of the flat plate for an hemicylindrical leading edge fin at an angle of attack of α = 4, 8 and 12deg. As can be seen the isobar lines are independent of α until the plateau region and are nearly symmetric with respect to the x axis. In the windward side the location of the second peak pressure point moves downstream with the increase of α. In contrast to the windward side the effects of the angle of attack is small in the leeward side. It is to be noted that the bow shock shape in x-y plane is almost independent of α.

In figures 7(a) and (b) pressure distributions on typical rows are compared for different values of α. Figure 7(a) is for the leeward side and at y/d = -0.6 the difference in pressure appears after the second peak point and the pressure decreases with

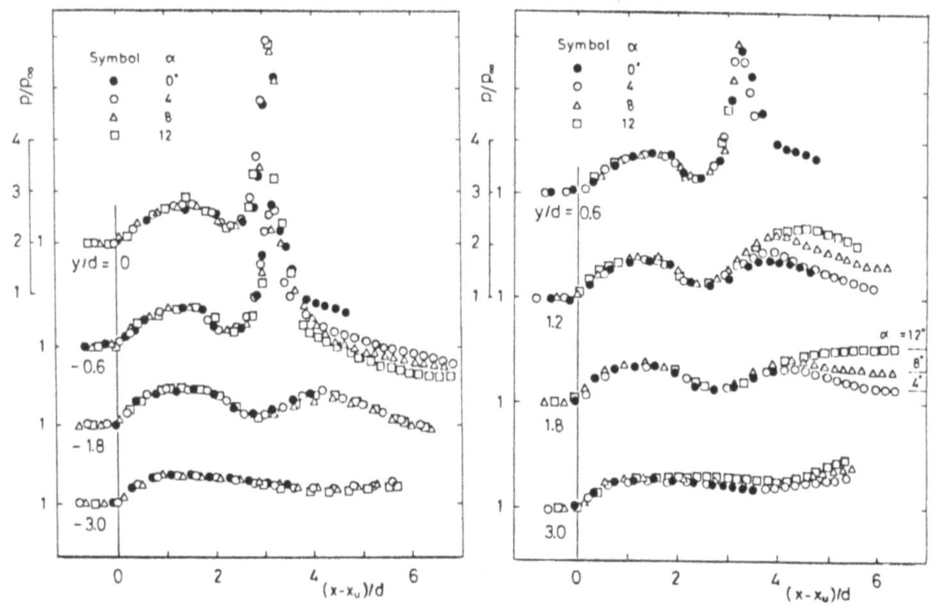

(a) leeward side (b) windward side

Fig.7. Effect of angle of attack on streamwise pressure distribution

the increase of α. This may be due to the effect of the expansion wave. Further outboard, the peak location moves downstream and at the same time the pressure difference weaken. At y/d = -1.8 the effect of α is no longer detectable. On the windward side as is shown in figure 7(b) the effects of α appear near the pressure trough and the second peak pressure increases with the increase of α. In downstream part the pressure tends to a constant value, which is close to the inviscid pressure rise for each deflection angle α (shown on the right of ordinate by a bar).

3. Pressure distribution of fin

Figure 8 shows pressure distributions along the wedge surface and the leading edge of hemicylinder and flat faced fins. The ordinate is the distance of a point measured normal to the flat plate z, normalized by d and the abscissa the surface pressure normalized by the free stream static pressure p_∞. It is clear that with the increase of Θ_W the separated shock wave shifts upstream and therefore the triple point moves up since the separated shock angle is independent of the leading edge shape. Near the root the surface pressures retain nearly the same shape but deviates as z/d increases and tends to a constant pressure level. As can be seen this pressure level increases with the increase of the bluntness. The pressure hump is near the shock triple point and is considered due to the impingement of jet [16].

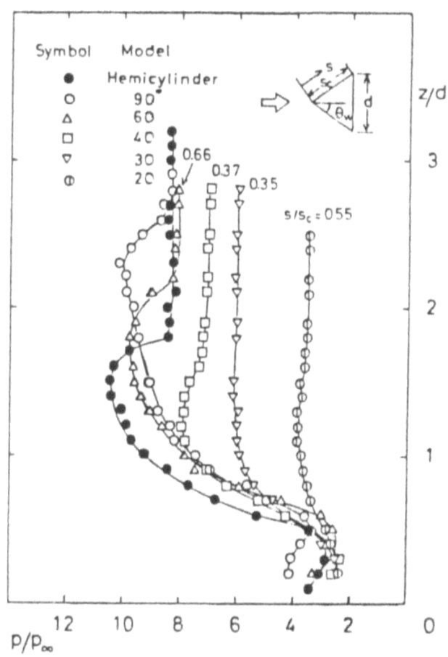

Fig.8. Pressure distributions along the wedge surface

Figures 9(a), (b) and (c) show the oil flow patterns on the surface of the hemicylindrically blunted fin. When compared with α = 0deg case with that of α = 12 deg cases the surface flow pattern is nearly the same in the leeward side

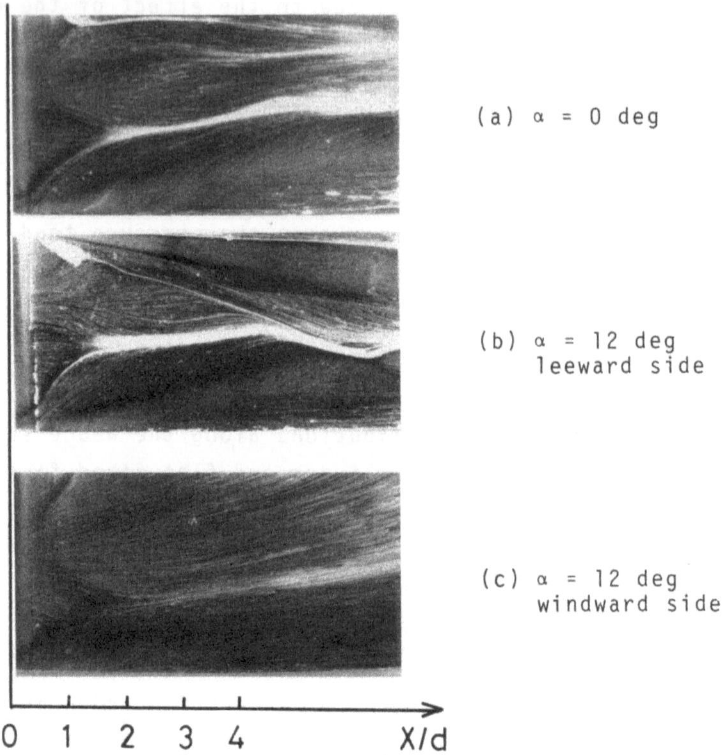

Fig.9. Oil flow photographs

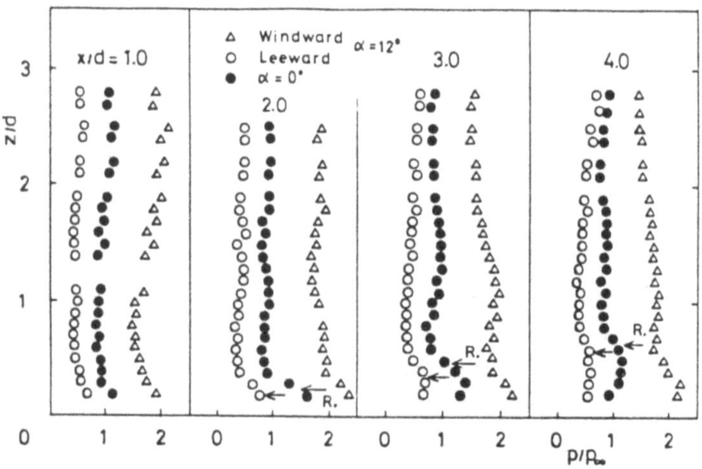

Fig.10. Effect of angle of attack on fin side face pressure distributions

but deviates in the windward side surface. It is to be noted that the herringbone type of surface flow pattern is not clearly visible on the windward side surface. In figure 10 corresponding fin's side face pressure ratio p/p_∞ at x/d = 1.0, 2.0, 3.0 and 4.0 is plotted against z/d. The arrows in the graph show the re-attachment point obtained from the oil film flow. As seen in figures 6, near the root the pressure must rise to the flat plate value. This can be seen at x/d = 2.0 and with the increase of x/d this pressure rise value decreases and at the same time this point moves up. Near this point corresponds to the re-attachment point and therefore the vortex would be expected to grow in downstream direction. The pressure fall above the re-attachment point may be considered due to the expansion waves generated from the separated region of the flat plate [17][18]. By taking the angle of attack, the pressure decreases in leeward side and at the same time the variation of the pressure in z direction decreases. On the other hand in windward side pressure increases and the pressure rise point moves to the flat plate as x/d increases.

Conclusions

Conclusions derived from the present experimental investigation are summarized as follows:

1. The separated shock angle is independent of the leading edge shape but depends on Mach number.
2. The first peak pressure is independent of the leading edge shape but the second peak pressure depends strongly upon the leading edge geometry.
3. The upstream influence length Lu increases linearly with the increase of θ_W.
4. For all cases, the separation line coincides with the p/p_∞ = 1.5∿1.6 isobar line and the distance between the upstream influence line and the separation line is almost constant in spanwise direction.
5. The effects of the angle of attack appear near the root of the fin and especially on the windward side downstream of the trough.

A part of this work was supported by the Ministry of Education under the Grant-in-Aid for Scientific Research.

References

1. Voitenko,D.M.;Zubkov,A.I.;Panov,Yu.A.: Supersonic Gas Flow Past a Cylindrical Obstacle on a Plate. Izv. AN. SSSR. MZhG. 1 (1966) 121-125.
2. Voitenko,D.M.;Zubkov,A.I.;Panov,Yu.A.: Existence of Supersonic Zones in Three-Dimensional Separation Flows. Izv. AN. SSSR. MZhG. 2 (1967) 20-24.
3. Price,A.E.;Stallings,R.L.: Investigation of Turbulent Separated Flows in the Vicinity of Fin Type Protuberances at Supersonic Mach Numbers. NASA TN D-3804 (1967).
4. Westkaemper,J.C.: Turbulent Boundary Layer Separation ahead of Cylinders. AIAA Journal, 6 (1968) 1352-1355.
5. Kaufman II,L.G.;Korkegi,R.H.: Shock Impingement Caused by Boundary Layer Separation Ahead of Blunt Fins. AIAA Paper No.73-236 (1973).
6. Sedney,R.;Kitchens Jr.,C.W.: Separation ahead of Protuberances in Supersonic Turbulent Boundary Layers. AIAA Journal 15 (1977) 546-552.
7. Dolling,D.S.;Cosad,C.D.;Bogdonoff,S.M.: An Examination of Blunt Fin-Induced Shock Wave Turbulent Boundary Layer Interactions. AIAA Paper No.79-0068 (1979).
8. Gillerlain,J.D.: Fin-Cone Interference Flow Field. AIAA Paper No.79-0200 (1979).
9. Dolling,D.S.;Bogdonoff,S.M.: Scaling of Interactions of Cylinders with Supersonic Turbulent Boundary Layers. AIAA Journal 19 (1981) 655-657.
10. Dolling,D.S.: Comparison of Sharp and Blunt Fin-Induced Shock Wave/Turbulent Boundary Layer Interaction. AIAA Journal 20 (1982) 1385-1391.
11. Dolling,D.S.;Bogdonoff,S.M.: Blunt Fin-Induced Shock Wave/ Turbulent Boundary-Layer Interaction. AIAA Journal 20 (1982) 1674-1680.
12. Saida,N.;Hattori,H.: Shock Wave-Turbulent Boundary Layer Interactions Induced by a Blunt Fin. Trans. JPN. Soc. Aeron. Space Sci. 27 (1984) 67-77.
13. Hung,C.M.;Kordulla,W.: A Time-Split Finite Volume Algorithm for Three-Dimensional Flow Field Simulation. AIAA Paper No. 83-1957 (1983).
14. Saida,N.: Interaction between Shock Waves and Boundary Layers Induced by Sharp Fin. JPN. Soc. Aeron. Space Sci. 32 (1984) 403-409 (in Japanese).
15. Chapman,D.R.;Kuehn,D.M.;Larson,H.K.: Investigation of Separated Flows in Supersonic and Subsonic Streams with Emphasis on the Effect of Transition. NACA TR.1356 (1958).
16. Edney,B.: Anomalous Heat Transfer and Pressure Distributions on Blunt Bodies at Hypersonic Speeds in the Presence of an Impinging Shock. Aero. Res. Inst. Sweden Rep.115 (1968).
17. Zheltovodov,A.A.: Properties of Two- and Three-Dimensional Separation Flows at Supersonic Velocities. Izv. AN. SSSR. MZhG. (1979) 42-50.
18. Kubota,H.;Stollery,J.L.: An Experimental Study of the Interaction Between a Glancing Shock Wave and a Turbulent Boundary Layer. J. Fluid Mech. 116 (1982) 431-458.

Interactions in Three-Dimensional Steady Flows

Some Observations of Three-Dimensional Shock-Wave/Turbulent Boundary-Layer Interactions

Seymour M. Bogdonoff

Gas Dynamics Laboratory
Mechanical and Aerospace Engineering Department
Princeton University
Princeton, NJ 08544

Summary

The Gas Dynamics Laboratory at Princeton University has carried out a series of studies of three-dimensional interactions of shock waves and turbulent boundary layers which have resulted in the generation of several key data sets, a classification of structures generated by three-dimensional interactions, and many observations which appear to differ from the classical projections from two-dimensional flows. The early papers revealed the general features for weak shock interactions. A series of subsequent papers have extended these studies to a family of geometries and to shocks which are quite strong. The present paper draws on much of this work to make some specific comments in three areas where there are difficult conceptual problems which have to be resolved if these flows are to be really understood: surface flow visualization, flowfield structure, and unsteadiness.

INTRODUCTION

The complex three-dimensional interaction of a shock wave with a turbulent boundary layer has been the subject of research for many years. Each investigation has provided some additional piece of new data, but much of the proposed structure of this flow has been based on concepts which have been difficult to support by detailed measurements. Some of the earliest efforts to consider three dimensional flow were detailed by Lighthill[1] and Maskill[2] and were concerned with concepts of separation. Korkegi[3], Kubota & Stollery[4], and others have proposed flowfield models. The group in the Gas Dynamics Laboratory of Princeton University has been concerned with detailed experimental studies in an attempt to define the flowfield and to work closely with those developing the new capability for computations in this difficult area. A general observation is that this complex flow has been studied with only rather limited success with

Turbulent Shear Layer/Shock Wave Interactions
IUTAM Symposium Palaiseau 1985
Editor: J. Délery

regards to generating data which directly supports or refutes many of the concepts which are the basis for the flow models which have been proposed.

The present paper attempts to briefly review the current state of the art and specifically focuses on three areas: surface flow visualization, proposed flowfield structure, and the steadiness of these interactions. The three topics are closely inter-connected and have to be examined in detail in the light of our current experimental, theoretical, and computational efforts.

EXPERIMENTAL DATA BASE

A turbulent supersonic three-dimensional flowfield is not an easy one to investigate experimentally, and only rather limited data has been obtained thus far. One technique to examine these flows has been the use of surface flow visualization. For many of the experimental studies, wall static pressure measurements have been obtained using the classical orifice, tubing, and a pressure indicator. These two surface studies have been supplemented by very limited flowfield surveys and visualization. Flowfield surveys with physical probes have included total pressure and yaw angle as the primary measurements. The difficulty of flowfield measurements has resulted in only a limited data base. In contrast to two-dimensional flows, the standard optical techniques for flowfield visualization can be applied in only a very limited sense for three-dimensional interactions. Some efforts have been made using light scattering, but the complexity of the flowfield and the difficulty of experimentally tracing streamlines has provided limited data which is difficult to interpret. Finally, all of the above measurements have been made using instrumentation or techniques which provide mean measurements in a flowfield which is turbulent and for which the steadiness of the flow has been assumed and not measured. Only a few high frequency measurements have been made over the years, primarily concerned with wall static pressure determination. It is the use of the above rather sparse data base in the construction of the complex flowfield model which the present author believes is a key limitation in our full understanding of this phenomena.

GENERAL OVERVIEW

The author, for simplicity, will limit his considerations to a specific set of data generated within the Gas Dynamics Laboratory, but the comments are applicable to a wide range of studies. Although a general class of geometries has been studied (Fig. 1), we will limit our discussions to two geometries, a sharp fin and a swept wedge at a Mach number of three, to concentrate on the concerns of the present paper. Let me provide a data set from previous work[5] for a specific configuration of the sharp fin. Fig. 2 is a typical surface flow visualization. For weak interactions, (less than 7°), the surface streaks, initially parallel to the main flow direction, are slowly deflected throughout the interaction. Above 8° a surface feature of converging streaks starts to be noticeable and is clearly defined by 10°.

Wall static pressure distributions for a range of fin angles in a plane parallel to the original stream direction is presented in Fig. 3. An example of the three-dimensionality of the flow is shown by the yaw surveys, Fig. 4., made through the 10° interaction. The distortion of the original two-dimensional profile is complex, with high yaw angles limited to the lower part of the interaction. An accompanying pitch angle contour plot is shown in Fig. 5. The data presented in the foregoing figures is a demonstration of what has usually been available in only a few cases and for few interactions.

The following sections will focus on specific details associated with the surface flow visualization, the use of the surface flow and other information to construct the general flow structure, and the specific questions of flowfield unsteadiness.

SURFACE FLOW VISUALIZATION

The simplicity and ease of application of surface flow visualization has made this technique one of the most widely used to examine three-dimensional flows. It is from this information that much of the early modelling concepts were developed. Converging and diverging streaks were associated with "separation" and "reattachment" and the conjecture that such flows generate the "reverse type" flows of two-dimensional separation with its associated vortex. The shape of the upstream influence line and these other features permitted a general classification of three-dimensional interactions. There was clearly a region near the apex where the upstream influence region appeared to be highly curved. Outboard of the apex, the general features appeared to be conical although with different vertix locations. Similar surface flow visualization and static pressure data for swept wedges, tested under the same conditions, seem to show that the original two-dimensional or cylindrical flow became concial as the wedge sweep was increased[6].

Since the assumption that converging surface streaks are an indication of an important physical phenomena, separation, let us examine the surface flow data using the oil flow picture of Fig. 2. There is a convergence of the surface streaks, but there appears to be no single line of "separation" and only an asymptotic approach of adjacent streamlines to each other. In the middle of the "separated" region, the streaks are parallel and the region is wedge shaped as the flow further out along the shock direction is examined. The result of computation is shown in Fig. 6. These surface traces were calculated by Horstman (NASA-Ames)[7]. It is clear that the convergence or actually meeting of surface streaks does not occur. Several other surface flow features have been observed and described (without experimental verification) as "secondary sepation"[8], "reattachment", or vortices. The inclusion of these results in the flowfield structure section which follows are important in trying to understand the connection between these surface observations and the flowfield.

DETERMINATION OF FLOWFIELD STRUCTURE

Models of three-dimensional flows of the type considered have been proposed in the past by several authors, including Korkegi[3], Kubota and Stollery[4], using concepts of Lighthill and Maskill. These works proposed general flowfield structures which the author has been unable to support on the basis of the current experimental data.

Let us try first to connect the primary surface details with the flowfield information that is available. The upstream influence line, as determined from the surface oil flows, is in excellent agreement with the first indication of a change in wall pressure. The rest of the surface flow features, the line of convergence called "separation", secondary "separation" found in strong interactions, lines of "reattachment", are all features which cannot be clearly identified with surface pressure distribution features or with flowfield surveys. Of greatest concern to the author is the concept of "separation" and the use of the surface oil flows to determine this phenomena. Two-dimensional separation has been widely examined. When a signficant pressure rise is imposed on the boundary layer, the flow separates from the surface. The usual modelling is that there is a stagnation point from which a surface stream line diverges from the surface at some finite angle. Although this is a widely accepted classical picture, recent work has shown that the phenomena is highly unsteady and three dimensional. The "mean" flows noted are only a crude approximation of a complex phenomena. Two-dimensional separated flows are highly dissipative and, thus far, cannot be calculated. In three-dimensional flow, the physics proposed by Korkegi suggests that boundary layers subjected to three-dimensional interactions "separate" at considerably lower pressure ratios than two-dimensional boundary layers. It is clear from the surface flow visualization, and also from the flowfield yaw surveys, that the layer at the bottom of the original boundary layer does not have to negotiate the imposed pressure rise. It simply turns and flows outward along the interaction. It does not seem physically reasonable that the turbulent boundary layer, minus its low energy part, would "separate" at lower pressure rise than that experienced in two dimensions where the low energy part of the layer must negotiate the pressure rise. A detailed examination of the surface flows has been based, in large part, on the continuity argument of Peake and Tobak[8], Fig. 7. The surface flows show a convergence of streamlines, but there does not appear to be a coalescence on a single line which would be a clear indication of strong vertical velocities. The proposed continuity model must take into account several factors - the original thickness of the observed layer, the effect of compressibility, and the effect of the "skewing" of the strongly three-dimensional flow. In the author's view, the flow "separation" has not been supported by any experimental data to date. In contrast to the two-dimensional case, computations have been quite successful. There is, thus far, no feature of the flow field which can be associated with the surface feature of streamline convergence. Many attempts have been made to make this connection, but to no avail. Computations, which has been able to describe much of the details of these flows, has shown no indication of vertical velocities close to the surface above the convergence region. It is clear that the limiting streamlines on the surface are converging. There must be some very small increase in thickness of what was originally a very thin layer on the surface, but this phenomena alone is not adequate to be considered separation. It does not appear to be an important factor in determining the flow structure. In the outer part of the interaction one can see, from the flowfield surveys and from streamline tracing from the computations of Horstman and Knight, that a vortical flow is generated. The author believes that this vortical flow is a rearrangement of the original vorticity in the initial two-dimensional boundary layer which must be reoriented in the three-dimensional flow. It is not generated by a "separation" and "reattachment" phenomena which has been proposed by the general models to date.

The final resolution of this question will depend on further detailed mesurements and perhaps a very careful examination by computations which are now being carried out. The many features of the surface flow which have been found by many investigators, for example, Zeltovodov[9] clearly have something to do with the complex flowfield that has been generated, but the concept of three-dimensional "separation" so widely used does not have a firm base for acceptance as a physical model.

From the surface flow visualization and from pressure distributions, the general flowfield characteristics for various geometries and conditions have been categorized as conical or cylindrical. Recent detailed examinations, however, have brought some question about the detailed applicability and accuracy of the proposed model. Settles[6] carried out the first extensive tests which seemed to divide the flowfield for a swept wedge into areas where the flow was cylindrical or concial, Fig. 8. Recent more detailed analyses now seem to indicate that the boundary is not as clean-cut as originally conceived. Cylindrical flows have been found in the area labelled conical and conical flows have been found in much of the area defined as cylindrical. Studies of the inception region and the subsequent regions outboard or downstream of the apex seem to indicate that there is at least another interpretation. A proposal by Shuyi Wang of our laboratory is that there is no conical region per se.

The region previously labeled as the conical region is part of what we now call the "near" flowfield which has grown, as the sweep increases, to such an extent that it covers the entire flowfield under observation. The limited range of the current available experiments are not definitive nor of sufficient detail to support the conical/cylindrical boundary concept except as an approximation which is locally applicable. The author believes that the inception region, particularly the area close to the apex of the three-dimensional models, holds the key to the development of the entire three-dimensional flow and wave system. Our present ability to understand this flow is far from adequate because of the lack of experimental information. Computations, in the long run, may be the only way to reveal the details.

STEADINESS

Much of the work on shock wave turbulent boundary layer interactions has relied on mean flow measurements. The fact that flow is turbulent was treated, as in most turbulent boundary layer studies, with the assumption that the mean values measured revealed most of the key physics. High frequency surface pressure studies of the three-dimensional flow around a blunted fin and separated 2-D flows revealed that the flows were highly unsteady[10] and the general mean flow picture usually assumed and computed were clearly not indicative of the physics. Considering that a large part of the data base for three-dimensional flows is surface flow visualization and static pressure, both with only low frequency response, these techniques give no indication of when flows may be unsteady. The flowfield information which has been obtained to date also uses techniques which are mean measurements. In trying to construct the flowfield details, knowledge of flow steadiness is crucial if any of the modelling is to be believed. Some early studies of steadiness of these flows were those of Tan et. al[11] and Tran et. al [12]. Their results are summarized in Fig. 9 where high frequency surface pressure measurements through three-dimensional interactions have been carried out for a range of shock strengths. The non-steady wall static pressures ahead of the interaction reflect the influence of the undisturbed turbulent boundary layer and free stream. At the point where the surface oil flow and mean pressure distribution show the beginning of the interaction, the wall pressure fluctuations show a significant increase. In general, this behavior would be expected on the basis of other work which indicates the amplification of turbulent fluctuations by pressure rises. As the shock strength increases, a peak in the rms fluctuations occurs. The peak is rather narrow and is limited to the region which, on the surface flow visualization, would lie between the upstream influence line and the line of convergence. The peak is followed by a region of constant value of the rms fluctuations. Downstream of the position of the externally imposed shock, the fluctuations start to decrease towards the values that might be expected for a undisturbed boundary layer. The sharp peak in the rms fluctuations has been, in the two-dimensional case, associated with rapid and large motions of the shock structure[10]. There is much to be studied in this nonsteady area and the mechanisms which are the result of or cause the surface pressure fluctuations are not at all clear. To a first approximation, it appears as though the three-dimensional interactions of this type studied thus far are considerably more steady than the two-dimensional case. We do not have sufficient information to determine whether the details of the flow can be reasonably approximated by our mean measurements. There are still questions of whether the surface flow directions, given by the surface flow visualization traces, are some sort of an average of a highly fluctuating flow, or provide an indication which is difficult to quantify in the light of the detailed modeling of the flowfield.

CONCLUSION

The comments have been limited to discussions of experimental investigations where it is clear that the instrumentation is inadequate to answer many of the questions which must be resolved. Multiple instrumentation and new techniques are required to connect surface and flowfield features. Some aspects of high frequency measurements must be included to settle the question of the steadiness of the flow and the validity of mean flow measurements. Although much has been learned by the studies to date, the author believes that many of the fundamental concepts are not very firm and that three-dimensional interactions still suffer from application of two-dimensional concepts that may not be valid. Computations hold special promise for aiding in the understanding of these complex flows. The Navier-Stokes solutions of Horstman and Knight[12] are providing valuable insights into the phenomenon and the computation techniques, once validated, provide a way to examine flowfield details that are probably not accessible to experimental techniques. The combination, a close inter-relationship of computation and experiment with new instrumentation, provides, in the author's view, a requirement for future understanding of this important and complex flow phenomena.

ACKNOWLEDGEMENT

The work reported herein was primarily supported by the Office of Scientific Research of the United States Air Force, monitored by Dr. J. Wilson.

REFERENCES

1. Lighthill, M.J., "On Boundary Layers and Upstream Influence, II. Supersonic Flows Without Separation", Proc. of the Royal Society, A, Vol. 217, pp. 46-109, 1953.

2. Maskell, E.C., "Flow Separation in Three Dimensions", RAE Report No. Aero. 2565, 1955.

3. Korkegi, R.H., "On the Structure of Three-Dimensional Shock-Induced Separated Flow Regions", AIAA Journal, 14, No. 5, May 1976, pp. 597-600.

4. Kubota, H., and Stollery, J.L., "An Experimental Study of the Interaction Between a Glancing Shock Wave and a Turbulent Boundary Layer", Journal of Fluid Mechanics, 116, 1982, pp. 431-458.

5. Oskam, B., Vas, I.E. and Bogdonoff, S.M., "Mach III Oblique Shock Wave - Turbulent Boundary Layer Interactions in Three Dimensions", AIAA Paper 76-336, 1976.

6. Settles, G.S. and Teng, H.-Y., "Cylindrical and Conical Flow Regimes of Three-Dimensional Shock/Boundary-Layer Interactions", AIAA Journal, 22, No. 2, February 1984, pp. 194-200.

7. Horstman, C.C., Private Communication.

8. Peake, D.J. and Tobak, M., "Topology of Three-Dimensional Separated Flows" , NASA Tech. Memo. 81294, April 1981.

9. Zheltovodov, A.A., "Regimes and Properties of Three-Dimensional Separated Flows Initiated by Skewed Compression Shocks", Zhurnal Prikladnoi Mekhaniki i Teknicheskoi Fiziki, No. 3, May-June 1982, pp. 116-123.

10. Dolling, D.S. and Bogdonoff, S.M., "An Experimental Investigation of the Unsteady Behavior of Blunt Fin-Induced Shock Wave/Turbulent Boundary Layer Interactions", AIAA Paper 81-1287, 1981.

11. Tran, T.T., Tan, D.K.M. and Bogdonoff, S.M., "Surface Pressure Fluctuations in a Three-Dimensional Shock Wave/Turbulent Boundary Layer Interaction at Various Shock Strengths", AIAA Paper 85-1562, July 1985.

12. Knight, D., Horstman, C., Shapey, B. and Bogdonoff, S.M., "The Flowfield Structure of the 3-D Shock Wave-Boundary Layer Interaction Generated by a 20° Sharp Fin at Mach 3", to be presented at the AIAA 24th Aerospace Sciences Meeting, Reno, Nevada, January 6-8, 1986.

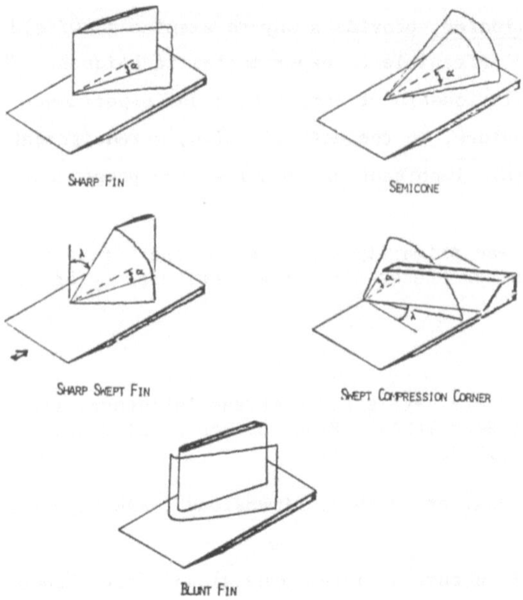

Figure 1
Model Configurations

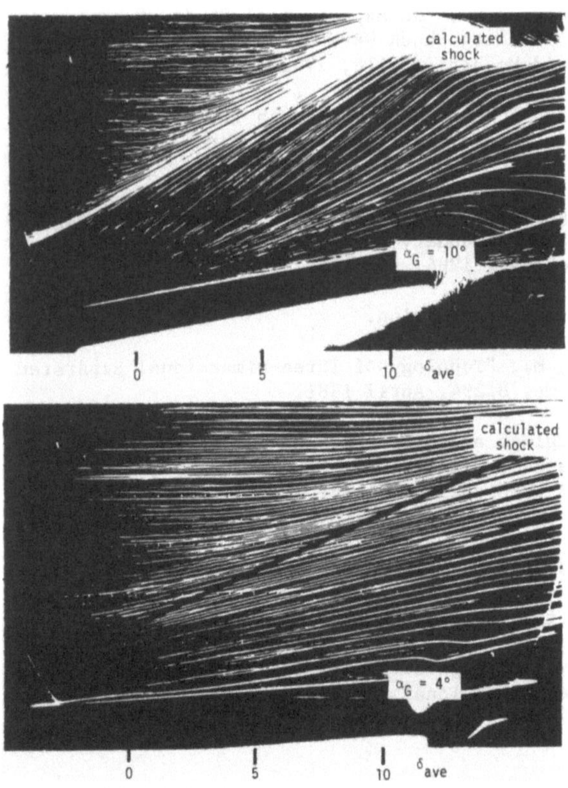

Figure 2
Surface Oil Flow Patterns
for 4° and 10° Fins

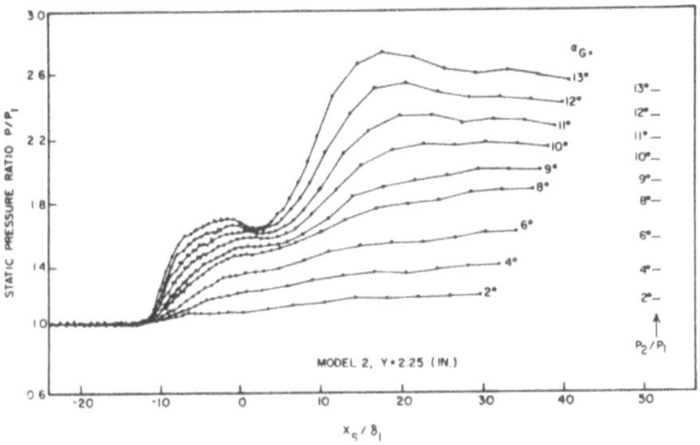

Figure 3
Static Pressure Distributions
for Variable Fin Angles

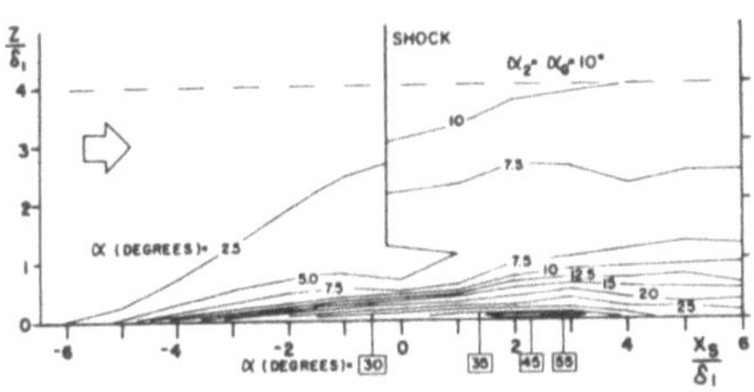

Figure 4
Yaw Angle Contour plot,
Fin Angle 10°

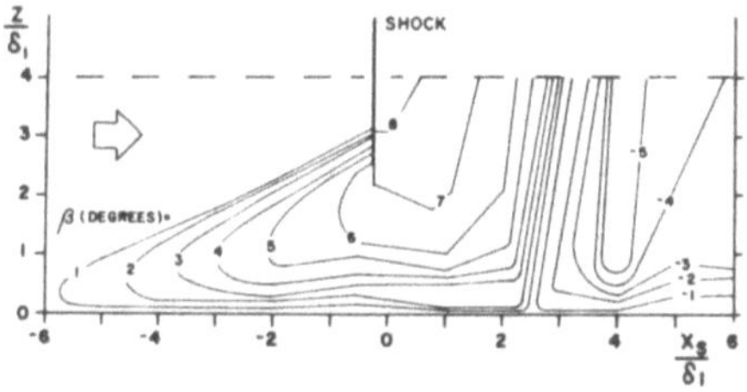

Figure 5
Pitch Angle Contour Plot,
Fin Angle 10°

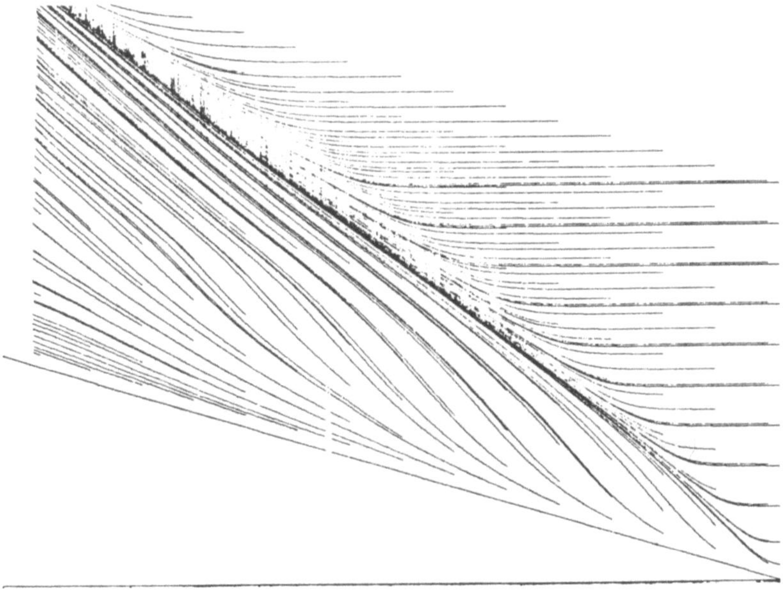

Figure 6
Computed Surface Streaks

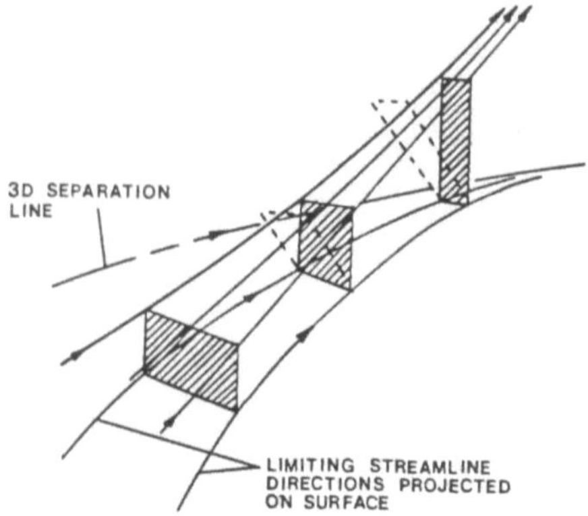

Figure 7
Model of Peake and Tobak

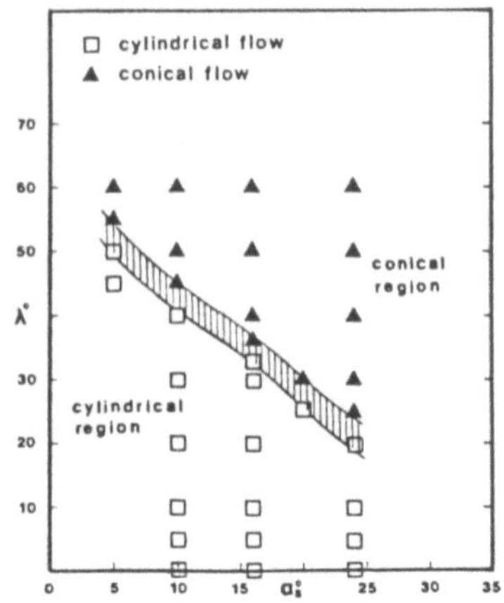

Figure 8
Conical-Cylindrical Boundary
for Wedge Flows

3-D SHARP FIN

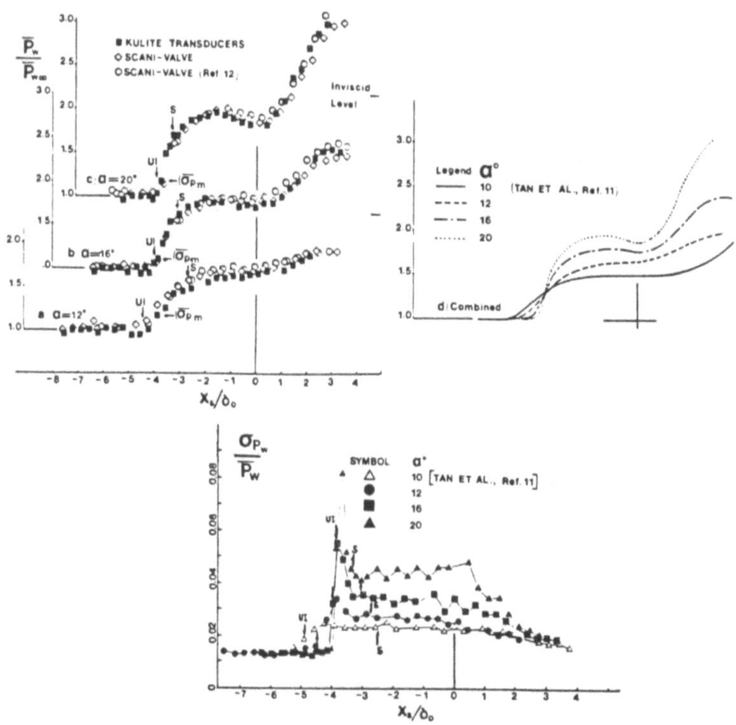

Figure 9
High Frequency Data for
Various Fin Angles

Experimental Study of Shock-Wave/Boundary-Layer Interaction in a Three Dimensional Channel Flow

R. BENAY and T. POT

ONERA - 29, Avenue de la Division Leclerc - 92320 CHATILLON (France)

SUMMARY

The paper presents the first results of a thorough experimental analysis of a strong 3D shock-wave/turbulent boundary-layer interaction occurring in a channel. Many field measurements have been performed by using a 3 component LDV system. The results thus obtained allow a clear description of the complex flow structure

1. INTRODUCTION

The aim of the present study is to provide a well documented set of experimental data on a 3D shock-wave/turbulent boundary-layer interaction. For this purpose , we have made an extensive use of Laser Velocimetry which allows the measurements of the mean as well as of the turbulent properties of the flow. The results thus obtained can be used to help in the understanding of the physics of complex interacting flows. They can also be used as bench-mark to test computational methods. This study is an extension to 3D flows of research already made at ONERA on 2D transonic channel flows [1]

2. EXPERIMENTAL ARRANGEMENT

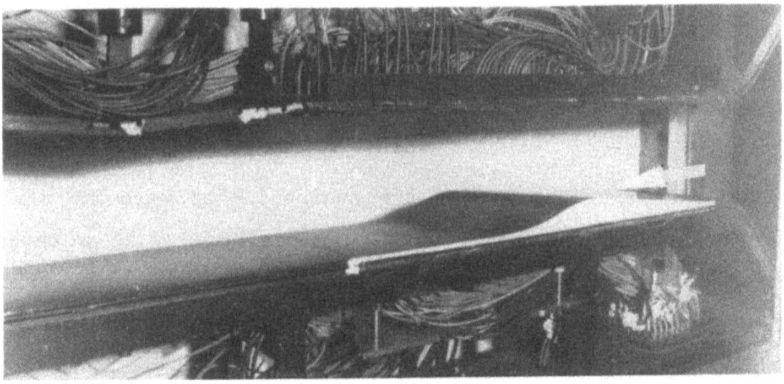

Fig. 1 - Photograph of experimental arrangement

Turbulent Shear Layer/Shock Wave Interactions
IUTAM Symposium Palaiseau 1985
Editor: J. Délery

The general arrangement of the test set-up is shown in Fig. 1. Basically, it consists in a bump mounted on the lower wall of a small supersonic wind tunnel test section. The shape of the bump consists in a rectilinear front part followed by a convex circular part; the rear part being a concave circular arc insuring smooth transition with the plane downstream wall of the channel . As shown in Fig 2, the contoured portion of the bump is at an angle of 60° with the upstream flow.

The incoming subsonic flow accelerates in the converging part of the channel and reaches sonic conditions at the top of the bump which constitutes a throat. Downstream, the flow is supersonic and due to the deflexion imparted by the rear part of the bump, shock-waves form in the channel giving rise to strong interactions with the boundary-layer developing on the four channel walls. The flow thus obtained is highly three dimensional and far more complex that 3D interactions already studied [2,3].

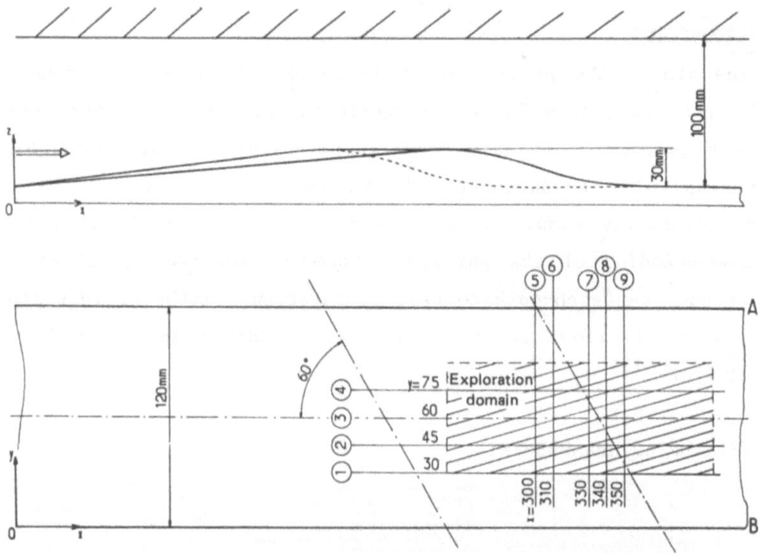

Fig. 2 - Definition of exploration domain

The upper and lower walls are equipped each with 250 static pressure orifices. Field measurements have been performed with a three color LDV system allowing the simultaneous measurement of the three velocity vector components [4]. Statistical quantities of interest (e.g., mean velocity components, Reynolds tensor components, histograms) were computed from samples of 2000 triplets constituted by the three velocity components recorded at each probing point.

The shaded area in Fig. 2 shows the projection on the bump surface of the surveyed domain. The flow has been probed along 75 vertical stations contained in this area, each of these stations comprising 50 measurement points. In what follows, only results concerning the mean velocity field will be presented.

3. PRESENTATION OF RESULTS

Photographs of surface flow patterns are shown in Figs, 3, 4 and 5. Figure 3, relative to the flow on the bump, reveals a nearly 2D incoming flow which is turned by a shock forming slightly downstream of the top of the bump. In the outer part of the flow, the Mach number on the upstream face of this shock is equal to 1.65. At the shock foot, a strong interaction with the boundary-layer takes place entailing separation. There, the skin friction lines pattern shows the existence of a separation line starting from a location close to side B of the channel (see Fig. 2).

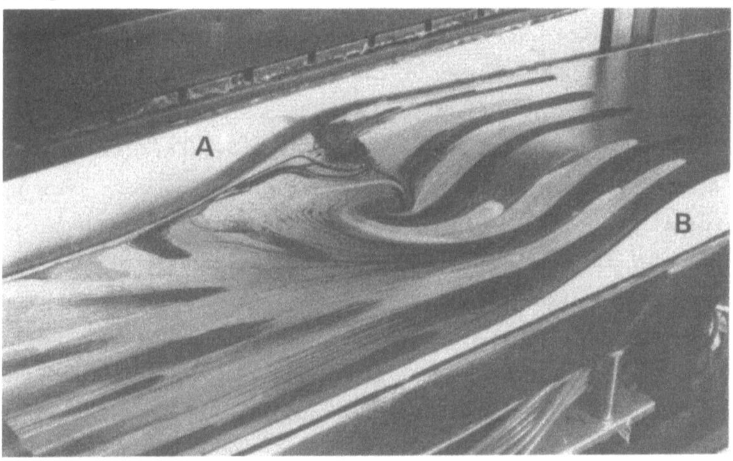

Fig. 3 - Surface flow pattern on the bump

Fig. 4 - Surface flow pattern on side A

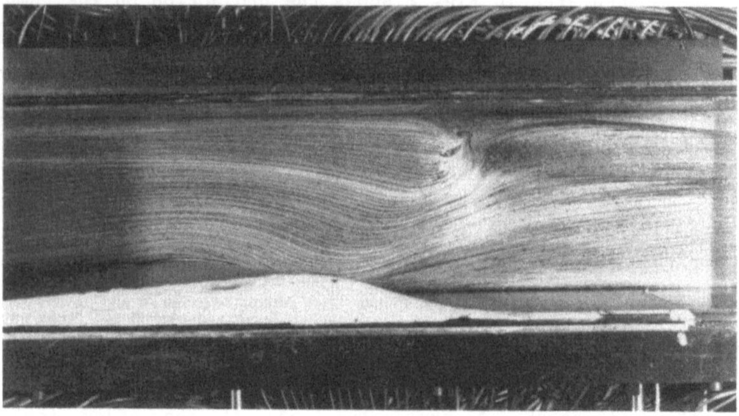

Fig. 5 - Surface flow pattern on side B

This line separates skin friction lines constituting a vortical structure from skin friction lines streaming downstream and which becomes parallel to the channel side walls. The surface flow pattern on the vertical side A is shown in Fig. 4. It reveals the existence of separation along with a very complex structure specially in the vicinity of the channel upper wall. On the other hand, as shown by Fig. 5, there is no separation on side B, the skin friction lines streaming continuously from upstream to downstream [5].

In order to have some idea about the outer flow structure let us now examine the tracings of iso-Mach lines shown in Figs. 6 and 7 which correspond to the longitudinal vertical planes 3 and 4 respectively (see Fig. 2). These tracings reveal the existence of a lambda shock system constituted of a front oblique shock and a quasi normal rear shock. This system is associated with strong interactions generally entailing separation. It is observed both in 2D flows and in transonic flows over swept wing.

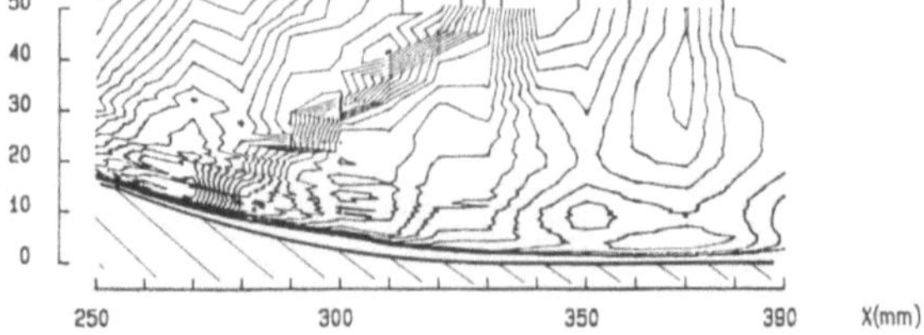

Fig. 6 - ISO - Mach lines in vertical plane 3 of Fig. 2

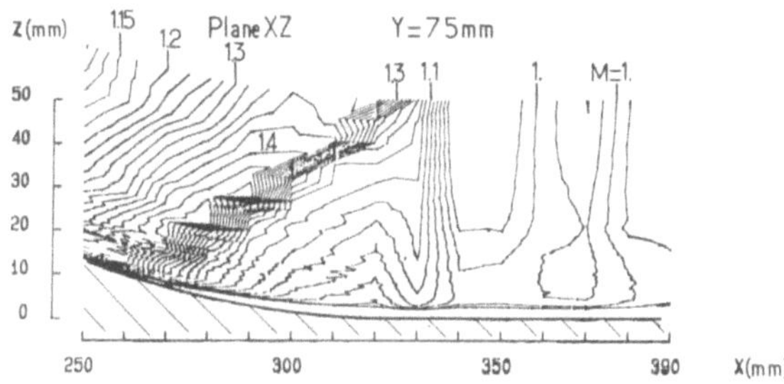

Fig. 7 - ISO - Mach lines in vertical plane 4 of Fig. 2

More instructive information on the complex flow structure is provided by vector plots of the mean velocity field. Thus, Figs. 8 to 11 show the vector field in longitudinal vertical planes (x,z) (see Fig. 2).

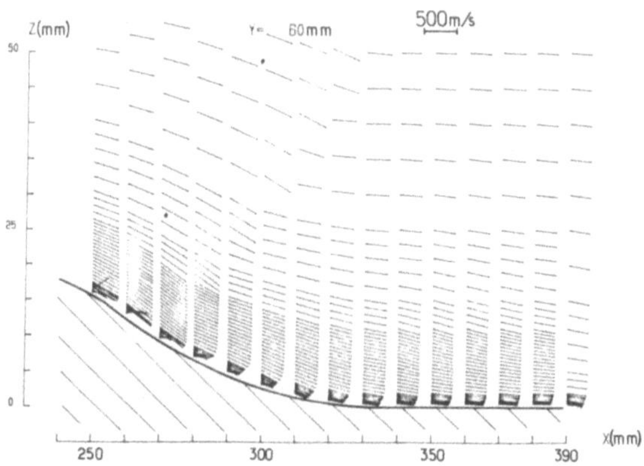

Fig. 8 - Vector plot in vertical plane 3 of Fig. 2

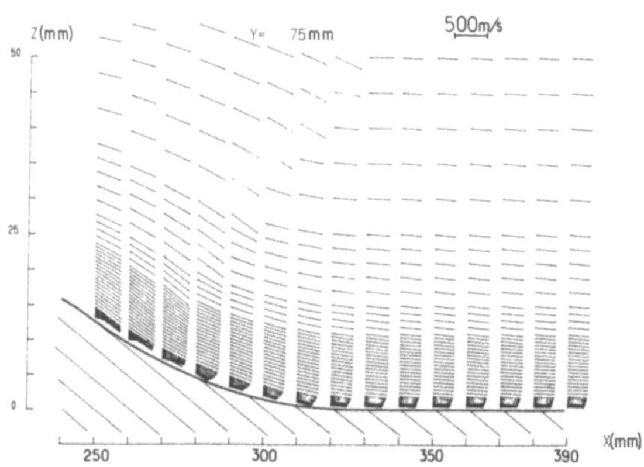

Fig. 9 - Vector plot in vertical plane 4 of Fig. 2

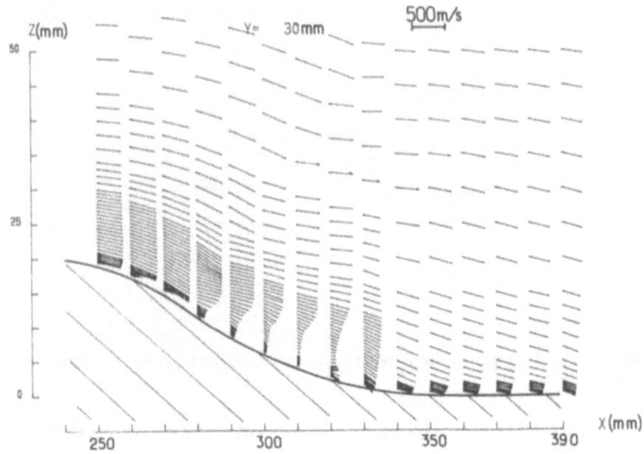

Fig. 10 - Vector plot in vertical plane 1 of Fig. 2

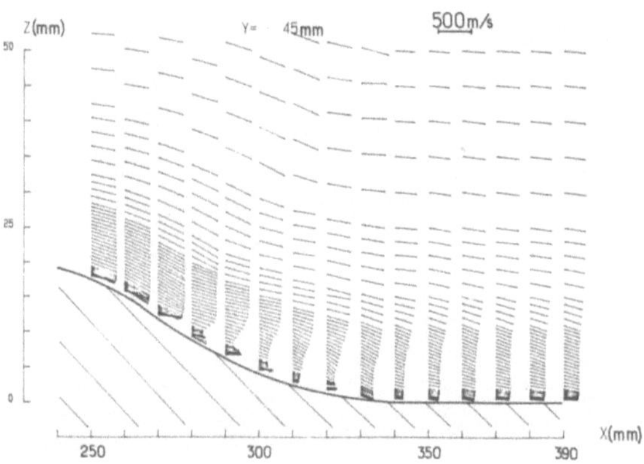

Fig. 11 - Vector plot in vertical plane 2 of Fig. 2

In planes 3 and 4, the boundary-layer remains relatively thin throughout the interaction domain. However, on approaching side A of the channel, one observes in planes 1 and 2 a region of strong decrease of the longitudinal velocity along with a rapid thickening of what appears to be the viscous layer. Thus, in plane 2, the velocity profiles present a minimum located at an appreciable distance from the wall. On approaching the wall, the longitudinal velocity component increases again and very close to the wall a classical boundary-layer profile is obtained. Figures 12 to 15 show the projections of the velocity vector in transverse vertical planes (y,z) (see Fig. 2). These plots display clearly the skewing of the flow in the so-called separated region (note the change in velocity scale). This skewing is moderate along the vertical lines close to side B. It increases on approaching side A and it is specially important along the line located at x = 45 mm. There is even a change in the sense of the transverse velocity component which runs from A to B in the outer part of the flow and from B to A in regions close to the wall. These plots exhibit the existence of a strong vortex which forms in the shock foot region and whose axis tends to become parallel to the general flow direction when going downstream of the interaction region. The "birth" of this vortex is evidenced by Fig. 16 which shows lines of constant velocity modulus in a (y,z) plane located at x = 290 mm. At this location, the vortex visualized by closed lines is flattened and still very close to the wall. The main difference with a 2D shock-wave/boundary-layer interaction is that here, in a 3D situation, the flow can take a tranverse, or crosswise direction, and consequently when separation occurs we generally do not observe recirculating bubble in the streamwise direction.

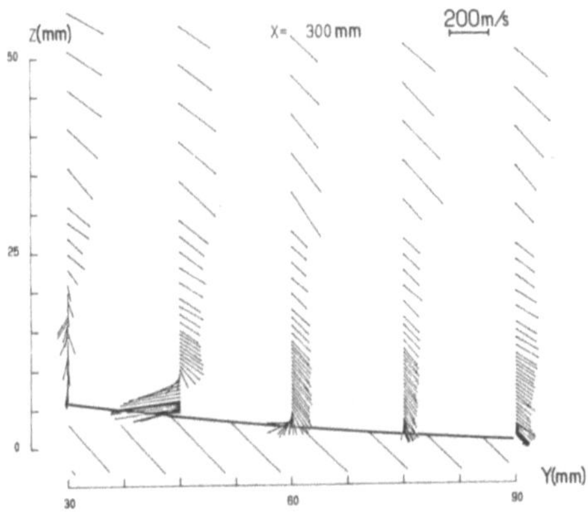

Fig. 12 - Vector plot in the transverse vertical plane 5

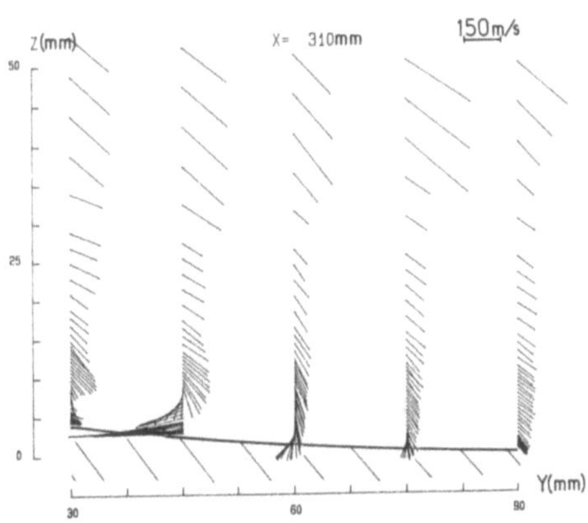

Fig. 13 - Vector plot in the transverse vertical plane 6

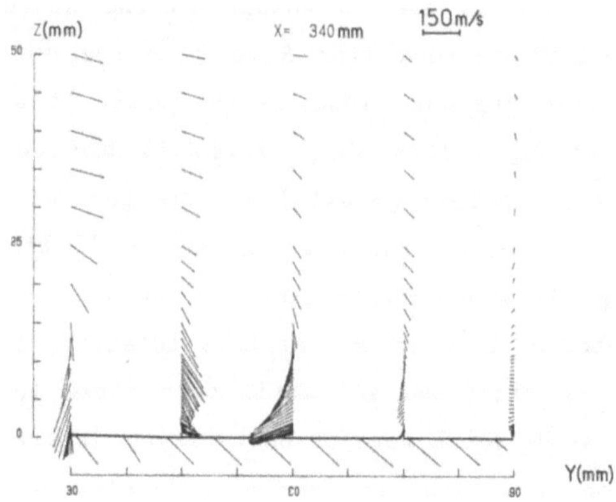

Fig. 14 - Vector plot in the transverse vertical plane 8

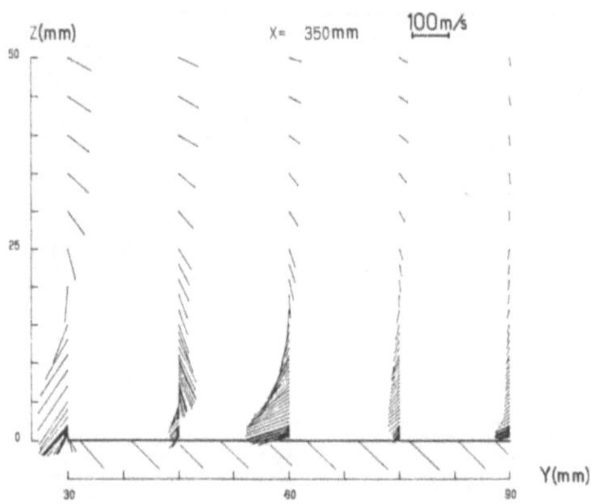

Fig. 15 - Vector plot in the transverse vertical plane 9

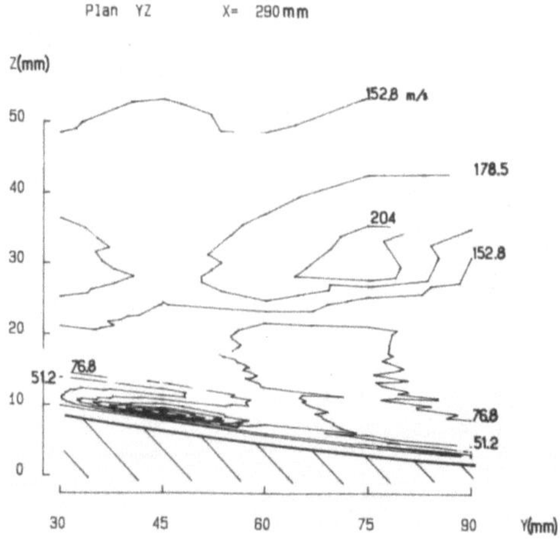

Fig. 16 - Lines of constant velocity modulus in a transverse vertical plane

4. CONCLUSION

The experiments performed with a 3D LDV system provide an exhaustive set of data allowing a detailed description of a 3D supersonic interaction occurring in a channel. More processing work is still necessary to obtain a clear understanding of the mean velocity field whose structure is very complex. Also, laser velocimetry has given many results on fluctuating properties whose processing is under way and which will be of great help for the modeling of turbulence in such complicated flows.

REREFENCES

[1] DELERY J. - "Experimental investigation of turbulence properties in transonic shock/boundary-layer interactions" - AIAA Journal, Vol. 21. n° 2 - Feb. 1983

[2] SETTLES G.S. - PERKINS J.J. and BOGDONOFF S.M. - "Investigation of three dimensional shock/boundary-layer interactions at swept compression corners" - AIAA Journal, Vol. 18. n° 7 - July 1980

[3] KUSSOY M.I. - VIEGAS J.R. and HORSTMAN C.C. - "An experimental and numerical investigation of a 3D shock separated turbulent boundary-layer" - AIAA Journal, Vol. 18, n° 12 - Dec. 1980

[4] BOUTIER A. - d'HUMIERES C. and SOULEVANT D. - "Three dimensional laser velocimetry : a review". - 2nd International Symposium on Application of Laser Anemometry to Fluid Mechanics, Lisbon, 2-5 july 1984

[5] OSKAM B. - VAS I.E. and BOGDONOFF S.M. - "Mach 3 oblique shock-wave/turbulent boundary-layer interactions in three dimensions" -AIAA Paper n° 76-336 - July 1976

Numerical Experiment with Inviscid Vortex-Stretched Flow around a Cranked Delta Wing: Transonic Speed

Arthur RIZZI
FFA The Aeronautical Research Institute of Sweden,
S-161 11 BROMMA, Sweden
and
Charles J. PURCELL & J. Thomas McMURRAY
ETA Systems, Inc., St. Paul, MN 55112, U.S.A.

Abstract

A numerical method that solves the Euler equations for compressible flow is used to study vortex stretching. The particular case simulated is transonic flow M_∞=0.9 α=10 deg. around the twisted and cambered cranked-and-cropped TKF delta wing of MBB. This geometry induces multiple leading-edge vortices in a straining velocity field that brings about flow instabilities but the result is a state of statistical equilibrium. The discretization contains over 600,000 cells and offers sufficient degrees of freedom in the solution to exhibit the onset of chaotic vortex flow that could well lead to turbulence. The simulated results are compared with wind-tunnel measurements.

Introduction

Today's practical airplane designer relies on wind-tunnel data in order to predict flight characteristics of transonic and supersonic aircraft. This paper explores the question if a numerical method can simulate the nearly inviscid vortical flowfield which develops around a wing at the very high Reynolds number of free-flight conditions for use in conjunction with wind-tunnel measurements. Realistic numerical simulations require three-dimensional discrete models of significant size. The recent construction of a 16-million-word memory for the CYBER 205 now allows tests of one-million-grid-point models on a practical basis within reasonable elapsed times. Prior to this it was hoped that this type of computation could be performed with disk I/O but experience showed that heavy time penalties were encountered due to the demands of the flow-solving algorithm. Therefore good virtual-memory-management techniques within a sufficient working set of real memory

Turbulent Shear Layer/Shock Wave Interactions
IUTAM Symposium Palaiseau 1985
Editor: J. Délery

seem to be the only effective way to carry out such large-scale simulations. The other crucial requisite is a large band-width communication channel between the memory and the arithmetic unit.

Does the solution converge to some limit as the discrete model grows larger by adding more and more grid points in the discretization? It does as soon as the mesh size becomes smaller than the smallest feature in the flow. But if small-scale motions are continually being excited through an instability, then all of the available degrees of freedom in the solution may be activated, and mesh convergence will not be reached. In that case no matter how fine the mesh is, small-scale features will always saturate the resolution of the mesh. But with the densest meshes now available, it may be possible to study fundamental flow instabilities like the interacting of stretched vortices as described by the continuum equations. The advent of supercomputers with very large real memory just now offers the possibility to explore such phenomena numerically.

The type of vortex flowfield that we want to simulate develops when a delta-shaped wing meets an oncoming stream of air at a high angle of attack. The flow separates from the wing in a free shear layer along the entire sharp leading edge, and under the influence of its own vorticity coils up to form a vortex over the upper surface of the wing. The high velocities in the vortex create a low pressure region under it which gives the wing a nonlinear lift. The flow also separates in a shear layer from the trailing edge of the wing which forms into a wake vortex. The primary and wake vortices then interact with each other behind the wing. The appropriate model to describe the fluid mechanics of this vortex flow is the compressible Navier-Stokes equations. But if the Reynolds number of the flow is very large, say over ten million, the shear stresses and dissipation terms take effect only in very thin layers of the flow on the surface of the wing and across the shear layers that separate from the leading and trailing edges. For this type of flow it is these free shear layers, and not the boundary layer, that contribute the greatest amount of vorticity. Free-shear layers contain inflexional velocity profiles, and hence are subject to inviscid instabilities. Because of the inherent nonlinearity of the equations, it is very difficult to analyze these instabilities, and even more difficult to predict the resulting flow. We hypothesize that the instabilities are linked to the intensification of vorticity and that this mechanism is an agent in the transition to disordered motion, and finally turbulence. The intent of the paper is to test the hypothesis with a numerical experiment designed to study the dynamics and stability of free-shear layers by means of computer simulation. Because the layers are thin and not influenced by the no-slip boundary conditions, the shear stresses and dissipation terms do not have to be accurately represented. They can instead be modelled by simpler nonphysical expressions. In this way the equations we solve are the Euler equations with an artificial-viscosity model. The flow therefore is inviscid except across thin discontinuities like shock waves and vortex sheets that are admitted in the solution where irreversible processes can take place.

Vorticity and Flow Instability

How do the dynamics of vorticity lead to a flow instability? Let's consider first the simple case of an isolated vortex filament in an inviscid and incompressible fluid. If this single filament is straight or circular, the dynamics are well known. But give the filament a varying curvature, and immediately it induces a torsion. The coupling of these two phenomena, curvature and torsion, was studied by Betchov[1]. If the vortex filament is plane but of variable curvature, the different segments of the filament will move out of the plane at different velocities (Fig. 1). Thus it acquires torsion.

In general, if a curve moves under the influence of its own segments, two equations can be constructed specifying the evolution of the radius of curvature R and of the torsion T. Betchov studied these equations for the case of a helicoidal filament (Fig. 2) and found that it moves in space with a translation velocity V_T and a rotation producing a tangential velocity V_R. He found two opposing mechanisms in the system of equations, one an intensification of the torsion and the other a dispersion of the torsion, but the analysis indicates that for some range of conditions the filament may find a statistical equilibrium between the two.

This process describes the way a single filament may reach a distorted but balanced stated, and suggests a mechanism for the transfer of energy from large-scale to small-scale motion, the so-called energy cascade. In the more general situation the flow is compressible and contains many vortex filaments which may interact with each other. Much more difficult to analyze, the dynamics now are governed by the vorticity equation for compressible nonisentropic flow

$$\frac{D}{Dt}\frac{\vec{\omega}}{\rho} = \frac{\vec{\omega}}{\rho}\cdot \underset{\approx}{E} + \frac{1}{\rho^3}\,(\text{grad }\rho \times \text{grad }p) \tag{1}$$

where $\underset{\approx}{E}$ is the rate-of-strain tensor

$$\underset{\approx}{E} = \frac{1}{2}(\text{grad }\vec{V} + \text{grad}^T\vec{V})$$

The first term on the right side is the production of vorticity by stretching caused by a straining velocity field, and the second term is a production by nonisentropic processes like shock waves. If both are zero, the ratio $\vec{\omega}/\rho$ is convected unchanged with the fluid element. In the more complicated case they are nonzero, and under what conditions Eq.(1) may reach some kind of equilibrium is not generally known. Insight into this matter, however, can be achieved by observing the dynamics in numerical simulations. The key questions are: do the dynamics of vorticity in Eq.(1) describe a mechanism for the cascade of energy from large to small scales, and are these scales maintained in some statistical equilibrium?

The aim of this paper then is to carry out a numerical experiment where we seek to observe the evolution of vorticity according to Eq.(1). When a stream meets a conical delta wing, it separates from the straight leading edge. The vorticity is greatest right at the leading edge, and its direction, given by the leading edge, is primarily along conical rays starting from the apex. To a good approximation the core of the vortex also lies along one such ray. This fact together with the conical-flow condition that all radial gradients of velocity vanish tells us that the term $\vec{\omega}\cdot\underset{\sim}{E}$ is very nearly zero for such flows. A conical flow therefore is not a good candidate for our experiment. Since the direction of vorticity is so closely governed by the geometry of the leading edge of the wing, a better choice is a wing whose leading edge is curved. In this way we can impart curvature, and hence torsion, to the vortex filaments which may initiate the cascade. In addition to the practical interest in understanding its flowfield, a good geometry for our experiment is the MBB wing, termed TKF, whose leading edge is not a straight line, nor does it even lie in a plane, a so-called cranked-and-cropped delta with camber and twist. We select transonic flow conditions $M_\infty=0.9$ $\alpha=10$ deg. at which compressibility effects are strong and produce shock waves. For this case then both of the vorticity production terms on the right side of Eq.(1) should be substantial.

Numerical-Simulation Procedure

We use first Eriksson's method of transfinite interpolation[2] to construct a boundary-conforming O-O type mesh with nearly 640,000 cells around the TKF cranked delta wing (Fig. 3). The Euler equations can be expressed as an integral balance of the conservation laws

$$\frac{\partial}{\partial t}\iiint q \, dvol+\iint \underset{\sim}{H}(q)\cdot\underset{\sim}{n} \, ds = \iint T \, ds \qquad (2)$$

where q is the vector with elements of mass, momentum, and energy. The inviscid flux quantity $\underset{\sim}{H}(q)\cdot\underset{\sim}{n}$ represents the net flux of q transported across, plus the pressure p acting on, the surface S surrounding the volume of fluid. The term T is the artificial viscosity model. It has the property of an energy sink, i.e., $(d/dt)q^2<0$ summed over all cells including those at the boundaries. The finite-volume method then discretizes (2) by assuming that q is a cell-averaged quantity located in the center of the cell, and the flux term $\underset{\sim}{H}(q)\cdot\underset{\sim}{n}$ is defined only at the cell faces by averaging the values on each side. A more detailed description of the method is given in Ref. 3. With the appropriate boundary conditions we integrate this last equation with the two-level three-stage scheme that steps the solution forward in time.

Simulated Vortex Flowfield

The flow model that we have described dissipates phenomena whose wavelength is on the order of the mesh spacing and so it mimics the action of the real Newtonian viscosity which for high Reynolds number flows acts on the molecular scale. Therefore the smaller we make the mesh spacing by adding more grid points, the finer the scale-length of the phenomena that we can resolve before it is obliterated by the dissipation. If there is an energy cascade at work in the flow, it can only be seen in very dense meshes (Fig. 4). The standard mesh dimensions, say about 60,000 cells, can support only the steady large-scale features. Many such simulations have been carried out before. But what happens if we use a much finer mesh? We believe that, given a sufficient number of grid points to support the small scales, the Euler equations do model vortical flow instabilities, which under suitable conditions, can be statistically steady and consistent, at least in an average sense, with the coarser-grid solution. The mechanism for the excitation, as well as the probable instability of the small-scale flow features sustained by the transfer of energy from the large scales, is the stretching of vorticity caused by the external straining of the free shear layers in this flow. The following computed results for flow around the TKF wing at M_∞=0.9 α=10 deg. support the argument.

Figure 5 presents isograms of the simulated flow on the upper surface of the wing and in one spanwise plane over it. These are results obtained upon a standard mesh size of 80 cells around the chord, 40 across the span, and 24 outward, and they indicate a laminar vortex flow much like the ones simulated previously. The isobars and isoMach contours (Fig. 5a and 5b) show that a vortex does form from the free-shear layer shed from the leading edge, but the footprint of the vortex weakens just beyond the crank, where we might have expected another vortex to form, or at least the core to curve. Losses in total pressure (Fig. 5b) are mostly found in the shear layer, for reasons which we now understand[4], shock waves do not contribute substantially to the vorticity in this case. The lift and drag coefficients are C_L=0.655 and C_D=0.120. We present this standard-mesh solution in order to portray the large-scale motion and to provide a basis for comparison with the fine-mesh results.

The solution obtained on a mesh having double the number of cells in each of the coordinate dimensions (160×48×80) as the standard mesh is given in Fig. 6. The most striking difference between Fig. 5 and Fig 6 shows up in the isobars and isoMach contours of the vortex footprint in the vicinity of the crank. Instead of very smooth flow, the contours in Fig. 6 suggest a wave character not seen in the standard mesh. At the crank a second vortex forms from the leading edge, interacts with the inboard vortex, and produces many small eddies. We believe that this small-scale structure comes from the torsional forces brought about by the mutual interaction of the vortex filaments coming from the leading edge. It is only possible to support scales like these on a fine mesh. A further refined mesh may well reveal even smaller ones. The destabilizing forces that bring about this motion, however, do seem to be in equilibrium with the overall flow. The integrated forces C_L=0.646 and C_D=0.115 are only slightly different from the standard-mesh values.

Figure 7a, contains the chordwise comparison of the surface pressures computed at 4 span locations in the standard and fine meshes with those measured in the wind tunnel. The physical model includes a fuselage, but the numerical one does not. On the lower surface the agreement of all three results is quite good, and although there are significant differences on the upper surface, the correlation is good enough to establish confidence in the realism of the numerical simulations. A spanwise comparison, for which there are no measurements, brings out the differences in the standard and fine mesh solutions.

Final Remarks

Vortical flows are among the most baffling for the fluid dynamicists to understand. They are susceptible to instabilities, they can develop local regions of extreme velocity and vorticity, and they are inherently nonlinear. Hence they are prime subjects for numerical study. Here we have focussed on one aspect of vortical flow, the bending or curving of vortex filaments brought on by the geometry of the boundary that determines the flow. With sufficient degrees of freedom given by the mesh size to the solution, we have observed small-scale torsional-wave effects due to a cranked leading edge in a flowfield which overall is in equilibrium. Wind-tunnel measurements verify that our particular simulation bears a certain degree of realism. Whether further mesh refinement would show even smaller scales and eventually chaotic turbulent flow for this case is still a matter of speculation. At higher angles of attack it is likely that the torsional wave grows in strength so that the flow no longer is in equilibrium. If so the wave may well be the harbinger of vortex bursting.

References

1 Betchov, R.: On the Curvature and Torsion of an Isolated Vortex Filament, J. Fluid Mech., Vol. 22, 1965, pp. 471-479.

2 Eriksson, L.E.: Generation of Boundary-Conforming Grids around Wing-Body Configurations using Transfinite Interpolation, AIAA J., Vol. 20, Oct. 1982, pp. 1313-1320.

3 Rizzi, A.W. and Eriksson, L.E.: Computation of Flow Around Wings Based on the Euler Equations, Journal Fluid Mechanics, Vol. 148, Nov. 1984, pp. 45-71.

4 Powell, K., Murman, E., Perez, E., and Baron, J.: Total Pressure Loss in Vortical Solutions of the Conical Euler Equations, AIAA Paper 85-1701, 1985.

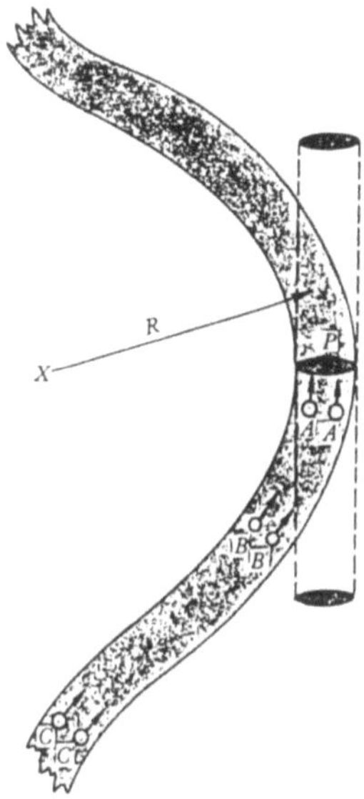

Fig. 1 The vortex filament. The short arrows indicate the vorticity. At point p, the particles b and B´ induce a velocity perpendicular to the page. Particlea A and A´ induce velocities of opposite directions which cancel out. Particles c and C´are too far away to be effective.

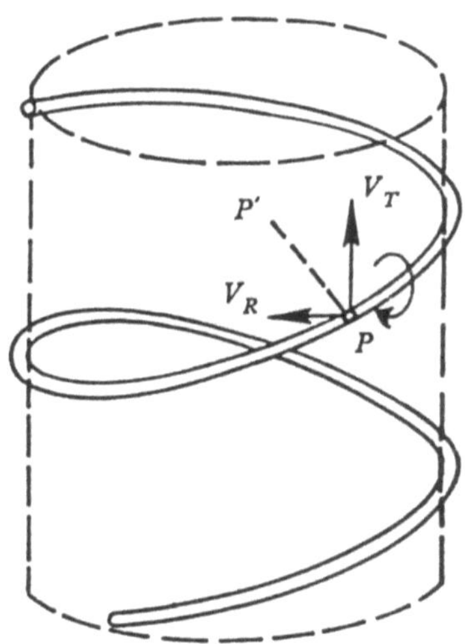

Fig. 2 Motion of a helicoidal fialment: point P moves toward P´.

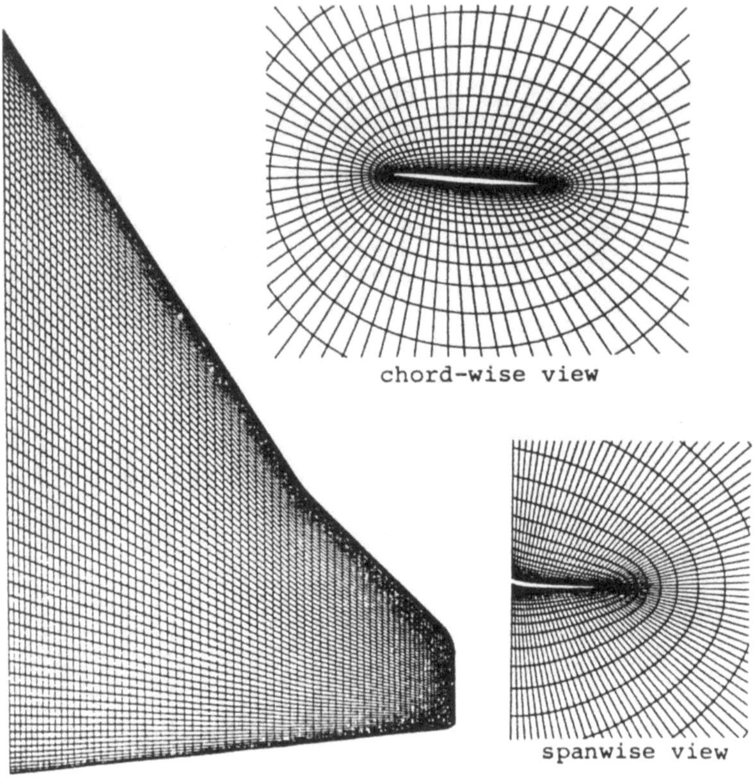

Fig 3 O-O mesh with 160x48x80 cells around cranked delta wing.

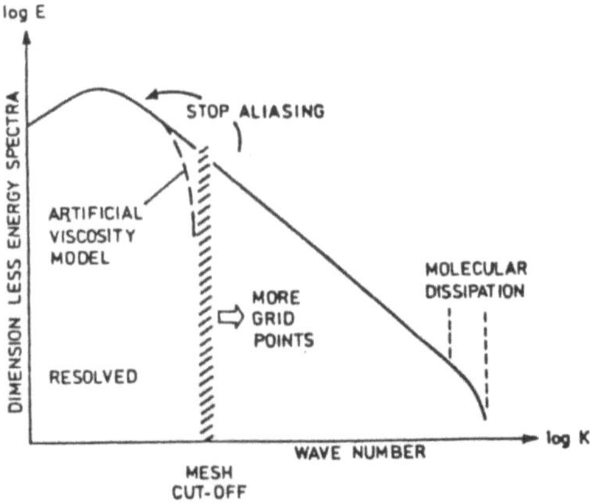

Fig. 4 **Energy cascade of a high-Reynolds-number turbulent flow. If the mesh number is large enough, it may be possible to approximate the broad-scale effects of turbulence.**

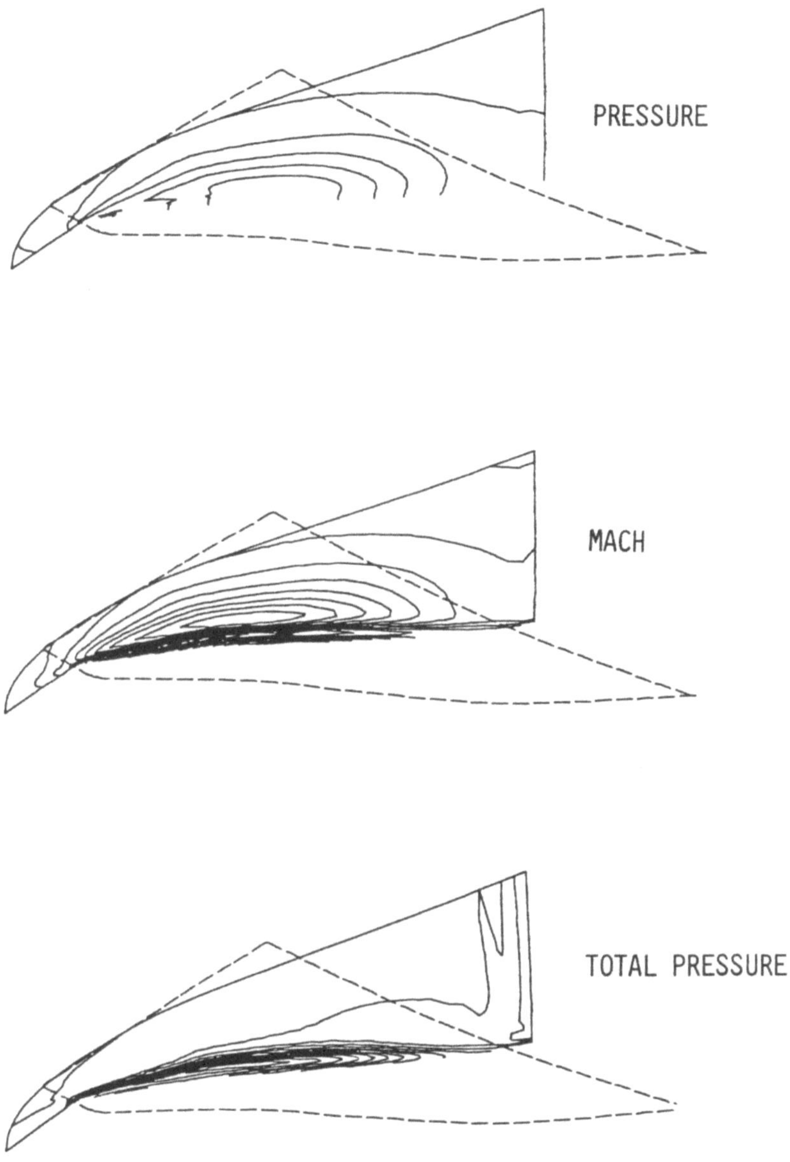

Fig. 5a Flow field computed with the standard mesh of 80x24x40 cells. $M_{\infty} = 0.9$ $\alpha = 10$ deg. Contours over the wing.

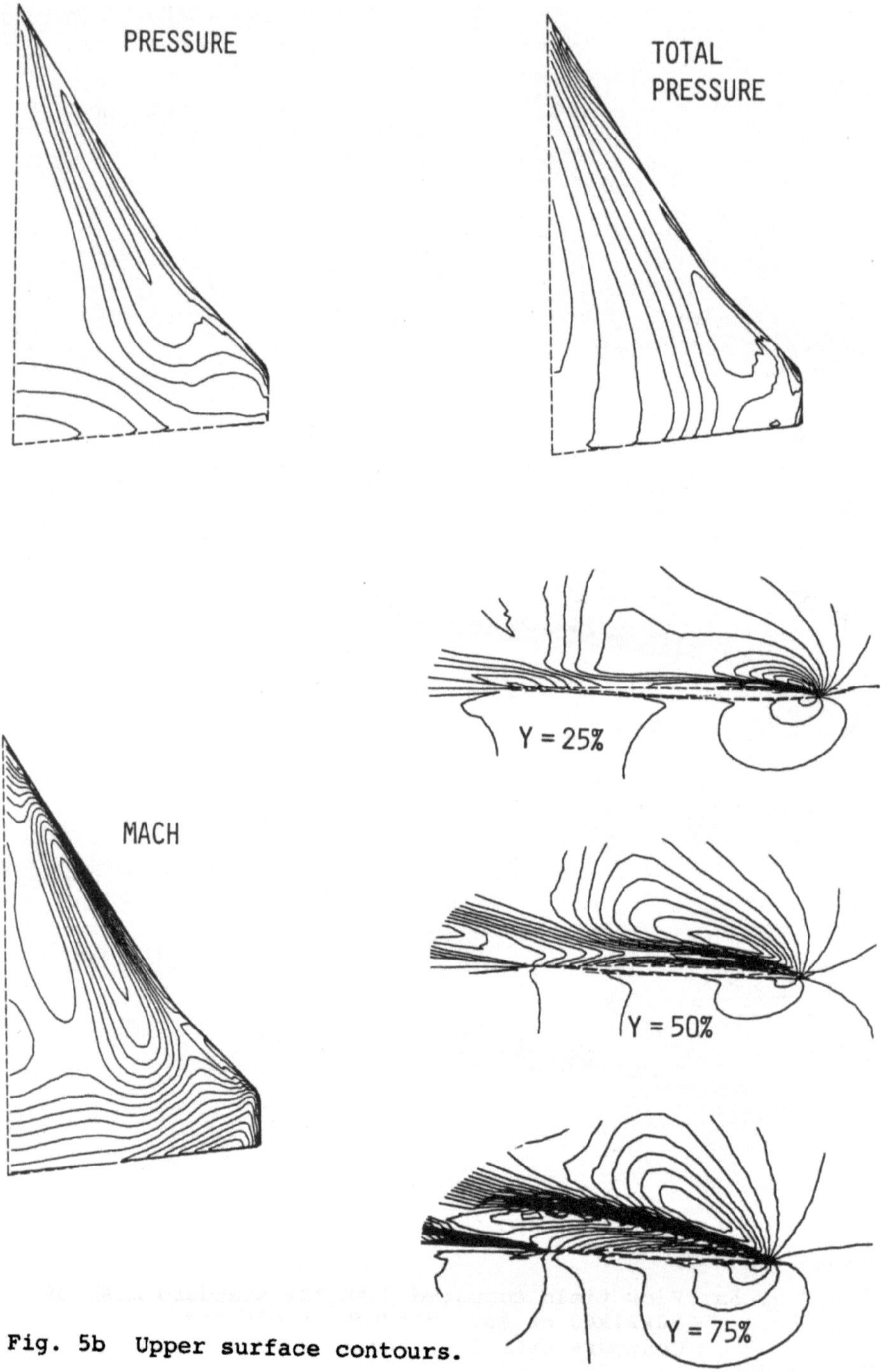

Fig. 5b Upper surface contours.

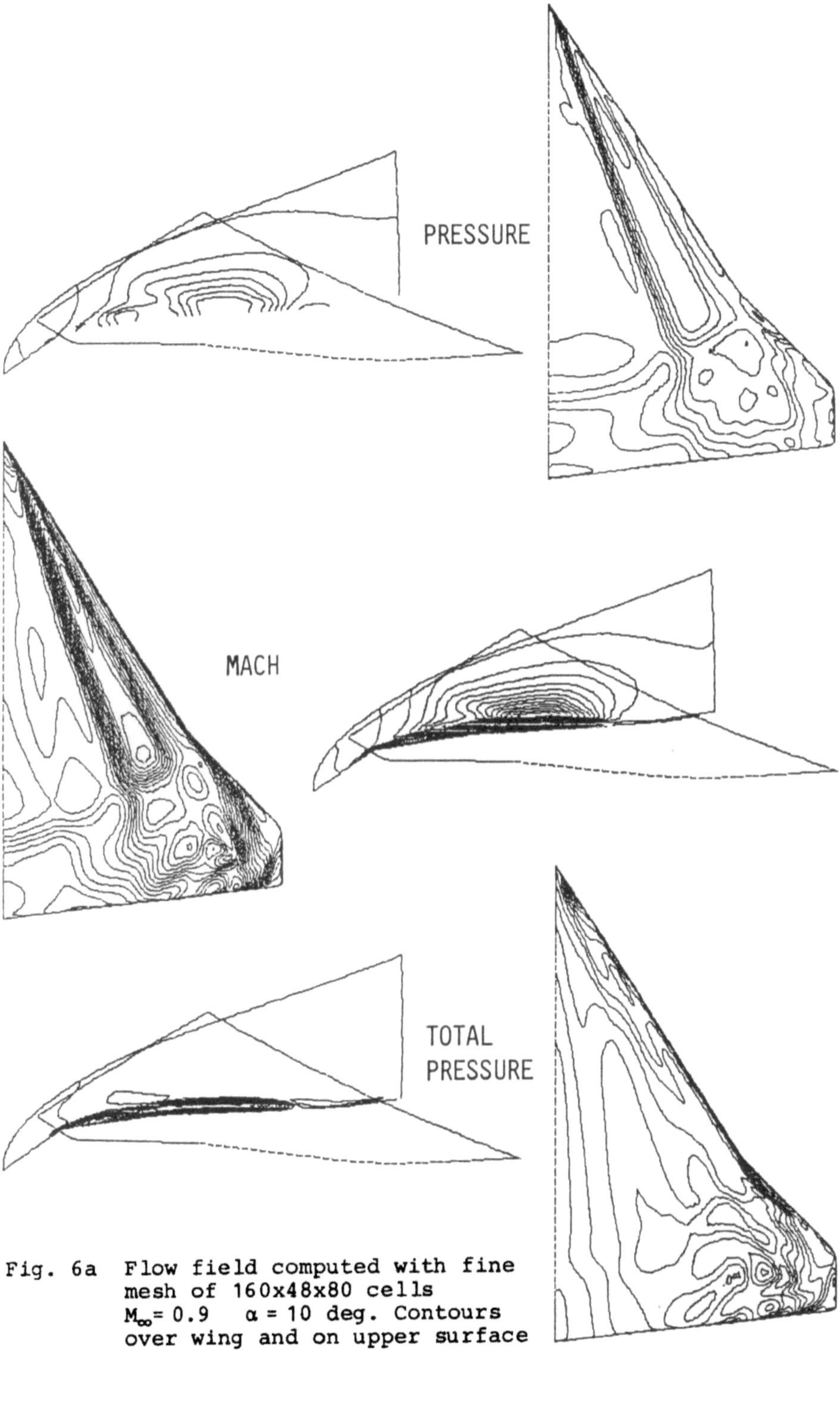

Fig. 6a Flow field computed with fine mesh of 160x48x80 cells $M_\infty = 0.9$ $\alpha = 10$ deg. Contours over wing and on upper surface

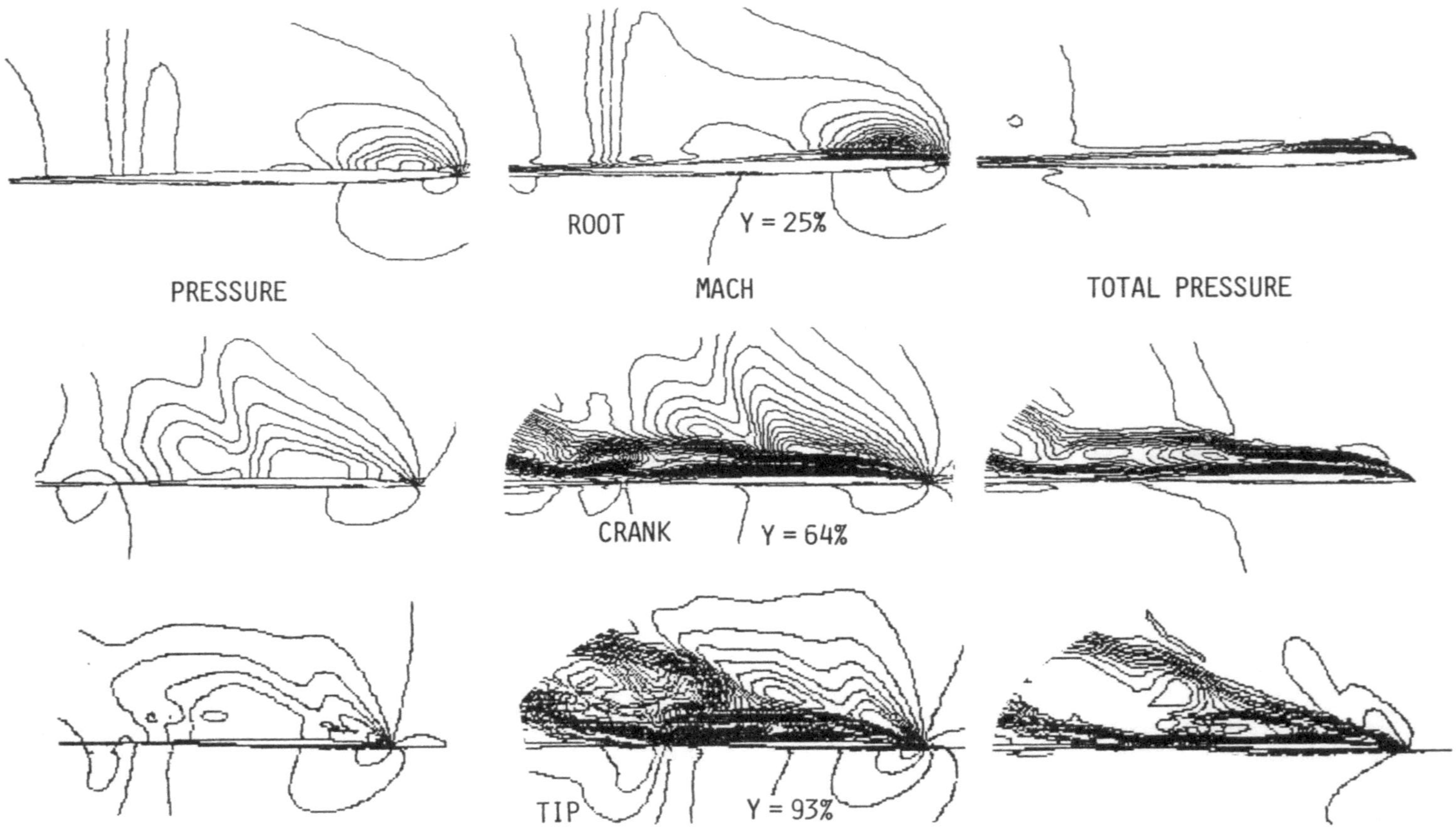

Fig. 6b Chord section contours.

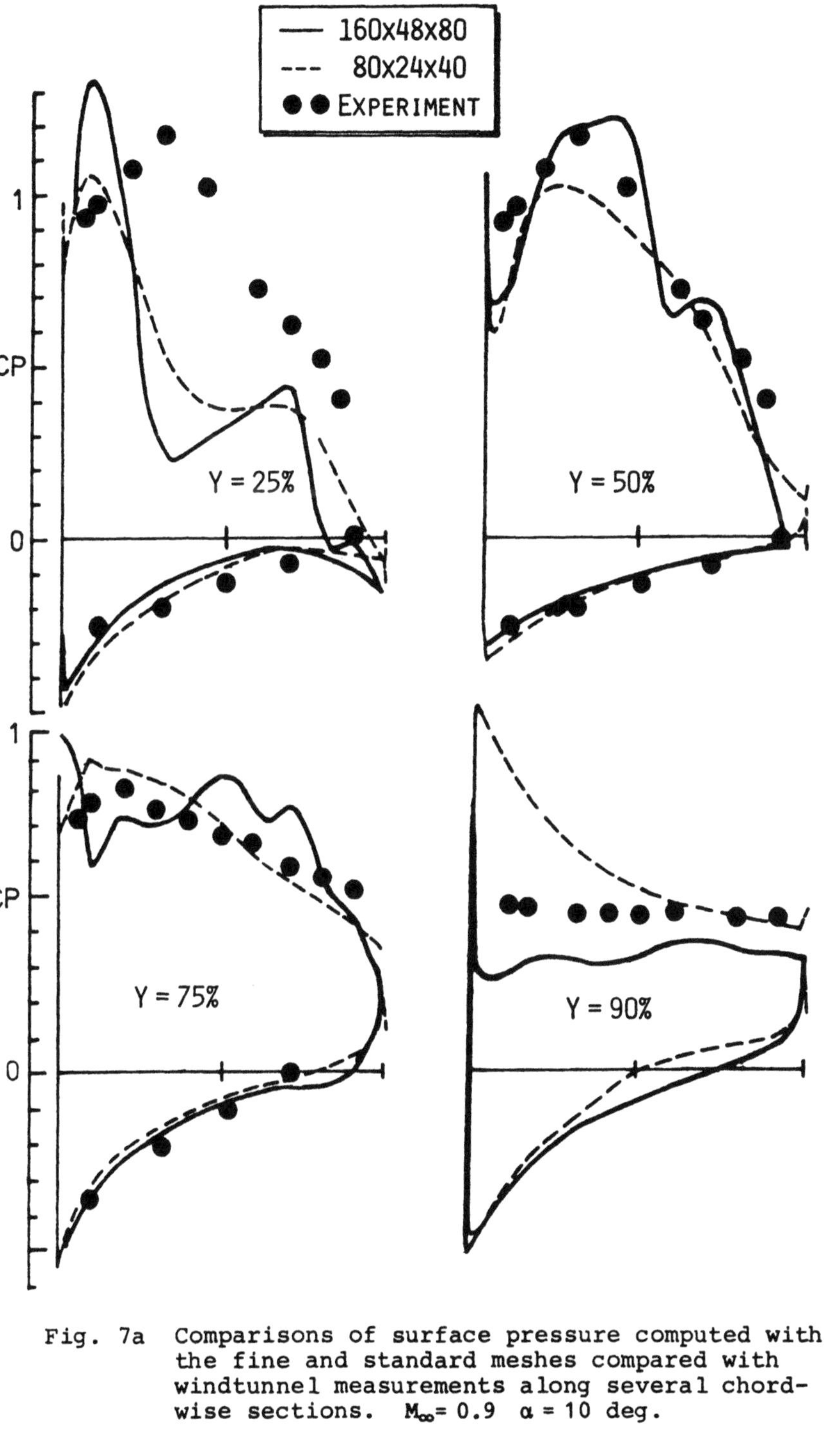

Fig. 7a Comparisons of surface pressure computed with the fine and standard meshes compared with windtunnel measurements along several chordwise sections. $M_\infty = 0.9$ $\alpha = 10$ deg.

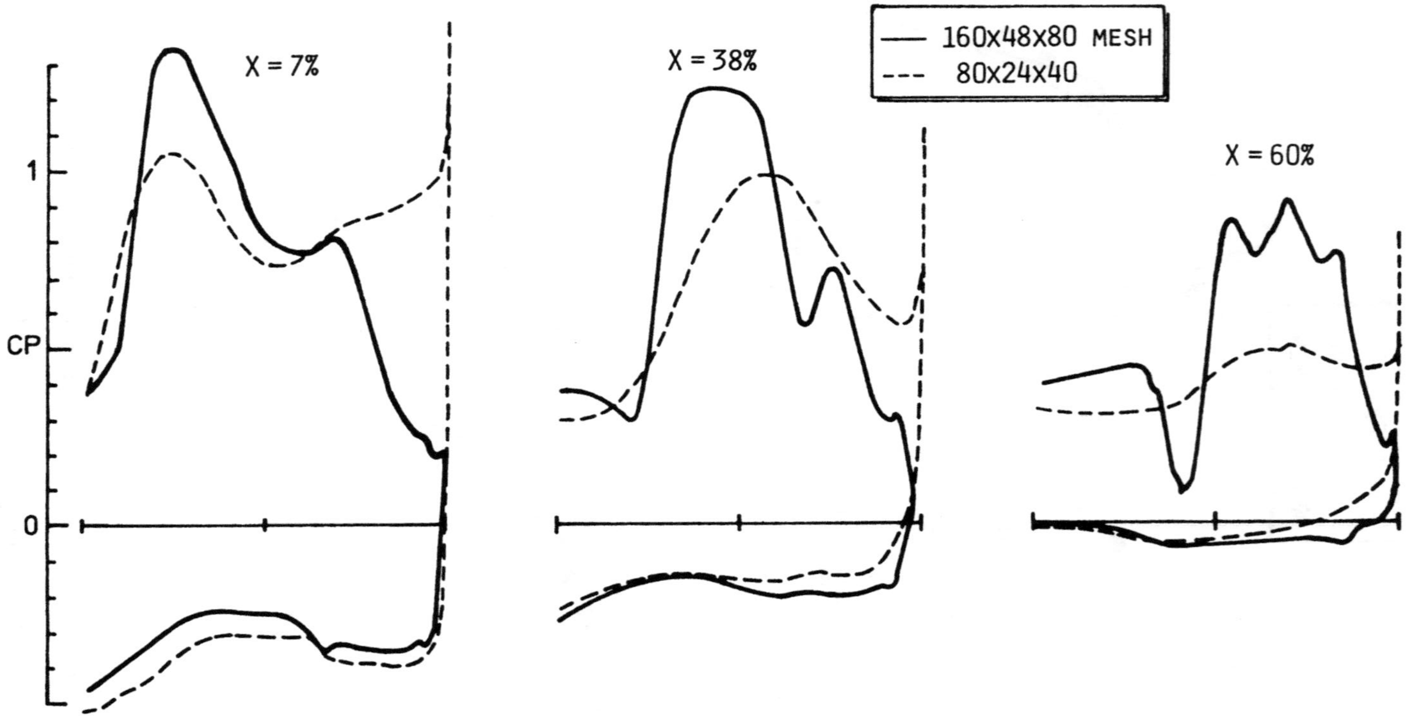

Fig. 7b Several spanwise comparisons of the standard-mesh and fine-mesh results for surface pressure.

Supersonic Viscous-Inviscid Interaction of a Swept Compression Ramp with a Turbulent Boundary-Layer

G. R. INGER

Department of Aerospace Engineering
Iowa State University
Ames, IA, USA

Summary

This paper analyzes the supersonic interaction of a swept infinite compression ramp with an adiabatic turbulent boundary layer by means of a non-asymptotic triple deck approach combined with the non-isotropic eddy viscosity model of Rotta. Fundamental analytical descriptions of the 3-D disturbance flow field are obtained for a wide range of Mach number, Reynolds number, boundary layer shape factor and sweep angles. In particular, the following aspects are treated: (1) prediction of the dependence of the upstream influence on sweep angle; (2) assessment of the validity of the so-called independence principle, including deviations due to combined compressibility and turbulence effects plus the cross-flow component of the interactive turbulence, and (3) examination of the shear stress disturbance region close to the wall to obtain basic predictions for the 3-D skin friction line geometry and the onset of separation.

Outline of the Nonasymptotic Interaction Theory

Referring to the planview of Fig. 1 we consider the idealized situation where the ramp is straight such that interactive disturbance flow gradients parallel to the ramp may be considered negligible ($\partial/\partial s = 0$). For non-separating flows at high Reynolds numbers, a triple-deck approach can be applied to treat the very rapid, short-ranged changes occurring normal to the ramp (Fig. 2) as follows: (1) an outer region of potential inviscid flow above the boundary layer, which contains a compression shock and the interactive wave systems; (2) an intermediate deck of frozen shear stress, rotational inviscid disturbance flow occupying the outer 90% or more of the incoming boundary layer thickness; (3) an inner shear-disturbance sublayer, lying within the Law of the Wall of the incoming boundary layer, which accounts for the interactive skin friction perturbations (and hence any possible incipient separation) plus most of the upstream influence of the interaction. Over

Turbulent Shear Layer/Shock Wave Interactions
IUTAM Symposium Palaiseau 1985
Editor: J. Délery

the short interaction zone, changes in the undisturbed boundary-layer properties are neglected. Here we emphasize the non-asymptotic treatment of the inner shear-disturbance and middle decks at realistic Reynolds numbers. The former, although exceedingly thin, contributes significantly to the upstream influence that is a significant feature of interactions, and contains the skin friction disturbances and thus is indispensible in 3-D flow to determine the skin friction line geometry and the onset of separation. The non-asymptotic treatment, which can be shown to be fully equivalent to a formal second order asymptotic-analysis approach [2], provides an experimentally well-verified [3,4] basic account of the interactive pressure gradient along the inner deck plus the normal pressure gradient, streamline divergence, and boundary layer profile shape effects across the middle deck (while these are excluded in the infinite Reynolds number limit, they are important in turbulent interactions at practical high Reynolds numbers $Re_{\delta*} \sim 10^3$ to 10^6).

The incoming two-dimensional boundary layer profile has the classical Law of the Wall/Law of the Wake structure, as modeled by Walz's [5] composite analytical expression with an adiabatic wall reference temperature-method correction for compressibility and arbitrary non-equilibrium values of its shape factor H_{i_1}. Along the interaction it is known that the turbulence effects in the boundary layer remain virtually frozen (i.e., the eddy viscosity remains unperturbed) except within the very thin Law of the Wall region. We use the well-regarded 3-D eddy viscosity theory of Rotta [6]. Detailed study of this model shows that its complicated non-isotropy aspects make a negligible contribution to the inner deck solution.

Formulation of Outer and Middle Deck Disturbance-Flow Problems. Assuming a weak ramp-shock with an isentropic inviscid small disturbance motion imposed upon the undisturbed incoming flow, we write $p = p_o + p'(N,y)$, $u_N = U_{N_o} + u_N' = U_{o_x}(y)\cos\Lambda + u_N'(N,y)$, $\rho = \rho_o(y) + \rho'(N,y)$, $u_s = U_{s_o} + u_s' = U_{o_x}(y)\sin\Lambda + u_s'(N,y)$ and $V = V'(N,y)$ where the primes denote the perturbations then above the inner deck we obtain for $\partial/\partial s = 0$ that

$$\left[1-M_{0N}^2(y)-2\frac{u_N'}{U_{0N}}M_{0N}^2\right]\frac{\partial^2 p'}{\partial N^2}+\frac{\partial^2 p'}{\partial y^2}-2\frac{dM_{0N}}{dy}\frac{\partial p'}{\partial y}=0 \tag{1}$$

$$\frac{\partial(V'/U_{0N})}{\partial y}=\left[\frac{1-M_{0N}^2(y)}{M_{0N}^2}\right]\frac{\partial p'/\gamma p_1}{\partial N} \tag{2}$$

$$\frac{\partial(u_N'/U_{0N})}{\partial N}=-\frac{\partial}{\partial N}\left[\frac{p'/\gamma p_1}{M_{0N}^2(y)}\right]-\frac{V'\,dU_{0N}/dy}{U_{0N}^2(y)} \tag{3}$$

plus the tangential-disturbance momentum equation

$$\frac{\partial}{\partial N}\left[\frac{u_s'}{U_{0s}(y)}\right]=-\frac{V'}{U_{0N}}\frac{dU_{0s}/dy}{U_{0s}} \tag{4}$$

where $U_{0_S} = U_0(y)\sin\Lambda$, $M_{e_N} = M_e\cos\Lambda$ and u_N' term in Eq. 1 accounts for possible mixed transonic-flow effects when $M_{e_N} \to 1$ [8]. This generalized Lighthill pressure perturbation equation for non-uniform flows applies throughout the boundary layer except at the singular point $M_0 \to 0$, which we avoid by consideration of the inner deck as shown below. Eq. (1) is solved for all N and $y > 0$ subject to the far-field condition as $y \to \infty$ that the disturbances ahead of the ramp shock vanish while the shock pressure jump becomes equal to the inviscid pressure rise associated with the swept compression ramp. The inner boundary condition is

$$\frac{\partial p'}{\partial y}(N, y_{w_{eff}})=-\rho_{0w}U_{0N}^2(y_{w_{eff}})\frac{d\theta_{wN}}{dN} \tag{5}$$

where θ_{w_N} is the ramp angle measured normal to the ramp line and $y_{w_{eff}}$ is the displacement height of the inner deck below.

It is interesting to note from Eq. (4) that a small tangential disturbance flow is induced because of the interactive upwash in the presence of the tangential boundary layer vorticity with sweep; it vanishes both at the boundary layer edge where $dU_0/dy \to 0$ and at the bottom of the middle deck where $v' = 0$ and so has a maximum somewhere deep down in the boundary layer profile. It can also be seen that whereas p', u_N' and V' are each independent of the corresponding tangential <u>disturbance</u> flow u_s', they are not independent of sweep because of the Δ-dependence of inner condition (5).

<u>3-D Flow in the Inner Shear-Disturbance Deck</u>. The original works of Lighthill [10] and Stalker [11] treated this problem by neglecting the turbulent stresses altogether and considering only the laminar sublayer effect; while this yields an elegant analytical solution, the results can be significantly in error at very high Reynolds numbers and cannot explain the ultimate asymptotic behavior that must pertain in the infinite Re_{ℓ} limit. The present work avoids this by including the interactive turbulent-stress effects, so that the resulting shear-disturbance sublayer theory yields a non-asymptotic treatment of the three-dimensional disturbance flow which encompasses the complete range of Reynolds numbers.

To facilitate a tractable theory for the main physical effects with weak non-separating shocks, the sub-layer flow is assumed to be a small perturbation upon the incoming boundary layer; we retain however, the physically important streamwise pressure gradient, streamwise and vertical acceleration, and both laminar and turbulent disturbance stresses. For adiabatic flows at low-to-moderate external Mach numbers, the density perturbations are neglected and the corresponding compressibility effect on the Law of the Wall portion of the undisturbed profile treated by the Eckert reference temperature method. Applying the basic conservation equations with $\partial p/\partial y \cong 0$ and the aforementioned assumptions including the isotropic turbulence model, we obtain (after some cross differentation to eliminate the continuity equation) in the leading small disturbance approximation the following pair of equations governing the 3-D motion in s and N coordinates:

$$\frac{\partial}{\partial N}\Big[U_{0N}\frac{\partial^2 V'}{\partial y^2} - V'\frac{d^2 U_{0N}}{dy^2} + \underbrace{\sin^2\Lambda\,\frac{\partial^2}{\partial y^2}\Big(\nu_{t_0}\frac{\partial U_s'}{\partial y}\Big)}_{\text{CROSS-FLOW TURBULENCE}}\Big] = \frac{\partial^2}{\partial y^2}\Big[(\nu_{w_0}+2\nu_{t_0})\frac{\partial^2 V'}{\partial y^2}\Big] \quad (6)$$

$$U_{0N}\frac{\partial U_s'}{\partial N} - \rho_{0w}^{-1}\frac{dp'}{dN} = \frac{\partial}{\partial y}\Big[(\nu_{w_0}+\nu_{t_0}\frac{\partial U_s'}{\partial y}\Big] \quad (7)$$

where we have introduced the new sidewash variable $U_s' = u_s' \times \cot\Lambda - u_N'$ and where

$$\nu_{t_o} = (.41y)^2 (1 - e^{-y/A})^2 \sec\Lambda \frac{dU_{o_N}}{dy} \tag{8}$$

$$A = 26\,\nu_{w_o} (\rho_{w_o} / \tau_{w_{o_N}} \sec\Lambda)^{1/2} \tag{9}$$

$$(\mu_{w_o} + \rho_{w_o} \nu_{t_o}) \frac{dU_{o_{N,S}}}{dy} = \tau_{w_{o_{N,S}}} \tag{10}$$

Equation (6) is to be solved subject to the initial condition $v'(-\infty, y)$ = and the wall boundary conditions $v'(N,0) = 0$ and $\partial v'/\partial y(N,0) = 0$ for $N < 0$ or $\theta_{W_N}(dU_{O_N}/dy)(0)$ for $N > 0$. A third condition involving v' is obtained by satisfying the N-differential at the N-momentum equation right at the wall. The fourth boundary condition is the outer inviscid matching requirement $v'(N,\delta_{SL}) = v'_{INV}(N,\delta_{SL})$ with the inviscid solution governed by the inviscid form of Eq. (6) and δ_{SL} pertaining to $\partial^2 v'/\partial y^2 = 0$ (i.e., vanishing total disturbance shear) somewhere within the Law of the Wall region.

A significant feature of this solution is the displacement thickness $y_{w_{eff}}$ that emerges from the asymptotic behavior of v' far from the wall, i.e., the value of y where the "back projection" of this behavior vanishes. Physically, $y_{w_{eff}}$ thus represents the effective wall position "seen" by the overlying middle deck disturbance flow, and serves to couple the inner and middle deck solutions by providing non-singular inner slip flow boundary condition (5).

Solution by Fourier Transformation

Solutions of the foregoing equations were obtained by introducing a Fourier transformation w.r.t. N and the non-dimensional inner deck variables defined by Inger [1,3]. The resulting ordinary differential equations and boundary conditions are in a "universal" form that can then be straightforwardly integrated (in fact, solved analytically for the quasi-laminar case where $\nu_{t_o} \ll \nu_{w_o}$) and the results neatly tabulated (see Ref. 9 for details).

Along the wall, the upstream disturbance pressure in the leading large scale approximation is found to be

$$\frac{p'_w}{\Delta P} \simeq \frac{T_e}{T_w} \frac{U_{0N}^2(y_{w_{eff}})}{U_{e_0}^2} C_{0_\ell} e^{N/\ell_{U_N}} \tag{11}$$

where $\Delta P \equiv \rho_{o_e} U^2_{e_N} \theta_{WN}/\beta_N$ is the overall inviscid pressure rise, C_{o_ℓ} is a Reynolds/Mach number-dependent coefficient of order unity and ℓ_{U_N} is the upstream influence distance normal to the ramp line

$$\ell_{U_N} \simeq \frac{M^2_{e_N}/\beta_N}{[M'_{0N}(0)]^2 \, y_{w_{eff}}} \qquad (\beta_N \equiv \sqrt{M^2_{e_N}-1}) \tag{12}$$

The corresponding result for the inner deck displacement thickness is found to be

$$y_{w_{eff}} = .776\left(\frac{\mu^2_{w_0}\ell_{U_N}}{\rho_{0w}\tau_{w_{0N}}}\right)^{1/3}\Big[H_1(T) - \underbrace{H_2(T)\sin^2\Lambda}_{\text{CROSS-FLOW TURBULENCE}}\Big] \tag{13}$$

where the functions $H_1(T)$ and $H_2(T)$, plotted in Fig. 3, represent the relative influence of the turbulent eddy viscosity in terms of sweep angle and the turbulent interaction parameter

$$T \equiv (.41)^2 \frac{\rho_{0w}\tau_{w_{0N}}}{\mu^2_{0w}} \left(\frac{\mu^2_{0w}\ell_{U_N}}{\rho_{0w}\tau_{w_{0N}}}\right)^{2/3} \sec\Lambda \tag{14}$$

The simultaneous solution of Eqs. (12) and (13) for ℓ_{U_N} and $y_{w_{eff}}$ constitutes the matching of the inner and middle decks.

The solution for the normal component of the disturbance skin friction can also be expressed in analytic form, as follows. Focusing attention on the large scale aspects that likely control observable skin friction line behavior, we obtain upstream of the ramp

$$\tau'_{w_N} \simeq -1.372\, f_p\, p'_w \left(\frac{\rho_{w_0}\tau_{w_{N0}}\ell^2_{U_N}}{\mu_{w_0}}\right)^{-1/3}\left[S_1 - S_2\sin^2\Lambda\right] \tag{15}$$

where S_1 and S_2 are the interactive turbulence effect-shear functions given in Fig. 3 and f_p is a non-dimensional interactive history function of order unity. The detailed analysis also yields expressions for the cross flow and hence the x- and z-components of the skin friction.

Figure 3, a central result, gives a unified account of the inner interactive physics over the entire Reynolds number range from quasi-laminar behavior at $T << 1$ (lower Reynolds numbers) to the opposite extreme of wall turbulence-dominated behavior at $T >> 1$ pertaining to asymptotic theory at very large Reynolds numbers where the inner deck thickness and its disturbance field become vanishingly small. Moreover, the additional effect of the cross-flow turbulence terms ($\sim \sin^2 \Lambda$) in reducing the interactive disturbance effects below their values based on the independence principle is clearly shown.

Discussion of Results

The Independence Principle for Small-to-Moderate Sweep. When the explicit $\sin^2 \Lambda$ terms are neglected, inspection reveals that the u'_N, v', p' fields are now explicitly independent of the tangent wash U'_S and influenced by sweep only through (1) the undisturbed flow eddy viscosity ν_{t_o} that depends on the total resultant wall shear and not on the normal component, and (2) the compressibility effect on the fluid properties ρ_{o_w} and μ_{o_w}, which depend on the resultant flow Mach number M_e rather than M_{e_N}. The first influence is moderate and completely accounted for by using modified Von Karman and Van Driest constants $K_3 = .41 \sec^{1/2} \Lambda$ and $A_3 = 26 \cos^{1/2} \Lambda$. The second influence can be accounted for by $T_w/T_e = 1 + .5 (\gamma - 1) r_3 M_{e_N}^2$ where $r_3 = .89 \sec^2 \Lambda$ is an equivalent "3-D" recovery factor.

Upstream Influence with Sweep. This important property is found from the above as

$$\ell_{u_N}^{4/3} = C \left(\frac{M_{e_N}^2}{\tau_{w_{o_N}}^{5/3} \beta_N} \right) \left\{ \frac{\left[1 + .178 M_{e_N}^2 \sec^2 \Lambda\right]^{2\left(\frac{1+2\omega}{3}\right)}}{H_1(T) - H_2(T) \sin^2 \Lambda} \right\} \quad (16)$$

where $C = a_e^2 \mu_e^{5/3} / .776 \nu_e^{1/3}$. This shows that the sweep effect modifies the upstream influence by the product of two terms: the first involves only the flow components normal to the ramp and thus embodies Stalker's [11] sweepback principle, while the second contains an adiabatic wall compressibility effect plus the influence of sweep on the interactive eddy viscosity, both of which do not obey the sweepback principle and increase ℓ_{U_N} over the value based on this principle.

Comparing with the zero sweep value for a given resultant incoming Mach number M_{e_o}, noting that $\tau_{W_N} = \tau_{W_o} \cos \Lambda$, Eq. (14) in the quasi-laminar approximation ($T << 1$, $H_1 = H_2 = 0$) yields the variation of $\ell_{U_N}/(\ell_{U_N})_{\Lambda=0}$ with sweep angle shown in Fig. 4. It is seen that this theory predicts a significant relative increase in ℓ_{U_N} with Λ, especially at lower Mach numbers. This result agrees with the earlier laminar works of Stalker and others, and is due to the dependence on normal component of the Mach number embodied in the sweepback rule. In the fully turbulent interaction case where $H_1 \sim 0.326T^{-1/2}$ and $H_2/H_1 \cong 1 - 0.58 \tan^2 \Lambda$, Eq. (14) predicts an even stronger relative increase of upstream influence with sweep, in accord with the experiments of Settles [12] (Fig. 4).

Skin Friction Lines and Incipient Separation. The planform skin friction line geometry is obtained from the relation (Fig. 5)

$$\frac{dZ_{SF}}{dX} = \frac{\tau_{wz}}{\tau_{wx}} = \frac{\tau'_{wz}}{\tau_{wx_o} + \tau'_{wx}} \quad (17)$$

where τ'_{W_N} is negative, implying an upward lateral deflection in the +z direction. A typical set of skin friction curves inferred from the streamwise integration of Eq. (17) using our solution values for τ'_{W_z} and τ'_{W_x} into the rising interactive pressure are illustrated in Fig. 5 for a given sweep angle; it may be seen that they closely resemble the experimentally observed behavior cited by Green [13]. Note that because $\partial/\partial s = 0$, all the skin friction lines in the present problem must deflect the same amount for a given distance normal to the ramp.

The foregoing also allows assessment of the flow behavior close to the surface at the onset of three-dimensional separation. Following Lighthill [14], we tentatively take this to occur when the skin friction lines merge into a single common line, corresponding to the condition $(dZ_{SF}/dx) \rightarrow \cot \Lambda$, Fig. 5 (any larger deflection with $\partial/\partial s = 0$ would cause the lines to cross). Thus from Eq. (17) we obtain $(\tau'_{W_N})_{i.s.} + \tau_{W_{O_N}} = 0$ (vanishing of the total normal shear stress component) which indeed is the physically expected result for a straight shock surrounded by a $\partial/\partial s = 0$ disturbance field. Moreover, this conclusion is independent of the turbulence model and carries all the more weight for that reason. Thus we have formally established the equivalence of the two conditions and lent basic support to Stanbrook's suggestion [15] that 3-D separation is imminent when the skin friction lines become parallel to the ramp.

References

1. Inger, G. R., "Upstream Influence and Skin Friction in Non-Separate Shock/Turbulent Boundary Layer Interactions," AIAA Paper 80-1411, July 1980 [see also article in Numerical and Physical Aspects of Aerodynamic Flows (T. Cebeci, Ed.), Springer-Verlag, N.Y. 1982].

2. Melnik, R., "A Survey of Turbulent Viscid-Inviscid Interaction Theory in Aerodynamics," AIAA Paper 84-0264, Reno, Nevada, January 10.

3. Inger, G. R., "Some Features of a Shock-Boundary Layer Interaction Theory in Transonic Flow Fields," in AGARD CP-291 ("Computation of Viscous-Inviscid Interactions"), September 1980, pp. 18-1 to 18-66.

4. Nandanan, M., E. Stanewsky and G. R. Inger, "A Computational Procedure for Transonic Airfoil Flow Including a Special Solution for Shock-Boundary Layer Interaction," AIAA Journal 19, December 1981, pp. 1840-1846.

5. Walz, A., Boundary Layers of Flow and Temperature, M.I.T. Press, 1969, pp. 113-115.

6. Rotta, J., "A Family of Turbulence Models for Three-Dimensional Boundary Layers," Proc. First Symposium on Turbulent Shear Flows, Penn State University, 1977.

7. Cebeci, T., K. Kaups and J. Rumsey, "A General Method for Calculating Three Dimensional Compressible Laminar and Turbulent Boundary Layers on Arbitrary Wings," NASA CR-2777, 1977.

8. Murman, E. M. and J. D. Cole, "Calculation of Plane Steady Transonic Flow," AIAA Journal 9, January 1977, pp. 114-121

9. Inger, G. R., Application of the Fourier Transform Method to the Solution of Swept Shock/Turbulent Boundary Layer Interaction Problems," AFOSR Report, Iowa State University, Ames, September 1985.

10. Lighthill, M. J., "On Boundary Layers and Upstream Influence: Part II--Supersonic Flows without Separation," Proc. Roy. Soc. A217, 1953, pp. 478-507.

11. Stalker, R. J., "Sweepback Effects in Turbulent Boundary Layer/Shock Wave Interaction," Jour. of the Aero. Sci. 27, May 1960, pp. 348-356.

12. Settles, G. S., J. Perkins and S. M. Bogdanoff, "Investigation of Three-Dimensional Shock/Boundary Layer Interactions at Swept Compression Corners," AIAA Journal 18, July 1980, pp. 779-785.

13. Green, J. E., "Interaction Between Shock Waves and Boundary Layers," Progress in Aeronautical Sciences, Vol. 11, 1970, Pergamon Press.

14. Lighthill, M. J., "Attachment and Separation in Three-Dimensional Flow," in Laminar Boundary Layers (L. Rosenhead, Ed.), Oxford U. Press, 1963, 72-82.

15. Stanbrook, A., "An Experimental Study of the Glancing Interaction Between a Shock Wave and a Turbulent Boundary Layer," British ARC CP-555, 1961.

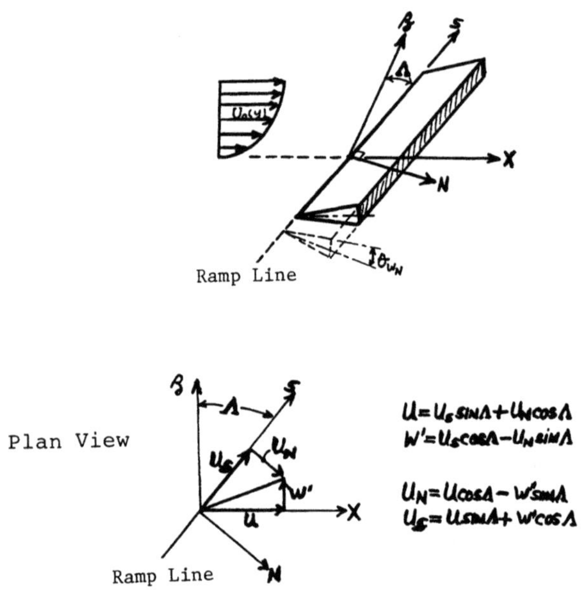

Fig. 1 Swept Ramp Flow Configuration

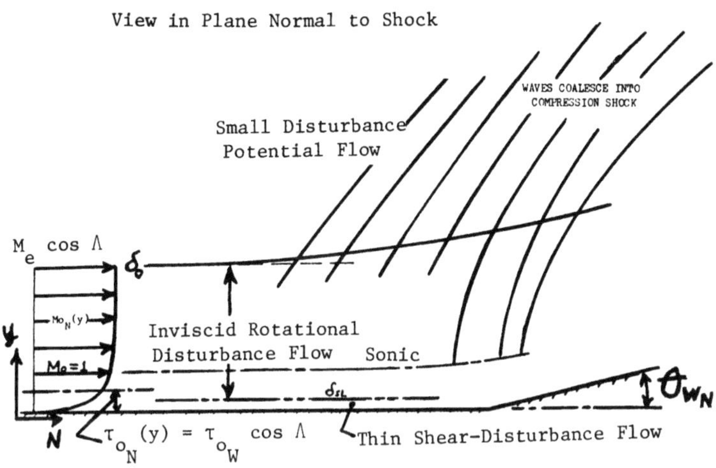

Fig. 2 Triple-Deck Structure of the Interaction

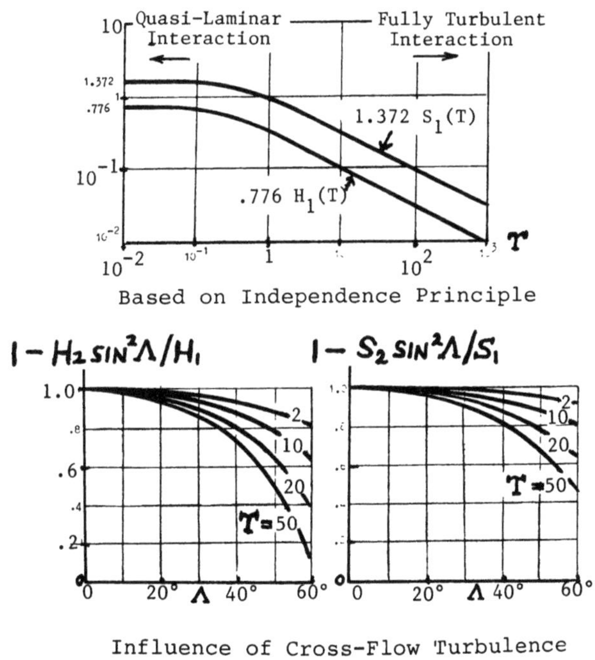

Fig. 3 Interactive Turbulence-Effect Functions

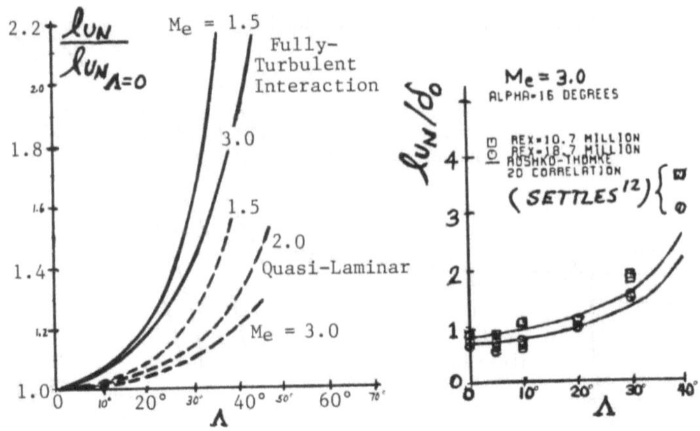

Fig. 4 Relative Sweepback Effect on Upstream Influence

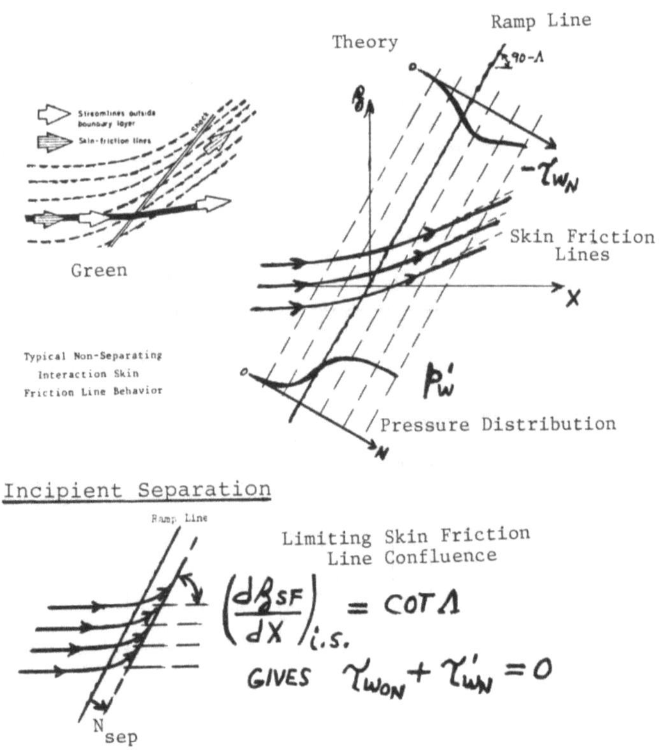

Fig. 5 Near Wall Skin Friction Lines

Shock/Boundary-Layer Interaction Model for Three-Dimensional Transonic Flow Calculations

J. D. McLean and T. K. Matoi

Boeing Commercial Airplane Company

Introduction

Several widely used computer programs for calculating viscous transonic flow in both two and three dimensions use a classical direct-iteration scheme to couple an inviscid-flow solution for the outer flow with a solution to boundary-layer equations for the viscous flow near the surface. In weak solutions to the inviscid-flow equations, shocks represent discontinuities that are inappropriate as input to the boundary-layer equations. In practice, however, the inviscid solutions are generated numerically, and the shock discontinuities are smeared over several grid cells. It is a common practice to feed the inviscid solution directly into the boundary-layer equations, allowing the boundary-layer solution to respond to the smeared shock jumps just as it would to any other pressure gradient. For very weak shocks the thickening of the boundary layer through the shock predicted in this way should be very nearly correct, but for stronger shocks there are three major drawbacks to this approach:

1) The prediction of the viscous flow development through the shock is inaccurate because the assumptions inherent in the simple boundary-layer equations are violated in such a strong-interaction region,

2) the coupled solution has an unrealistic dependence on the grid spacing used in the inviscid solution. In the real flow, the smearing of the pressure rise through the shock results from the interaction of the shock with the boundary layer, while in coupled calculations the smearing depends strongly on the grid spacing of the inviscid solution, and

Turbulent Shear Layer/Shock Wave Interactions
IUTAM Symposium Palaiseau 1985
Editor: J. Délery

3) the boundary-layer solution will often predict separation for cases in which the shock is moderately strong, but in which no separation appears in the real flow.

The alternative approach explored in this paper differs from the simple scheme described above in two basic areas: the treatment of the effects of the shock on the boundary layer, and the treatment of the effects of the boundary layer on the shock in the outer inviscid flow. First, the treatment of the the effects of the shock on the boundary layer is improved by replacing the boundary-layer equations in the shock zone with a set of semi-empirical jump conditions for the changes in the boundary-layer quantities through the shock. The boundary-layer equations are still used upstream and downstream of the shock zone. Several investigations (References 1, 2, 3) have shown that models of this type offer an improvement over the boundary-layer equations, and Inger (Reference 4) has used such a model in coupled transonic flow calculations for 2-D airfoils. Second, the effect of the boundary layer on the outer flow is modeled by defining the displacement thickness in the shock zone as to produce the "viscous-wedge" effect proposed by Yoshihara (Reference 5), i.e., to encourage the inviscid solution to produce a perpendicular Mach number slightly less than unity downstream of the shock.

In this paper, we describe the extension of this basic approach to three dimensions and its application to coupled viscous-inviscid three-dimensional transonic flow calculations. The approach has been incorporated into two computer programs that carry out such coupled calculations automatically. One program treats wing-body and wing-body-strut-nacelle flows using a transonic full-potential method (Reference 6) for the inviscid flow, and the other program treats nacelle-center-body flows using an Euler equation method (Reference 7). Both programs use a three-dimensional finite-difference boundary-layer method (References 8 and 9) for the boundary-layer calculations, modified to also carry out the calculations associated with the semi-empirical shock-jump conditions.

The Shock-Boundary Layer Interaction Model

Basic Operation

The interaction between the boundary layer and the inviscid flow is calculated by a classical direct-iteration scheme. In each cycle of the iterative procedure, the viscous flow is computed in the direct mode, and a weighted average of the new and old displacement thickness is used to modify the surface shape for the next cycle.

The calculation of the viscous flow for each cycle involves several steps. First, the curvilinear, orthogonal surface grid is generated and the inviscid-flow velocity vectors are appropriately transformed and interpolated onto this grid. Based on the inviscid-flow data, the intersections of the shocks with the chordwise boundary-layer grid lines are located. These shock locations are actually defined by upstream and downstream boundaries z_1 and z_2 of "shock zones" over which the shock is smeared in the inviscid-flow solution, as shown for a typical wing boundary-layer grid in Figure 1.

This determination of shock locations is based on the behavior of the shock-perpendicular Mach number. We make the simplifying assumption that the shock sweep is perpendicular to the pressure gradient at a point in the shock zone where the magnitude of the gradient is a maximum. The shock sweep can then be calculated on each chordline. The boundaries z_1 and z_2 are determined by applying a sequence of criteria based on the value and gradient of the shock-perpendicular Mach number.

The boundary-layer solution is determined in a marching sequence in which the program logic "visits" each station only once, determining the solution there as a function of the solution at upstream stations. The basic marching direction is from the leading-edge attachment line to the trailing edge, and from root to tip along each spanline.

When the program logic "visits" the first stations associated with a shock zone, the empirical shock-jump conditions for that zone are computed. This consists of computing the integral parameters of the velocity profile perpendicular to the shock/wing-surface intersection upstream of the shock, $\delta^*_\perp$, $H_\perp$, and $C_{f\perp}$ at $z = z_1$, and the corresponding downstream parameters, using the jump conditions described below.

The computation of the velocity profiles at each station associated with the shock zone takes place as follows. First, the local values of the shock-perpendicular integral parameters are interpolated from the upstream and downstream values using interpolation functions that depend on the local value of the perpendicular Mach number. Three-dimensional boundary layer velocity profiles consistent with these integral parameters are computed by an algebraic transformation of the velocity profiles at the station immediately upstream on the same chordline. With the velocity profiles thus determined, the three-dimensional displacement thickness is computed by integration of the continuity equation, using a procedure that has been modified so that the resulting surface provides the desired "viscous-wedge" effect on the outer flow. When all of these operations are completed, all of the same solution variables are defined as would have been provided by solving the boundary-layer equations.

Shock-Jump Conditions and Separation Criterion

Our model for the effects of a shock on a three-dimensional turbulent boundary layer is based on an existing two-dimensional model extended to three dimensions by the following physical arguments. When a transonic shock impinges on a solid surface, it represents a region of very strong pressure gradient concentrated in a relatively narrow band on the surface, which we will refer to as the "shock/solid-surface intersection." The surface isobars are closely spaced in the region of interaction and tend to assume the same orientation as the shock/solid-surface intersection itself. If we assume that the pressure gradient in the parallel direction is negligible, then we can expect that the flow will behave locally like the flow on an infinite span swept wing. The independence principle, which is exact only for laminar, incompressible boundary layers, states that the boundary-layer velocity profiles in the direction perpendicular to the isobars are independent of the spanwise flow. It has been found that the effects of turbulence do not cause large deviations from the behavior predicted by the independence principle (Reference 10), and we would not expect the effects of compressibility at transonic speeds to cause substantial deviations either.

To have a complete three-dimensional model, we must also model the development of the viscous layer in the shock-parallel direction. We also assume that the pressure forces and the viscous and turbulent stresses cause no substantial change in the shock-parallel velocity component through the shock zone. We, therefore, construct the parallel-direction velocity profiles assuming that the shock-parallel velocity component is unchanged along streamlines of the flow through the shock zone.

The jump conditions for $\delta_\perp^*$ and $H_\perp$ are the ones used by Bozatli (Reference 3) in his study of two-dimensional airfoil flows using an integral boundary layer method. The $\delta_\perp^*$ jump condition was derived from a numerical, non-asymptotic, triple-deck theory of transonic shock/boundary-layer interaction developed by Inger (Reference 1), making use of an algebraic curve-fit by Deane (Reference 11). To derive the jump condition for $H_\perp$ we used the integral conservation approach proposed by investigators at ONERA (Reference 2). For the shock-perpendicular skin-friction coefficient downstream of the shock, we use a simplified curve-fit of Deane's results. The dependence of $\Delta\delta_\perp^*$, $H_{\perp 2}$, and $C_{f\perp}$ on shock strength ($M_{\perp 1}$) is illustrated for typical conditions in Figure 2.

The criterion for shock-induced separation is based primarily on the upstream perpendicular Mach number $M_{\perp 1}$ and is applied in two stages. As $M_{\perp 1}$ increases, the first indication of shock-induced separation is given by the code when $M_{\perp 1}$ exceeds a threshold value that depends weakly on the shock-perpendicular, incompressible shape factor of the upstream boundary layer:

$$M_{\perp 1\ sep.} = 1.5 - 0.139\ H_{\perp 1\ inc.} \tag{1}$$

If $M_{\perp 1}$ exceeds this first threshold, the boundary-layer program continues to generate a solution in the shock zone and downstream of the shock unless a second threshold is reached at $M_{\perp 1} = 1.4$, when $C_{f\perp 2}$ vanishes. When this happens, the remainder of the chordline becomes a forbidden zone. Forbidden zones that can arise in solving the boundary-layer equations are discussed in Reference 8.

Generation of Velocity Profiles in the Shock Zone

At each station associated with the shock zone, the integral parameters of the shock-perpendicular velocity profile are determined by interpolation between upstream and downstream values. Shock-perpendicular velocity profiles consistent with these parameters are then calculated by an algebraic transformation of the profiles at the station immediately upstream on the same chordline.

The shock-parallel velocity profiles are calculated by interpolating the corresponding profiles at the upstream station, using the stream function as the independent variable.

Viscous Wedge Model

The viscous flow solution, including the shock/boundary-layer interaction model, affects the inviscid solution only through the distribution of the displacement thickness. Outside the shock zone, δ^* is determined from the three-dimensional boundary-layer velocity field through the integration of the continuity equation. Inside the shock zone the same integration is performed, but the resulting δ^* is then modified so as to have the desired "viscous-wedge" effect on the outer flow, i.e., to push $M_{\perp 2}$ toward a value slightly less than unity on each chordline. The δ^* derivative in the chordwise direction is changed by a factor that depends on the downstream perpendicular Mach number $M_{\perp 2}$. The factor provides the desired viscous-wedge behavior by increasing the growth of δ^* if $M_{\perp 2}$ is too low and by decreasing it if $M_{\perp 2}$ is too high. This dependence of the δ^* slope on $M_{\perp 2}$ has proven to be fairly effective in driving $M_{\perp 2}$ toward the desired range of values.

Three-Dimensional Wing Calculations

To illustrate the capabilities of the new model, we use the example of an experimental wing design tested in the Boeing transonic wind tunnel. Calculations were made over a range of flow conditions encompassing shock-induced separation; and for comparison purposes each flow condition was calculated both with and without the new model. The nominal experimental flight conditions (Mach number and Reynolds number) were used except that experimental values of α were increased by a constant increment of 0.32 to match the upper-surface roof-top pressure levels.

The predicted pressure distributions near the cruise point are compared with experiment in Figure 3. Both calculations give reasonably good agreement, especially near mid-semispan. The shock location and post-shock pressure distribution are predicted slightly better by the new model. The smearing of the shock in both calculations is greater than in the experiment. Outboard of mid-semispan both calculations underpredict the roof-top suction and the lift, and inboard of mid-semispan both calculations predict the shock too far aft.

At the higher Mach number of .79, all of the calculations without the new model prematurely predicted separation at the shock, and the pressure predictions are therefore meaningless. The pressure predictions for calculations with the new model display reasonably good agreement except that with increasing α, the calculations show an increasing tendency to predict the shock too far aft and with too low a post-shock Mach number.

At the highest angle of attack (α = 2.92; α = 2.6 experimental) the shock/boundary-layer interaction model's first separation criterion (eqn. 1) was exceeded over a large fraction of the span, and the final separation criterion was exceeded over most of the outboard wing. An examination of the experimental pressure data showed that the trailing-edge pressure divergence has just begun at α = 2.6; suggesting that in this case the onset of shock-induced separation in the experiment correlates roughly with the final separation criterion built into our model.

Three-Dimensional Nacelle Calculations

The new model is also being used for modeling flows about transport-aircraft fan-cowls. For one particular case, comparison calculations were made both with and without the new model.

In the calculations without the new model, the boundary-layer program encountered separation at the shock (at about 20 percent chord near the crown line of the external surface) on cycles 1 and 4. Thus, it is unlikely that the calculations would ever have converged, and the corresponding pressure predictions are therefore meaningless. In the calculations with the new model, the solution is not very well converged at cycle 4, but there is no separation and the solution is converging. In Figure 4 external-surface pressure distributions with the new model are compared with experiment. Already, agreement with the data is quite good, except that the forward suction peak is underpredicted at θ = 90.

Conclusions

A new computational model for three-dimensional shock-boundary layer interaction has been developed for use in viscous transonic flow calculations. In general, the new model has been found to be an improvement over the old method in which the boundary-layer equations are simply integrated through the shock. The new model has been tested in coupled viscous-inviscid interaction calculations for wings and for nacelles. From comparisons with measured pressure distributions, the following conclusions can be drawn:

1) For cases with weak-to-moderate shocks the new model gives a slight improvement in agreement over the old method.

2) For cases with strong shocks that caused the old method to fail by predicting shock-induced separation prematurely, the new model continues to converge, providing reasonable agreement with experiment.

3) The new model appears to be able to predict, at least crudely, the onset of shock-induced separation, provided the onset is not preceded by significant trailing-edge separation.

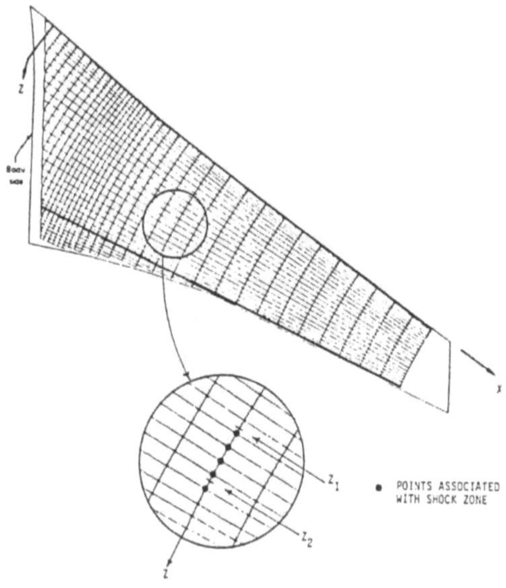

Fig. 1 Typical wing surface grid used by 3-D boundary layer program illustrating the definition of a shock zone on one chordline

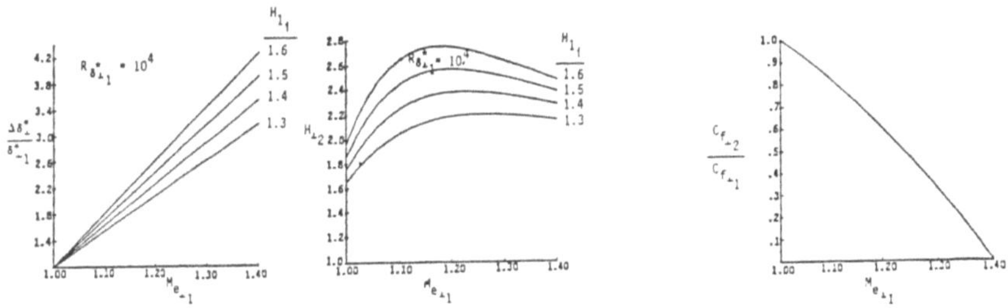

Fig. 2 Jump conditions for δ^* , H , and C_f

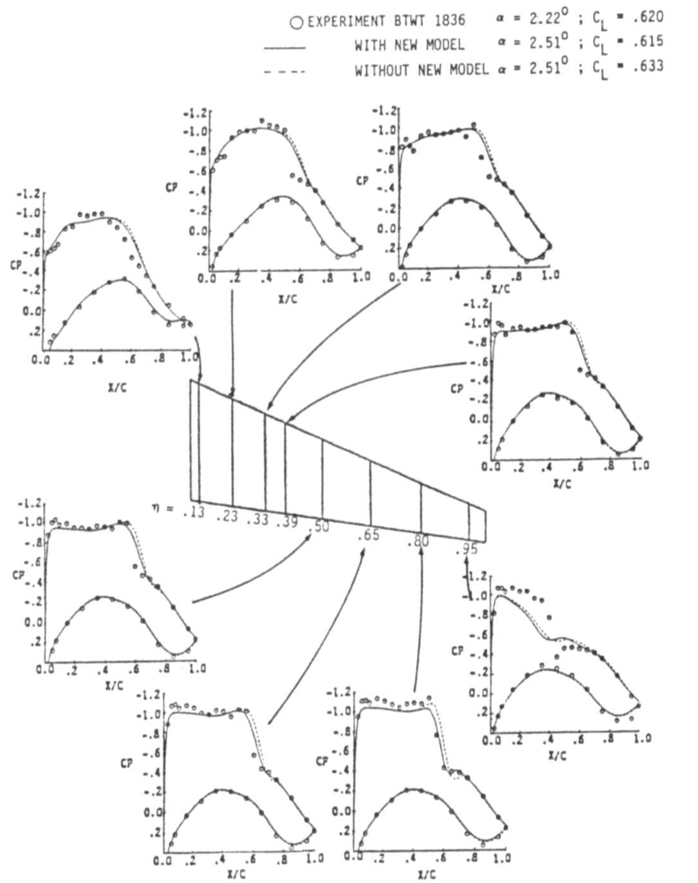

Fig. 3 Test - theory comparison for experimental wing (M = .77)

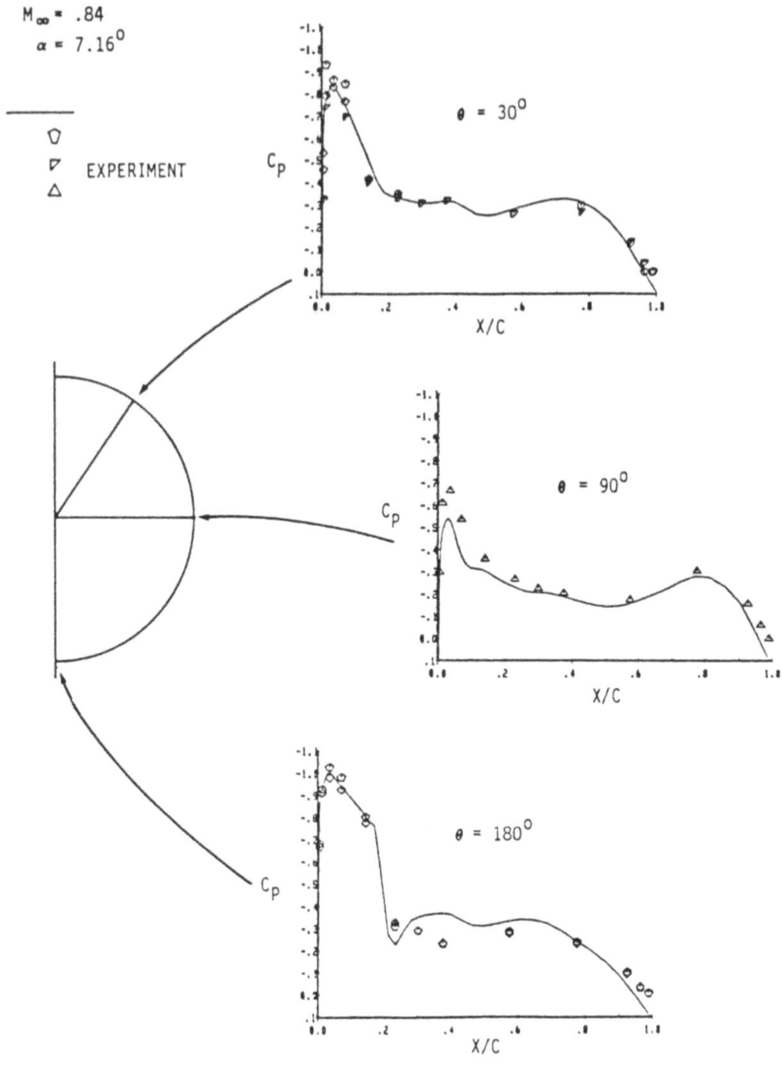

Fig. 4 Predicted pressure distributions for a nacelle

References

1. Inger, G. R., "Some Features of a Shock-Turbulent Boundary-Layer Interaction Theory in Transonic Flow-Fields," Computation of Viscous-Inviscid Interactions, AGARD Conference Proceedings No. 291, October, 1980.

2. Sirieix, M., Director, ONERA, private communication.

3. Bozatli, A. N., "Shock-Boundary Layer Interaction," unpublished Boeing Report, April, 1982.

4. Inger, G. R., "Application of a Shock-Turbulent Boundary Layer Interaction Theory in Transonic Flow Field Analysis," Transonic Perspective - A Critique of Transonic Flow Research, NASA Ames Research Center, February, 1981.

5. Yoshihara, H., and Zonars, D., " An Analysis of Pressure Distributions on Planar Supercritical Profiles with and without Jet Flaps at High Reynolds Numbers," GDCA-ERR-1684, December, 1971.

6. Yu, N. J., "Transonic Flow Simulations for Complex Configurations with Surface-Fitted Grids," AIAA Paper No. 81-1258, 1981.

7. Chen, H. C., Yu, N. J., and Rubbert, P. E., "Flow Simulations for General Nacelle Configurations using Euler Equations," AIAA Paper No. 83-0539, 1983.

8. McLean, J. D., and Randall, J. L., "Computer Program to Calculate Three-Dimensional Boundary-Layer Flows over Wings with Wall Mass Transfer," NASA CR-3123, May, 1979.

9. McLean, J. D., and Randall, J. L., "Three-Dimensional Boundary-Layer Calculations for Fuselages," 5th U.S. - German Data Exchange Agreement Meeting on Viscous and Interacting Flow Field Effects, 1980.

10. Nash, J. F., and Tseng, R. R., "Three-Dimensional Turbulent Boundary Layer on an Infinite Yawed Wing," The Aeronautical Quarterly, November, 1971.

11. Deane, A. E., "Reynolds Number and Shape Factor Effects on Transonic Shock-Turbulent Boundary Layer Interactions, including Incipient Separation," M. S. Thesis, Dept. of Aerospace Engineering, Virginia Polytechnic Institute and State University, Blacksburg, Virginia, December, 1980.

Aerothermal Problems Associated with Viscous/Inviscid Interaction over Hypersonic Flight Vehicles

Michael S. HOLDEN
Calspan/UB Research Center
Buffalo, New York 14225 USA

1. Introduction

In recent years, there has been a resurgence of interest in the development of hypersonic vehicles whose designs are significantly more sophisticated than the essentially ballistic re-entry vehicles which have been constructed and flown to date. In assessing the prediction techniques associated with the design of systems like the trans-atmospheric vehicle (TAV), the orbital transfer vehicle (AOTV), and a number of vehicles designed to be highly maneuverable at hypersonic speeds, it has become evident that there are a number of important aerothermal problems which must be addressed. Such problems include the prediction of low density non-equilibrium flows in the transition regime, the transition to turbulence at hypersonic speeds, turbulent boundary layer and shear layer development in strong pressure gradients, and the development of turbulent reacting boundary layers over ablating and transpiration-cooled surfaces. However, in both laminar and turbulent flows, the classes of problems associated with viscous/inviscid interaction or shock/boundary layer interaction represent those which create the most serious aerothermal load problems and, at the same time, are the most difficult to compute with accuracy.

While in supersonic flows, regions of shock wave/boundary layer interaction have their largest impact on the aerodynamic loads; in hypersonic flow, it is the large thermal loads and gradients generated in turbulent shock interaction regions that are of greatest concern. Of course, the large dynamic loads associated with the intrinsic unsteadiness of these flows also remain a principal concern. Since the severe heating loads on certain key components mandate the use of ablative- or transpiration-cooled thermal protection systems, the effects of surface roughness and blowing on the characteristics of shock/boundary layer interaction over such components is an important and, as yet, an unresolved problem. Although the aerothermal heating loads associated with regions of shock wave/turbulent

Turbulent Shear Layer/Shock Wave Interactions
IUTAM Symposium Palaiseau 1985
Editor: J. Délery

boundary layer interaction will present significant problems in hypersonic flight at low altitudes, it may be the laminar viscous/inviscid interaction and flow separation, which occurs in high altitude hypervelocity flight, which provides fundamental limitations on the performance of maneuvering hypersonic vehicles, and in particular those employing airbreathing propulsion systems. In the high altitude regime, the effectiveness of jet interaction and flap control systems, and intakes and combustors of hypersonic ram jets may be seriously reduced as the surfaces are in essence "faired-in" by viscous/inviscid interaction and flow separation as discussed in Reference 1. This paper reviews some of the important aerothermal problems associated with viscous/inviscid interaction over hypersonic flight vehicles.

2. Interacting Flows Over Transitional Nose Shapes

The shaping of ablative leading edges as boundary layer transition spreads over the surface can be strongly influenced by regions of shock wave/boundary layer interaction. The development of concave or idented nose shapes has been observed in both ground and flight test studies. The heating in the transition region near the sonic point causes the initial indenting (which is often three-dimensional in character) and the induced shock/boundary layer interaction subsequently enhances the indenting process. The large heating rates, which are generated at the base of the shocks formed on such indented nose shapes (Figure 1), coupled with roughness-enhanced heating, can induce heating many times those recorded at the stagnation point.

In hypersonic flows, the shock pattern and distribution of properties over blunt nose shapes are relatively independent of Mach number. However for flows over indented noseshapes, where embedded shocks and separated flows are generated in the flow field, this "Mach number independences principal" is invalid. This point is illustrated in Figures 2 and 3, where schlieren photographs and heat transfer measurements over an indented configuration are shown for Mach numbers of 11 and 13. It can be seen that the size of the separated region is reduced and the magnitude of the reattachment heating is nearly doubled as the Mach number is increased. Surface roughness, which is an intrinsic feature of an ablating nosetip, can also play an important role in controlling the flow field and the distribution of heating over indented nose shapes as illustrated in Figures 4 & 5.

It can be seen that adding surface roughness increased the size of the separated region and almost doubled the heating rate in the reattachment compression region. Notice that roughness does not have as great an effect upstream of the interaction, because it occupies a relatively smaller fraction of the thickness of the boundary layer in this region. While the configurations shown above are idealized axisymmetric representations of a transitional ablated nose shape, in reality, such a nose shape may be far from axisymmetric, as shown in Figure 1 (see Ref. 2 & 3).

The Nosetip Recovery Vehicle (NRV) was one of the few nosetips which has been recovered from a flight test during the part of re-entry where boundary layer transition was spreading over the nosetip. The nosetip is intrinsically three-dimensional in shape, and it is clear from the size of the indentations that large embedded separated regions were formed in the streamwise flutes. The schlieren photograph of the flow over the nosetip shown in Figure 1 clearly shows the flow features typical of embedded shock interaction regions over indented nosetips. The flow over the top of the model separates downstream of the spherical cap forming counterrotating longitudinal vortices in the flow between the grooves. A reattachment shock is formed as the compression waves formed at the back of the cavity coalesce. On the lower half of the model, the flow is basically attached. However, a recompression region is formed on the afterbody and a walljet is formed. The measurements of heat transfer shown in Figure 6 indicate that the boundary layer remains laminar over the spherical cap of the model with transition occurring in the free shear layer, downstream of the shoulder expansion. The heat transfer rates in the recompression region of the model can rise to three times the stagnation point heating value. It is clear that the prediction of these flows, which in reality are also influenced by surface roughness and blowing, is far from simple. However, unless such complex phenomena are modeled in some way in the prediction techniques, little progress will be made in describing such flows.

3. Hypersonic Interacting Flows Over Two- and Three-Dimensional Compression Surfaces

The complexity of the flow field in regions of shock wave/turbulent boundary layer interaction is such that it is unrealistic to expect to describe such regions in any detail within the framework of the boundary layer equations. Indeed their are some who would question whether the time or mass averaged Navier-Stokes equations capture the basic fluid mechanics associated with the intrinsically unsteady nature of separated regions. In hypersonic flows the effects of compressibility on the structure and development of turbulence must also be considered.

While there have been strenuous efforts to obtain predictions of two-dimensional and three-dimensional turbulent interaction regions, it is currently recognized that "successes" with "Navier-Stokes" code in describing some three dimensional turbulent interactions regions is a result of the dominance of the pressure and inertial terms in these flows. In these latter comparisons (References 4, 5, 6) it was found that the modeling of turbulence could be changed without significantly changing the numerical solution. For two dimensional interactions it appears the modeling of turbulence is more critical. To obtain good agreement for these latter flows, some very gross assumptions must be made in the turbulence model: Shang and Hankey[7], for example, chose to apply an empirical relationship (selected by matching the length of the separated region) to rapidly decrease the turbulent scale size through the interaction region, as shown in Figure 7. Horstmann[8], however, found the best agreement with Settles[9] measurements in wedge-induced separated regions using a two equation model for turbulence scale size and vorticity, as shown in Figure 8. Working with this same turbulence model, however, Horstmann was unable to predict the occurrence of separation on two incident shock/turbulent boundary layer configurations studied by Holden at Mach 11.2. As shown in Figure 9 these flow fields are clearly separated and yet the numerical solution fails to predict the characteristic plateaus in either the heat transfer or pressure distributions. The modeling of turbulence in separated interaction regions at hypersonic Mach numbers should account for the effects of compressibility and the generation of turbulence by the unsteady movement of the incident and induced shocks as they traverse and interact with a major region of the turbulent boundary layer. Clearly, further detailed experimental work on insightful theoretical modeling is required to develop numerical prediction techniques which are capable of describing turbulent interaction regions in detail.

Corner Interaction Studies

Studies with the emphasis on the heating in swept-shock interaction regions have been conducted by Newmann and Burke[10], Law[11], Token[12], and Scuderi[13] and Holden[14]. At low Mach numbers ($M = 2 \rightarrow 4$) and for adiabatic surfaces, a large body of data exists on the mean characteristics of swept-shock interactions. Strangely, this body of three dimensional data has been found to be in better overall agreement with the Hung and MacCormack[15] Horstmann[8] Shang and Hankey[7] Settles and Horstmann[6] solutions to the Navier-Stokes equations than the relatively less complex two-dimensional flow separation over a flat plate/wedge. These results are not as sensitive to the turbulence model and suggest that the gross features of the flows are controlled principally by inviscid effects. In a recent experimental study at Mach 11, Holden[14] used the incipient formation of a plateau in the heat transfer distribution, together with a marked increase in the fluctuation levels in the output of the thin film instrumentation, as marking the onset of flow separation. As shown in Figure 10, Holden's measurements indicated that in hypersonic flow over highly-cooled surfaces, the turbulent boundary is more tenacious in resisting boundary layer separation than predicted by the methods derived by McCabe[16] and Korkegi.[11] The measurements of the peak pressure ratio through the interaction and the plateau pressure rise are in better agreement with calculations based on an inviscid flow model in the two dimensional theory of Reshotko and Tucker[17] than the correlations of Scuderi[13], as shown in Figure 11. In was found that, as in the studies of two-dimensional separated interaction regions, the peak heating can be related to the overall pressure rise by a simple power law relationship as shown in Figure 12. Figure 12 shows that the maximum pressure rise through the interaction region can be calculated with good accuracy from inviscid flow relationships. While there appears to be merit for the development of simple prediction methods in describing the flow in terms of the normal flow Mach number, this is clearly a gross oversimplification and it should be noted that the plateau pressure measurements obtained in Holden's study were relatively independent of $M_0 \sin \theta$.

Swept Wedge and Skewed Shock/Boundary Layer Interaction

Another approach to exploring flow separation in regions of three-dimensional shock wave/boundary layer interaction is to begin with a two-dimensional or axisymmetric interaction and sweep this interaction (or introduce angle of attack for the axisymmetric case) to progressively introduce crossflow into the interaction region. Experimental studies of this type have been conducted by Ericsson, Reding and Guenther[18], Settles and Perkins[19], and Settles and Teng[20]. Settles, who studied the interaction region over swept and unswept flat plate/wedge configurations in an adiabatic Mach 3 airflow, found that introducing crossflow increased the scale of the separated interaction region. Considerable effort was expended in this latter study to determine the Reynolds number scaling, and the length from the upstream tip of the wedge for the flow to become quasi-two-dimensional. However, the effect of changing the overall spanwise scale of the model on the scale of the interaction was not examined explicitly. The measurements of surface and pitot pressure through the interaction regions were in good agreement with solutions to the Navier-Stokes equations obtained by Horstman[8]; however, some key features of the flow were poorly predicted. It is known that agreement with pressure data is not the most definitive of tests. More recently Holden[14] performed studies of crossflow effects on the size and properties of the interaction region induced by a swept-oblique-shock incident on a turbulent boundary layer over a flat plate at Mach 11 and $R_{e_L} = 30 \times 10^6$. Experiments were conducted for two strengths of incident shock, the first ($\theta_{SG} = 12.5^o$) to generate a separated condition close to incipient separation, and the second ($\theta_{SG} = 15^o$) to generate a well-separated flow. Distributions of heat transfer and pressure as well as schlieren photographs of the unswept or two-dimensional flow condition and the 30 degree swept condition are shown in Figures 14 and 15. It is clear from the well defined plateau regions in the distributions of pressure and heat transfer, as well as the well defined separation shock in the schlieren photograph, that a well separated region, extending two inches in length, is induced beneath the stronger incident shock. The measurements made of the distribution of heat transfer and pressure beneath the well separated flow induced by both the 12.5 degree and the 15 degree shock generators swept at angles of 0 and 30 (shown in Figures 14 and 15) indicated that the induced crossflow has little effect on the size and characteristics of the interaction regions. If there is a perceptible effect, it is a decrease in the length of the separated region with increased crossflow. The significant differences between Holden's and Settles' measurements of the variation of interaction length with sweep angle and those obtained in these studies are shown in Figure 16. While Settles finds an almost threefold increase in separation length at sweep angles of 40 degrees,

Holden found 10 percent reduction in this length. Further studies are required to resolve this issue.

4. Concluding Remarks

Boundary layer transition and transitional flows are ever present factors in hypersonic flows, and (as shown inadvertently in many experimental studies) have significant effects on the size and properties of separated interaction regions. The prediction of transition at hypersonic (or even supersonic) speeds is best describe as an art, and this, coupled with an inability to describe (even empirically) the characteristics of transitional interactions, probably represents the largest gap in our predictive capabilities.

While, in the absence of real gas effects, a number of the salient characteristics of shock wave/turbulent boundary interaction regions in hypersonic flow can be predicted using semi-empirical techniques. A numerical solution to Navier-Stokes equations has yet to be obtained to successfuly describe such flows. A fundamental problem associated with obtaining such a solution is, of course, the generation of a technique for modeling the turbulence in these flows; however, correctly gridding the interaction region is not a trivial matter. Correctly selecting a turbulence model which will describe the generation of turbulence at the base of the boundary layer as separation occurs, and its restructuring as the boundary layer undergoes radical thinning through the strong reattachment compression process, is central to the correct description of separation and reattachment in any speed range. Also, it remains to be determined whether the unsteady flow phenomena observed in separated flows at any speed can be described in the framework of the time- or mass-averaged Navier-Stokes equations. In hypersonic flow, however, the effects of compressibility on the generation and development of turbulence cannot be ignored. Only through carefully directed and conducted experimental studies can the insight and information be generated to resolve the issues raised above. An intrinsic problem which faces the experimentalist studying turbulent boundary layer separation in hypersonic flow is that the wall layer, within which separation first occurs and the properties of which are required to define the characteristics of both the attached and separated boundary layer, is very thin--typically five percent of the boundary layer thickness, making the accurate probing of this layer difficult. Thus, to obtain the required resolution, measurements must be made on very large models or on tunnel walls, with no intrusive techniques or miniature but robust flow field instrumentation. In the past, mostly because of difficulties in designing equipment robust enough to survive enough not to disturb interaction regions in high Reynolds number hypersonic flows, only mean surface measurements have been made. In many cases, only surface pressure measurements were obtained, and such measurements have little value in the evaluation of prediction techniques. In future experimental studies, every attempt must be made to obtain both mean and fluctuating flow field, and surface skin friction and heat transfer measurements to provide necessary to correctly model the development of turbulence generated in these flows.

References

1. Holden, M.S., "Studies of the Heat-Transfer and Flow Characteristics of Rough and Smooth Indented Noseshapes Part 1. Steady Flows," AIAA-86-0384, January 6-9, 1986, Reno, NV.

2. Holden, M.S., "A Review of Aerothermal Problems Associated with Hypersonic Flight," AIAA-86-0267, January 6-9, 1986, Reno, NV.

3. English E.A., "Nosetip Recovery Vehicle Postflight Development Report" SAND75-8059, Sandia Laboratories, Livermore, CA, January 1976.

4. Knight, D.D., "Prolems in Reconciling Computation and Experiment" 1985 Princeton University Workshop on the Structure of High-Speed Turbulent Boundary Layers.

5. Knight, D., Horstmann, C.C., Shapey, B., and Bogdonoff, S., "The Flowfield Structure of the 3-D Shock Wave Boundary Layer Interaction Generated by a 20 degree Sharp Fin at Mach 3" AIAA-86-343.

6. Horstmann, C.C. and Hung, C.M., "Computations of Three-Dimensional Turbulent Separated Flows at Supersonic Speeds," AIAA Paper 79-2, January 1979.

7. Shang, J.S., Hankey, W.L., and Petty, J.S., "Three-Dimensional Supersonic Interacting Turbulent Flow Along a Corner," AIAA Paper 78-1210, July 1978; also AIAA Journal, Vol. 17, No. 7, July 1979, pp. 706-713.

8. Settles, G.S. and Horstmann, C.C., "Flowfield Scaling of a Swept Compression Corner Interaction--A Comparison of Experiment and Computation," AIAA-84-0096, 22nd Aerospace Sciences Meeting, January 9-12, 1984.

9. Settles, G.S., Bogdonoff, S.M., and Vas, I.E., "Incipient Separation of a Supersonic Turbulent Boundary Layer at Moderate to High Reynolds Numbers," AIAA Paper 75-7, 1975.

10. Newmann, R.D. and Burke, G., "The Influence of Shock Wave-Boundary Layer Effects on the Design of Hypersonic Aircraft," AFFDL-TR-68-152, USAF Flight Dynamic Laboratory, 1968.

11. Law, H.C., "Three-Dimensional Shock Wave Turbulent Boundary Layer Interactions at Mach 6," ARL TR-75-0191, June 1975.

12. Token, K.H., "Heat Transfer Due to Shock Wave Turbulent Boundary Layer Interactions on High-Speed Weapon Systems," AFFDL-TR-74-77, April 1974.

13. Scuderi, L.F., "Expressions for Predicting 3D Shock Wave-Turbulent Boundary Layer Interaction Pressures and Heating Rates," AIAA Paper 78-162, January 1978.

14. Holden, M.S. "Experimental Studies of Quasi-Two-Dimensional and Three-Dimensional Viscous Interaction Regions Induced by Skewed-Shock and Swept-Shock Boundary Layer Interactions" Calspan Report No. 7018-A-2 Report Covering Period 15 January 1982 - 31 July 1984.

15. Hung, C.M. and MacCormack, R.W., "Numerical Solutions of Supersonic and Hypersonic Laminar Flows Over a Two-Dimensional Compression Corner," AIAA Paper 75-2, January 1975.

16. McCabe, A., "The Three-Dimensional Interaction of a Shock Wave with a Turbulent Boundary Layer," Aeronautical Quarterly, Vol. XVII, August 1966, pp. 231-252.

17. Reshotko, E., and Tucker, M., "Effect of a Discontinuity on Turbulent Boundary Layer Thickness Parameters with Application to Shock Induced Separation," NACA TN3454, 1955.

18. Ericsson, L.E., Reding, J.P., and Guenther, R.A., "Effects of Shock-Induced Separation," Lockheed Missiles and Space Co., Sunnyvale, CA, L-87-69-1, July 1969.

19. Settles, G.S., and Perkins, J.J., "Upstream Influence Scaling of 2D & 3D Shock/Turbulent Boundary Layer Interactions at Compression corners," AIAA-81-0334, 19th Aerospace Sciences Meeting, January 12-15, 1981.

20. Settles, G.S. and Teng, H.Y., "Flow Visualization of Separated 3D Shock Wave/Turbulent Boundary Layer Interactions," AIAA-82-0229, 20th Aerospace Sciences Meeting, January 11-14, 1982.

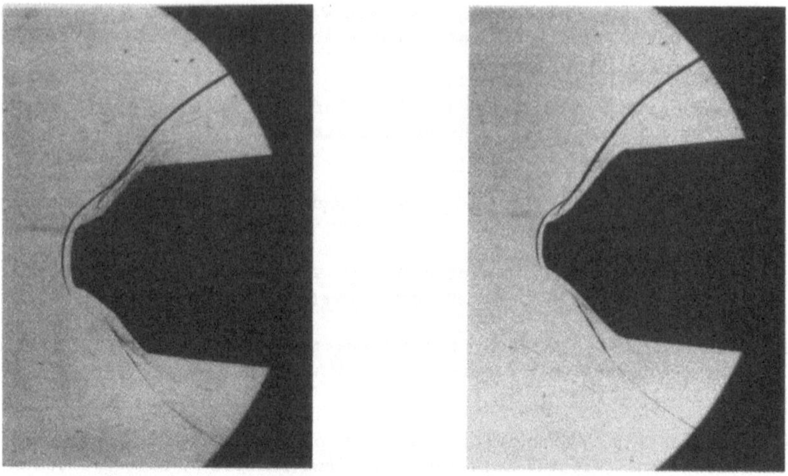

NRV NOSETIP RTE NOSETIP

Fig. 1 Schlieren photographs of the flow over the NRV and RTE nosetips exhibiting regions of body shock – boundary layer interaction.

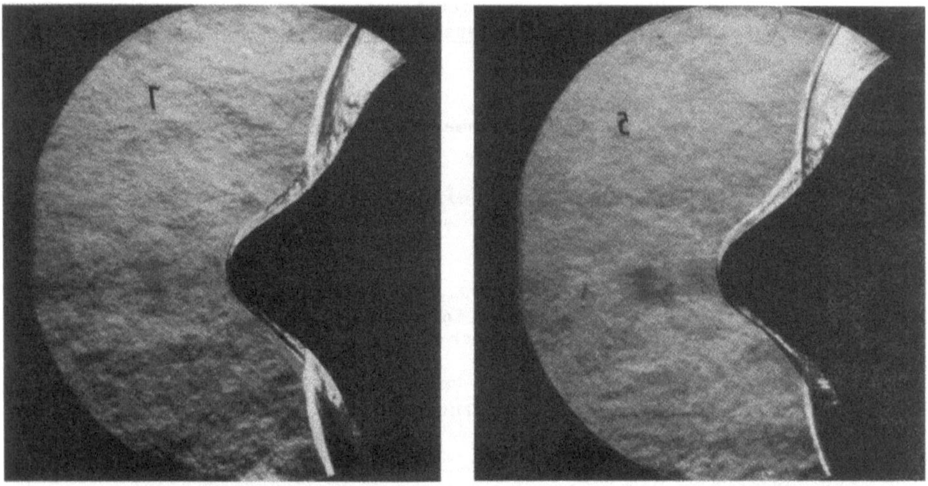

(a) MACH 11 (b) MACH 13

Fig. 2 Effects of Mach number on flow pattern over the indented nosetips (K = 15 MILS).

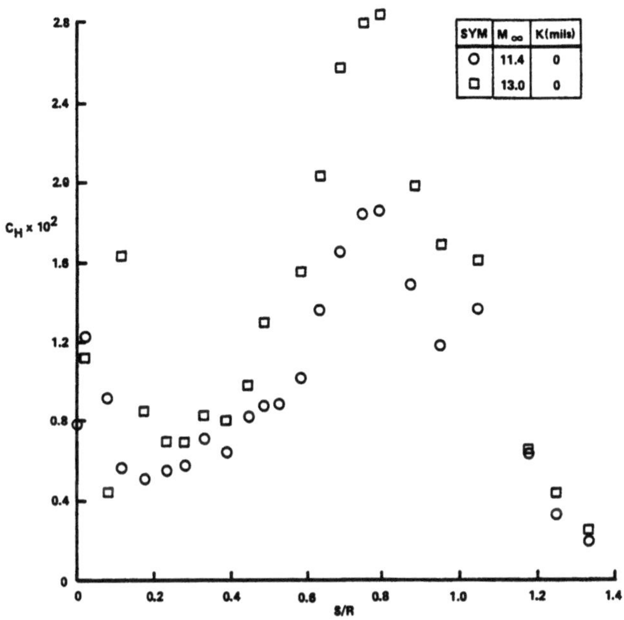

Fig. 3 Influence of mach number on the distribution of heating to the smooth model.

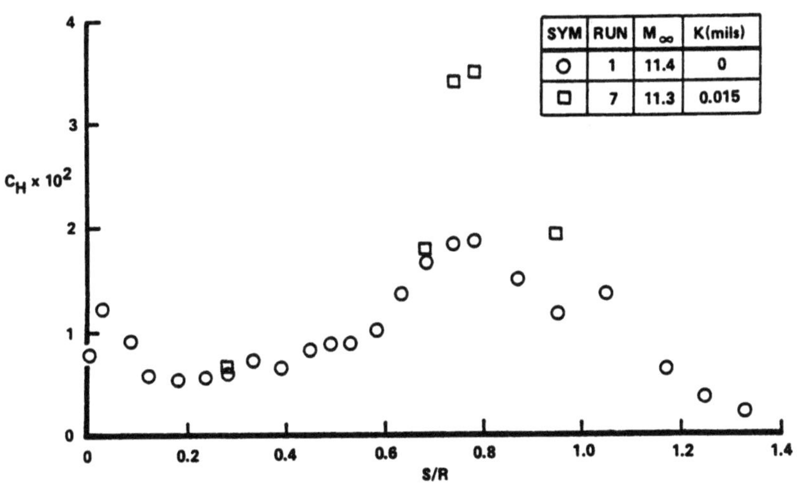

Fig. 4 Influence of surface roughness on the heating distribution. ($M = 11.4$ $Re_D = 12 \times 10^6$)

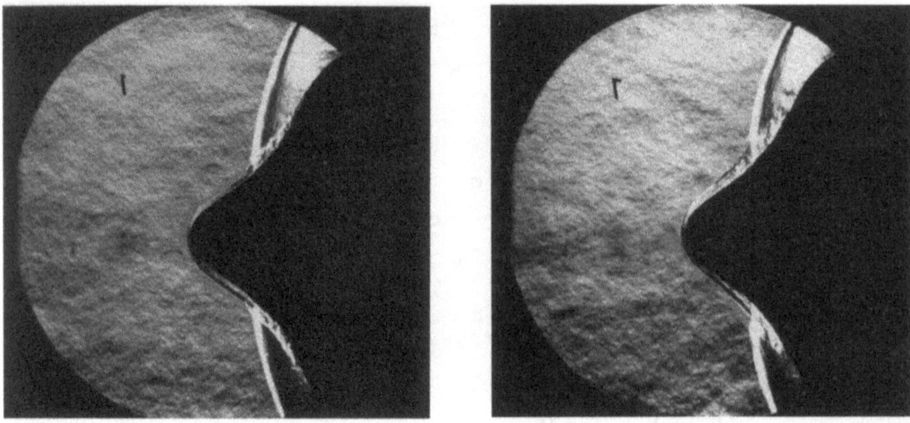

(a) K = 0 (b) K = 15 MILS

Fig. 5 Influence of surface roughness on the flow field on indented nosetips.

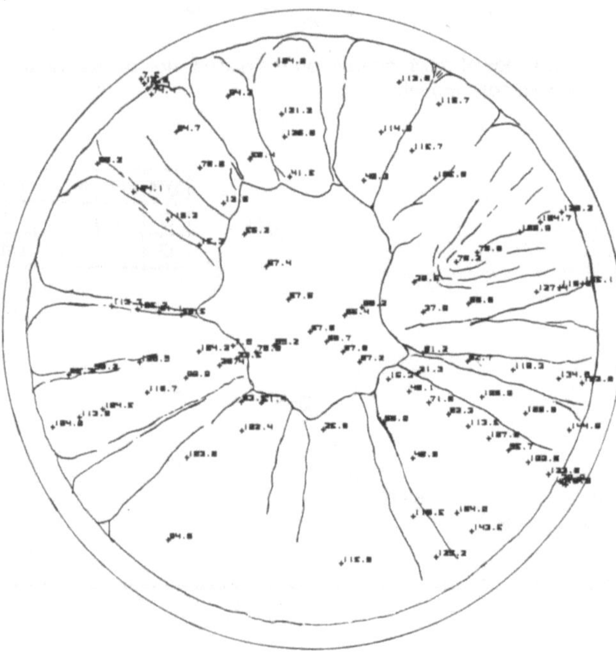

Fig. 6 Heat transfer distribution over the NRV configuration. M = 11 Re/Ft = 10×10^6

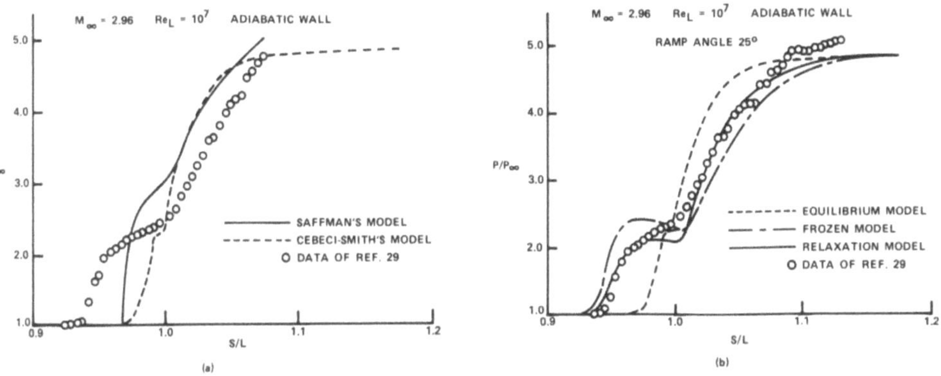

Fig. 7 Comparison of different turbulent models with experiment. Ref. 7.

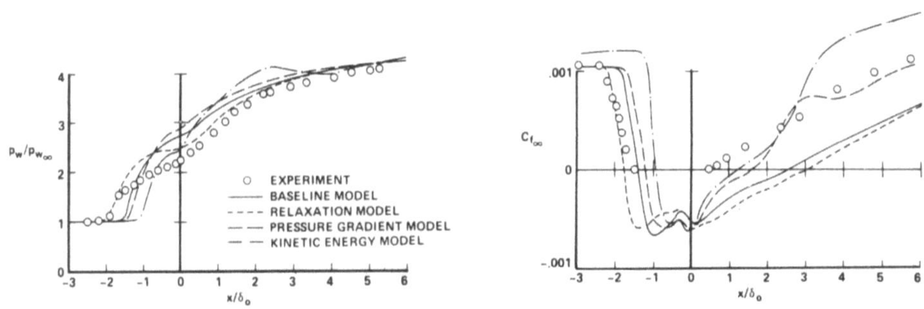

Figure 8 Comparison of computations and surface measurements; $\alpha = 24^o$, Ref. 8.

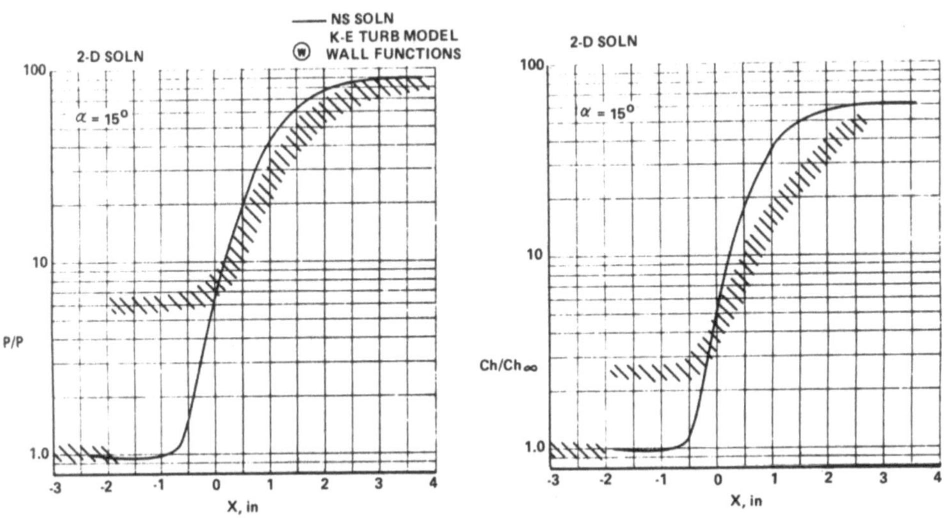

Fig. 9 Comparison between Navier-Stokes solution and experiment for a well separated flow.

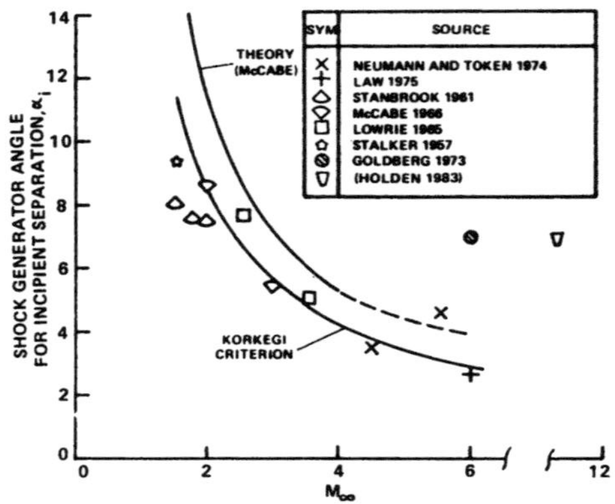

Fig. 10 Variation of shock generator angle to induce incipient separation with mach number.

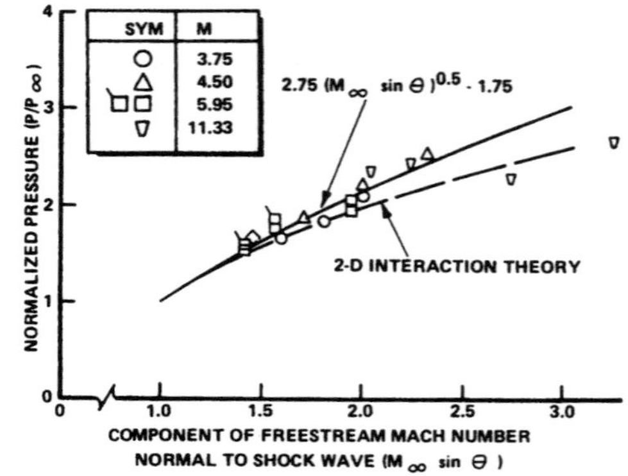

Fig. 11 Correlation of plateau pressure measurement from swept-shock interaction studies.

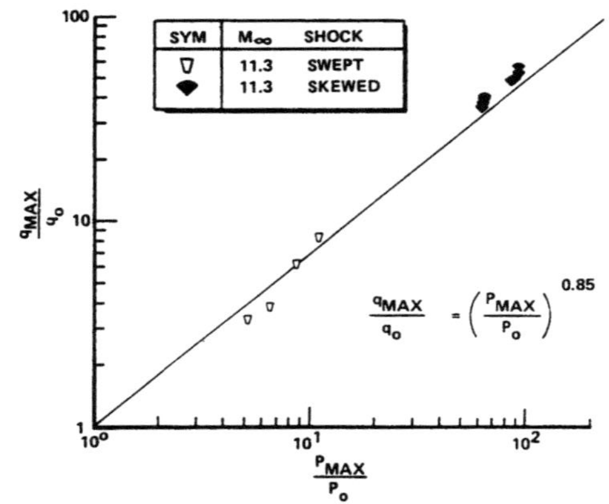

Fig. 12 Correlation of peak heating rates in skewed- and swept-shock interaction regions.

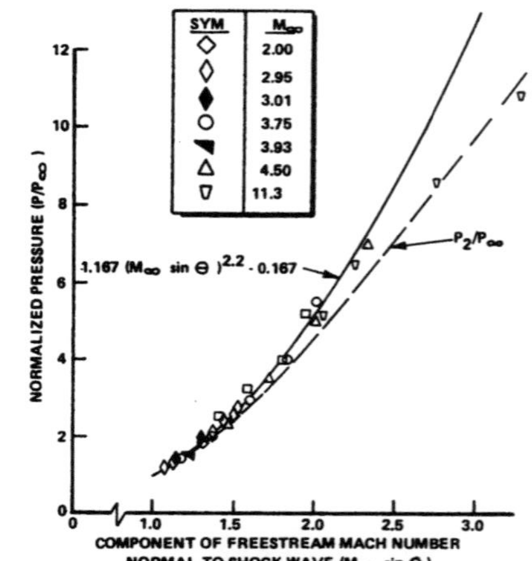

Fig. 13 Correlation of maximum pressures recorded in swept-shock interaction regions.

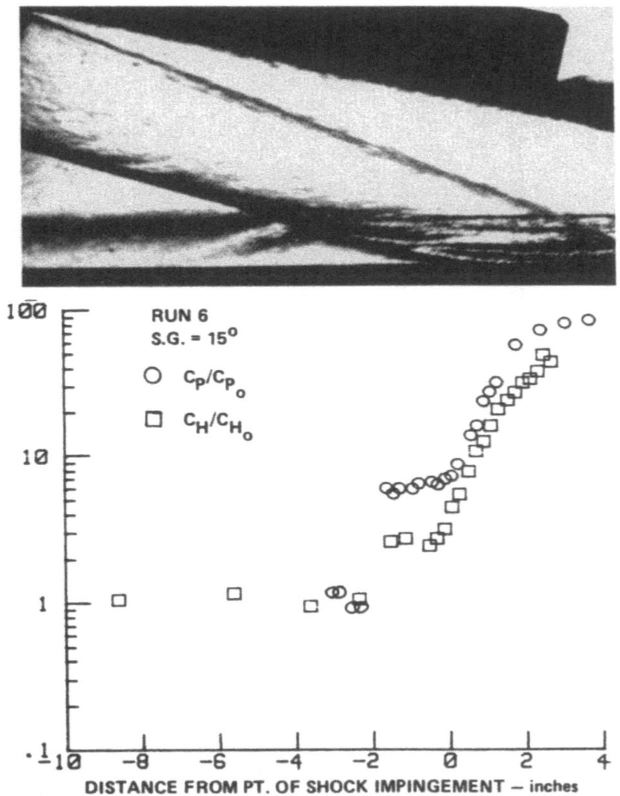

Fig. 14 Streamwise distributions of heat transfer and pressure through skewed-oblique-shock boundary layer interaction (θ = 15° ψ = 0°)

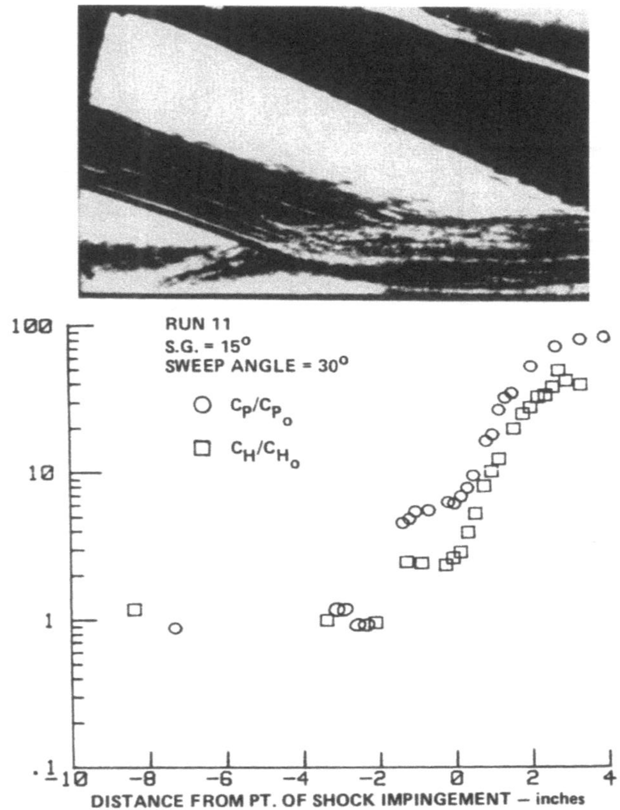

Fig. 15 Streamwise distribution of heat transfer and pressure through skewed-oblique-shock interaction (θ = 15° ψ = 30°)

SYM	GEN/WED	SOURCE	Re_x
○	16°	SETTLES, PERKINS AND BOGDONOFF	18.7×10^6
□	16°	↓ (M = 3)	10.7×10^6
◇	12.5°	HOLDEN STUDY	50×10^6
◺	15°	↓ (M = 11)	50×10^6

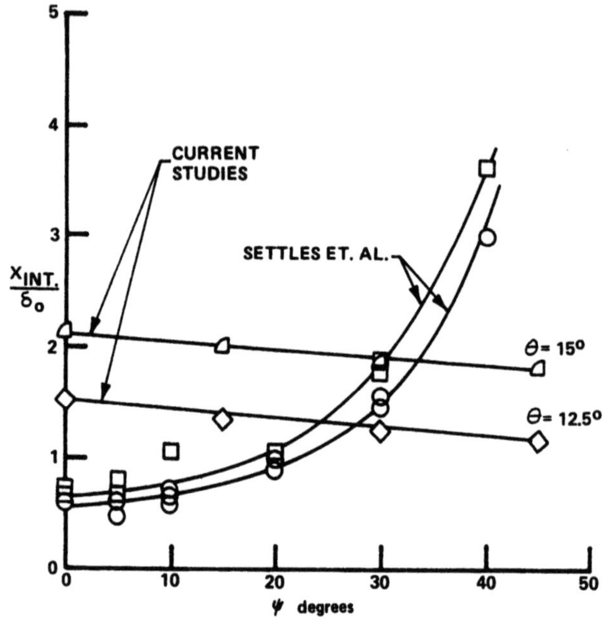

Fig. 16 Variation of streamwise extent of interaction ahead of shock impingement (or corner) with sweep angle.

Unsteady Aspects of Interaction Phenomena

Unsteadiness of Shock-Wave Induced Turbulent Boundary-Layer Separation. A Review

D. S. Dolling

Department of Aerospace Engineering and Engineering Mechanics
The University of Texas, Austin, Texas

Summary

This paper presents a brief review of the unsteadiness of shock-induced turbulent boundary layer separation. A large fraction of what is known about the unsteadiness comes from measurements of wall pressure fluctuations and, consequently, these data are the focus of the review. In the final discussion and conclusions, some suggestions are offered for experiments which would help improve our understanding of this phenomenon.

1. Introduction

It has been known since the 1950's that shock-induced turbulent boundary layer separation is an inherently unsteady process [1,2]. In general, distinguishing between steady and unsteady interactive flows is somewhat arbitrary since there are always random rippling motions of the wave structure. Here, unsteady implies large-scale motions of the separation shock wave structure. Although "large-scale" cannot be defined with precision, in this context it means that the motions "have a characteristic time scale many times larger than the typical turbulent large eddy time scale [3]."

In refs. 1 and 2, and many later studies, the unsteadiness was evident from randomly shot sequences of microsecond spark shadow and schlieren photographs, or high-speed cinema records. In general, optical methods readily reveal unsteadiness but can be difficult to interpret (section 2.3). Other than providing approximate estimates of the shock motion length scale, little quantitative information has been obtained from them.

A large fraction of the experimental work in shock wave boundary layer interaction has been mean flow studies. Surface tracer

Turbulent Shear Layer/Shock Wave Interactions
IUTAM Symposium Palaiseau 1985
Editor: J. Délery

techniques are widely used to determine so-called mean separation lines, particularly in three-dimensional (3-D) flows where conditions vary spanwise. Such methods are easy to use and generate sharply defined, repeatable separation lines. When the term separation (or abbreviation 'S') is used in this paper, it is these lines which are being referred to. However, what they represent physically and how they are generated in flows for which there is overwhelming evidence of unsteadiness is not satisfactorily resolved. Wall pressure fluctuation data suggest that they are at, or close to, the downstream boundary of a streamwise region of intermittent backflow, but there is no direct evidence to confirm this.

Much of the quantitative data on the unsteady separation process is from wall pressure fluctuation measurements. The earliest study known to the author is that of Kistler [4] in 1964, who made measurements in forward facing step flows at Mach numbers, M_∞ , of 3 and 4.5. In the late 1960's and early 70's, the work of Robertson [5,6], Speaker and Ailman [7], Coe et al. [8], and Chyu and Hanly [9] at speeds ranging from transonic to supersonic helped establish an extensive data base. Much of this early work was motivated by the need to predict the fluctuating pressure environment around high-speed vehicles. In recent years, there has been a resurgence of interest in unsteady aspects of interactive flows, with an emphasis on the separation process [3,10-17]. These studies have been driven largely by the realisation that physically accurate interpretations of mean flow data and an understanding of the physical mechanisms require detailed knowledge of the flow structure dynamics. Again, wall pressure fluctuation measurements are the majority of the new data. However, studies of the evolution of turbulence quantities through interactions using hot wires (i.e., refs. 18-20) and laser doppler velocimetry [21,22] have also contributed new results in this area.

As a consequence, the focus of the review is on what has been learned from wall pressure fluctuation measurements. Only flows which are naturally unsteady are included. Forced unsteadiness, induced by oscillating boundaries or time varying upstream or downstream boundary conditions, is excluded, as are unstable oscillations of shock patterns such as those in spiked blunt body flows.

2. Shock-Induced Turbulent Boundary Layer Separation

In supersonic flow, details of the separation process are sparse. In contrast, in incompressible flow it is well-established that separation of a 2-D boundary layer in an adverse pressure gradient does not occur at a point [23,24]. Backflow data show separation taking place over a streamwise region bounded by "incipient detachment" and "detachment" points respectively. Over this region, the fraction of the time that the near wall flow moves downstream varies from unity to zero. Further, the spanwise line of detachment does not move as a unit, but ripples spanwise.

In incompressible flow, it has also been shown that the traditional view that the small mean backflow in the separated zone comes from far downstream is not always correct. Flow visualisation shows that the near wall backflow is supplied intermittently by large eddy structures as they pass through the separated flow [23]. The difference in mechanisms is important. Simpson [23] points out that, "the Reynolds shearing stresses in this region must be modelled by relating them to the turbulence structure and not to local mean velocity gradients." The latter are the result of time-averaging the turbulence and are not related to the cause of the turbulence.

In transonic flows, the mechanisms appear to be similar. Délery [21,22] has used 2 color laser velocimetry to measure velocities and Reynolds stresses in attached and separated flows over interchangeable bumps in a 2-D channel at $M_\infty = 1.30$, 1.37 and 1.45. Mean velocity profiles in the separated flow cases compare well with their incompressible counterparts. Further, contours of percent probability of backflow in the separated flow support the view that the outer region of compressible flow supplies some of the near wall flow.

In supersonic flow, the picture is far less clear. Data obtained using hot film gages [25], buried wire shear stress gages [26] and pressure transducers [4-17] as well as optical methods have established without a doubt that the shock structure is unsteady. The wall pressure data suggest a qualitatively similar separation process as that in incompressible flow, but, without detailed backflow measurements, this cannot be confirmed. Bimodal probability distributions for the u' amplitude near separation in a Mach 3, 2-D oblique shock interaction, measured by Modarress and Johnson [27] using laser velocimetry, give some indication of intermittent backflow, but more work is needed. As far as is known, there is presently no detailed information on the origin of the backflow in 2-D separated flows.

Most of the quantitative data is of wall pressure fluctuations, P_w' , caused by the separation shock motion. Before discussing these results, the difficulties involved in making measurements of P_w' are mentioned briefly. A more detailed discussion, applicable specifically to the data of refs. 10-17, is given in ref. 3. In general, measurements of P_w' can be seriously influenced by the effects of spatial integration due to finite transducer size. Transducers which are too large will give rms fluctuation levels, σp_w , less than the 'true' value. In subsonic flows, Schewe [28] has shown that the transducer diameter (in terms of wall variables, $du\tau/\nu_w$) should ideally be less than 20. Assuming that similar scaling parameters are valid in supersonic flows, this requirement calls for transducers which are typically an order of magnitude smaller than the smallest commercially available. Accordingly, published values of σp_w in supersonic boundary layers are probably underestimated. However, since the major contribution to the results discussed below comes from the large-scale, low frequency (section 2.4) shock motion, the data should not be seriously in error.

2.1 Qualitative Features of the Pressure Signal

Kistler was probably the first to observe the intermittent character of the pressure signal upstream of separation. Recent studies, all at Mach 3, using 2-D compression ramps [3,13-15], hemi-cylindrically blunted fins [16] and sharp fins at angle of attack [11,12] all show similar results. Typical time histories for P_w' in a blunt fin flow are shown in Figure 1a. The values of $\bar{P}_w$ calculated from the time histories or measured using conventional wall pressure tappings agree well (Fig. 1b).

The unsteady shock causes a rapid increase in σp_w which, at its maximum, is a large fraction of $\bar{P}_w$ (Fig. 1c) P'_w increases across the shock wave (compare P'_w at times t_1 and t_2 in Fig. 1a), but this contributes relatively little to the overall σp_w in the intermittent region. Similar distributions of σp_w have been measured in interactions induced by circular cylinders [5,6], impinging shocks [7], and axisymmetric flare and steps [8,9] over a wide range of flow conditions. Sample pressure signals are not shown nor is the signal character discussed explicitly in these references, but the similarities in shape and magnitude of σp_w is strong evidence of shock motion and an intermittent wall pressure signal.

These data show clearly that the separation shock structure undergoes a large-scale motion. Further, $\bar{P}_w$ is the result of the superposition of large amplitude shock-induced fluctuations on the undisturbed boundary layer pressure signature. $\bar{P}_w$ increases in the streamwise direction because stations further downstream spend a larger fraction of the time behind the shock wave. The traditional view of $\bar{P}_w$ increasing through a continuous compression fan is clearly incorrect.

Shock oscillation and its effects may be important for accurate turbulence modelling. Theoretical studies by Anyiwo and Bushnell [29] show that it is one of three major mechanisms responsible for turbulence generation in a shocked region of a shear flow. Their analysis indicates that turbulence of intensity ranging from 150% of the shock oscillation intensity (at transonic speed) to about 80% (at supersonic speed and above) can be generated by the oscillating shock. Turbulence measurements by Muck and Smits [18] using normal and inclined hot wires in Mach 3 compression ramp flows (attached, incipient separation and separated cases) suggest that shock oscillation has an effect on the turbulence structure. As anticipated, significant increases in the turbulent normal and shearing stresses occurred in all three flows, but it was only in the separated case, where large-scale oscillations of the separation shock occur, that the dimensionless structure parameter $\overline{u'v'}/\overline{u'^2}$ was significantly reduced through the interaction. The authors tentatively attribute this to the tendency of the oscillating shock to tranfer mean flow energy more to the normal stresses than to the shear stresses.

2.2 Intermittency and Upstream Influence

By defining ϵ , the intermittency, as the fraction of the time that $P_w' > \bar{P}_{wo} + 3\,\sigma p_{wo}$ (where subscript 0 refers to undisturbed flow), the shock motion length scale, Ls , (or intermittent region length) can be determined. Distributions of ϵ for blunt fins and compression ramps are shown in Figure 2. They have the same shape as curves for intermittent backflow in incompressible flow. In all cases, the downstream boundary of Ls is close to, but upstream of, 'S.' Up to this point, there is no evidence that Ls correlates simply in terms of any incoming flow or geometric parameter.

Upstream influence, Lu , is defined conventionally as the distance from where $\bar{P}_w$ first increases above the undisturbed level $\bar{P}_{wo}$ (defined as station X_o) to some reference position downstream (i.e., ramp corner line, fin leading edge, etc.). The pressure signals show that upstream influence varies continuously and that the instantaneous value should be measured from the foot of the shock. Hence, the absolute maximum and minimum values of Lu correspond to the furthest upstream and downstream shock locations (i.e., the variation in Lu is equal to Ls).

The data show that ϵ and σp_w actually increase upstream of X_o . Both parameters are more sensitive than $\bar{P}_w$ to the presence of infrequent shock waves. Increases in higher order moments such as the skewness and flatness coefficients (which are sensitive to a few points in a data record being far from the mean value) are detectable even further upstream. In blunt fin flows where X_o is 2.8D upstream of the leading edge, increases in skewness have been detected as far upstream as 3.4D. High-speed (35kHz) schlieren photography by Degrez [30] confirms the presence of shock waves this far upstream.

The wall pressure signals show clearly the mechanism of upstream influence. It is not the result of upstream propagation of disturbances in a nominally steady flow, but is a direct result of shock-wave oscillation. Upstream influence defined in the usual way is, in this sense, somewhat arbitrary. It depends on the parameter selected for defining it and the sensitivity of the measuring tool. X_o (which is where $\bar{P}_w$ first increases) is where ϵ reaches a value of 0.03 - 0.05 . This is simply the station at which an increment in $\bar{P}_w$ is detectable using a pressure tap and conventional instrumentation.

2.3 Separation Shock Wave Structure

Muck et al. [3] have used a rotatable wall plug in which 4 transducers were installed in-line with spacing, $\xi = 0.23\delta_o$. Simultaneously sampled wall pressure signals upstream of 'S' in a Mach 3 compression ramp flow, with the transducers aligned longitudinally, are shown in Figure 3. When the shock is upstream of a given transducer (i.e., the second one at time t_1), the pressure indicated by it and the downstream transducers are essentially the same, providing convincing evidence that there is a single separation shock wave.

The pressure signals for the plug rotated 90° show that the shock front has spanwise ripples with wavelengths as small (if not smaller) than $0.23\delta_o$ and as large (if not larger) than $0.69\delta_o$. This may be connnected with the development of spanwise structures. Space-time correlations obtained by Muck and Andreopoulos [10] show that, unlike the curves in the upstream boundary layer, which are symmetric with respect to the $\tau = 0$ axis (as expected in 2-D flow), large assymmetries appear in the intermittent region. It is suggested by the authors that this deviation of the flow from its mean longitudinal direction is possible evidence of the cell structure of Taylor-Gortler vortices.

This same plug has also been used by Tran et al. [11] in an interaction generated by a sharp fin at $\alpha = 20°$. The freestream conditions were the same as the compression ramp studies of Muck et al. In this case, with the transducers aligned with the swept upstream influence line, the shock-induced pressure pulses were much more highly correlated with little evidence of any spanwise rippling.

Rippling motions in the 2-D case compound the difficulties of interpreting spark shadow or schlieren photographs. Both techniques integrate across the test section width. Since widths are generally large compared to a typical ripple wavelength, the varying streamwise positions of waves at different spanwise positions result in a photograph with a layered look, giving the impression of a series of waves. The layered look does not change radically from one photograph to another and can give the misleading impression of a nominally steady flow. In certain 3-D flows, such as those generated by blunt fins or circular cylinders, the problem is less severe and motion of the centerline shock can be seen much more clearly [30].

2.4 Separation Shock Motion Frequency

None of the flow visualisation studies has given any indication that the shock wave motion is periodic or has any preferred frequencies. Power spectra of the pressure fluctuations all have similar features (Fig. 4). They are broadband with a large fraction of the energy at low frequencies (i.e., time scales of order 1/10 of that of a typical large eddy). With increased ϵ , the overall power level progressively increases, reaching a maximum at $(\sigma p_w)_{max}$ where $\epsilon \simeq 0.7$. Generally applicable parameters for normalising the power and frequency axes and correlating data from different flows are not known. Robertson [5,6] was able to correlate spectra in blunt fin flows using U_∞/q_∞^2 Lsep and Lsep/U_∞ to normalise the power and frequency axes respectively. q_∞ and Lsep are the freestream dynamic pressure and separated flow length, respectively. Dolling and Or [13] tried the same parameters in compression ramp flows, but, because of the lack of data in cases with extensive separation, the attempt was somewhat inconclusive.

The characteristic frequency of the shock motion has been determined from measurements of the coherence function, $\gamma^2(\xi_1,\xi_2,f)$ and other techniques. Tran et al. [11] have measured γ^2

in the sharp fin flow at $\alpha = 16°$ and $20°$. The transducers were along the line of upstream influence. The results (Fig. 5) show that the signals are highly correlated at low frequencies. With increased transducer spacing, the coherence at high frequencies decreases rapidly since the high frequency motions have small spatial extents. In contrast, the coherence at low frequencies, which is associated with the shock motion, decreases slowly. For the two larger spacings, the maximum value of γ^2 occurs at about 2kHz. This frequency, which can be regarded as that characteristic of the shock motion at the interaction start, agrees well with estimates of shock frequency made by Dolling and Murphy [14] in compression ramp interactions under similar flow conditions. The frequency with which the shock moved upstream over the transducer was estimated from a large number of data records by counting the shock-induced pressure pulses. Results are shown in Figure 6. For the same δ_o as ref. 11, a maximum frequency in the range 1.8-1.9 kHz was obtained.

Muck and Andreopolous [10] have used conditional sampling techniques to determine probability density distributions for the separation shock wave frequency and velocity in a Mach 3 separated compression ramp flow. The wall pressure signal was simulated as an on-off function after conditioning on its level and time derivative. Based on this, the duration of a cycle, T_i (Fig. 7) and the streamwise velocity of the shock can be defined. The probability distribution for the periods and frequencies ($F_i = 1/T_i$) are highly skewed (Fig. 8). The results support the idea that "the shock wave motion may be due to the turbulence of the oncoming boundary layer rather than due to the shear layer formed over the separated region and to the recirculating region" [11]. This stems from the observation that the non-dimensional zero crossing frequency, $f_c\delta/U_e$, $(= \delta/U_e\bar{T})$, where $\bar{T}$ is the mean period, has a value of about 0.1 which is the same order of magnitude as the bursting frequency of the boundary layer. The probability distribution for the shock velocity was based on a rather low number of samples and is, therefore, only approximate. It indicates shock velocities ranging from $0.05\,U_e$ to $0.8U_e$ with the most probable value around $0.15U_e$.

2.5 Space-Time Correlations

Longitudinal space-time correlations, $R_{pp}(\xi,0,\tau)$ measured by Muck et al. [3] in the 24° separated compression ramp flow, show clearly the large-scale, in-phase motions resulting from the oscillating shock (Fig. 9). For small τ, R_{pp} in the intermittent region is lower than in the upstream boundary layer, while for large τ the opposite is true. Near the mean shock position at $X/\delta = -2.18$, R_{pp} for $|\tau/U_\infty\delta| \geq 1$ ($|\tau| \geq 4\times10^{-5}$s) is larger than in the upstream boundary layer and approaches zero for $\tau U_\infty/\delta = 15$, a value 3 times larger than in the upstream flow.

Tran et al. [11] have measured $R_{pp}(\xi,0,\tau)$ in the intermittent region of the sharp fin interaction at $\alpha = 20°$. The leading transducer (#1) was on the upstream influence line. For all transducer spacings (1-2, 1-3 and 1-4 combinations), the maximum value of R_{pp} occurred at negative τ indicating that the downstream transducer registers a pressure pulse before the upstream one. The results can be attributed only to some unknown motion of the shock wave since there is no evidence of any upstream convection by the mean flow.

3. Final Discussion and Suggestions

Results from 2-D and 3-D studies, at transonic and supersonic speed, show that large-scale motion of the separation shock wave is a characteristic feature of shock-induced turbulent flow. Up to this point, the data indicate the presence of a single shock rather than a distributed structure. Spanwise rippling, which may be connected with the development of spanwise structures, is seen in 2-D flows, whereas in 3-D flows it is either absent or less pronounced.

The shock motion generates an intermittent wall pressure signal accompanied by a rapid increase in rms fluctuation level. It oscillates over a streamwise region whose downstream boundary is close to the separation line (indicated by surface tracers). The characteristic frequency is low, with a time scale an order of magnitude lower than that of a typical large turbulent eddy. Conditional sampling experiments show highly skewed probability distributions for the period and frequency and suggest a link with the bursting frequency of the boundary layer. More work is needed to clarify this issue.

To understand the separation process better, flow direction measurements close to the wall are needed. Instantaneous backflow measurements are needed to show unambiguously the similarities and differences with the incompressible flow results. Simultaneous backflow measurements, in conjunction with wall pressure fluctuations, would show if separation occurs across a single shock, and how mean separation lines from surface tracer techniques are generated and what they represent. The data should also indicate the physical mechanisms responsible for the observed distribution of surface properties such as the skin friction and heat transfer coefficient.

Little is known about the separated flow structure itself. Even for the 2-D compression ramp, it is not known if the bubble structure ever resembles that of the mean flow model or if the latter is simply the time-average of different randomly varying repetitive structures. For evidently non-periodic phenomena where repetitive events or structures are masked by high turbulence levels or time-averaging, conditional sampling techniques are a powerful means of detecting and quantifying the structure. Space-time correlations, coherence and time-delay measurements, using strategically placed pairs or groups of sensors, would also provide critical data on physical mechanisms.

References

1. Bogdonoff, S. M., "Some Experimental Studies of the Separation of Supersonic Turbulent Boundary Layers," Aeronautical Engrg. Dept., Princeton Univ., Report 336, June 1955.

2. Chapman, D. R., Kuehn, D. M., and Larson, H. K., "Investigation of Separated Flows in Supersonic and Subsonic Streams with Emphasis on the Effect of Transition," NACA TN 3869, March 1957.

3. Muck, K. C., Dussauge, J. P. and Bogdonoff, S. M., "Structure of the Wall Pressure Fluctuations in a Shock-Induced Separated Turbulent Flow," AIAA Paper 85-0179, Jan. 1985.

4. Kistler, A. L., "Fluctuating Wall Pressure Under a Separated Supersonic Flow," J. of the Acoustical Soc. of America, 36, March 1964, pp. 543-550.

5. Robertson, J. E., "Characteristics of the Static and Fluctuating-Presure Environments Induced by Three Dimensional Protuberances at Transonic Mach Numbers," Wyle Laboratories Research Staff Report WR-69-3, June 1969.

6. Robertson, J. E., "Prediction of In-Flight Fluctuating Pressure Environments Including Protuberance Induced Flow," Wyle Laboratories Research Staff Report WR71-10, March 1971.

7. Speaker, W. V. and Ailman, C. M., "Spectra and Space-Time Correlations of the Fluctuating Pressures at a Wall Beneath a Supersonic Turbulent Boundary Layer Perturbed by Steps and Shock Waves," NASA CR-486, June 1969.

8. Coe, C.F., Chyu, W.J. and Dods, J.B., "Pressure Fluctuations Underlying Attached and Separated Supersonic Turbulent Boundary Layers and Shock Waves," AIAA Paper 73-996, Oct. 1973.

9. Chyu, W. J. and Hanly, R. D., "Power and Cross Spectra and Space-Time Correlations of Surface Fluctuating Pressures at Mach Numbers Between 1.6 and 2.5," NASA TN-D-5440, Sept. 1969.

10. Muck, K. C. and Andreopoulos,, J., Private Communication, July 1985.

11. Tran, T. T., Tan, D. K. M. and Bogdonoff, S. M., "Surface Pressure Fluctuations in a Three Dimensional Shock Wave/-Turbulent Boundary Layer Interaction at Various Shock Strengths," AIAA Paper 85-1562, July 1985.

12. Tan, D. K. M., Tran, T. T. and Bogdonoff, S. M., "Surface Pressure Fluctuations in a Three Dimensional Shock Wave Turbulent Boundary Layer Interaction," AIAA Paper 85-0125, Jan. 1985.

13. Dolling, D. S. and Or, C. T., "Unsteadiness of the Shock Wave Structure in Attached and Separated Compression Ramp Flowfields," Experiments in Fluids, 3, 1985, pp. 24-32.

14. Dolling, D. S. and Murphy, M. T., "Unsteadiness of the Separation Shock Wave Structure in a Supersonic Compression Ramp Flowfield," AIAA J., 21, Dec. 1983, pp. 1628-1634.

15. Murphy, M., "An Experimental Investigation of the Separation Shock Wave Unsteadiness in a Compression Ramp Flowfield," M.S.E. Thesis 1605-T, Mechanical and Aerospace Engineering Dept., Princeton University, May 1983.

16. Dolling, D. S. and Bogdonoff, S. M., "An Experimental Investigation of the Unsteady Behavior of Blunt Fin-Induced Shock Wave Turbulent Boundary Layer Interactions," AIAA Paper 81-1287, June 1981.

17. Dolling, D. S., "On Upstream Influence in Shock Wave Turbulent Boundary Layer Interaction," Royal Aero. Soc. J., Oct. 1983, pp. 324-327.

18. Muck, K. C. and Smits, A. J., "Behavior of a Turbulent Boundary Layer Subjected to a Shock-Induced Separation," AIAA Paper 84-0097, Jan. 1984.

19. Lee, D. -H., "Study of the Evolution of Turbulence in a Shock Wave Boundary Layer Interaction," Thesis (Docteur Ingenieur), Univ. of Poitiers, June 1979.

20. Muck, K. C. and Smits, A. J., "The Behavior of a Compressible Turbulent Boundary Layer Under Incipient Separation Conditions," Turbulent Shear Flows 4, Sept. 1983 (published by Springer-Verlag).

21. Délery, J. M., "Experimental Investigation of Turbulence Properties in Transonic Shock/Boundary Layer Interactions," AIAA J., 21, 2, Feb. 1983, pp. 180-185.

22. Délery, J. M., "Investigation of Strong Shock-Turbulent Boundary Layer Interaction in 2-D Transonic Flows with Emphasis on Turbulence Phenomena," AIAA Paper 81-1245, June 1981.

23. Simpson, R. L., "Some Features of Two Dimensional Turbulent Separated Flows," AIAA Paper 85-0178, 23rd. Aerospace Sciences Meeting, Jan. 14-17, 1985.

24. Sandborn, V. A. and Kline, S. J., "Flow Models in Boundary Layer Stall Inception," J. of Basic Eng., Trans. ASME 83, 1961, pp. 317-327.

25. Horstman, C. C. and Owen, F. K., "New Diagnostic Technique for the Study of Turbulent Boundary Layer Separation," AIAA J., 12, Oct. 1974, pp. 1436-1438.

26. Murthy, V. S. and Rose, W. C., "Wall Shear Stress Measurements in a Shock Wave Boundary Layer Interaction," AIAA J., 16, 7, July 1978, pp. 667-672.

27. Modarress, D. and Johnson, D. A., "Investigation of Shock-Induced Separation of a Turbulent Boundary Layer Using Laser Velocimetry," AIAA Paper 76-374, July 1978.

28. Schewe, G., "On the Structure and Resolution of Wall Pressure Fluctuations Associated with Turbulent Boundary Layer Flow," J. Fluid Mech., 134, 1984, pp. 311-328.

29. Anyiwo, J. C. and Bushnell, D. M., "Turbulence Amplification in Shock Wave Boundary Layer Interaction," AIAA J., 20, 1982, pp. 893-899.

30. Degrez, G., "Exploratory Experimental Investigation of the Unsteady Aspects of Blunt Fin-Induced Shock Wave Turbulent Boundary Layer Interactions," Mech. and Aero. Eng. Dept., MS Thesis 1516-T, Princeton Univ., June 1981.

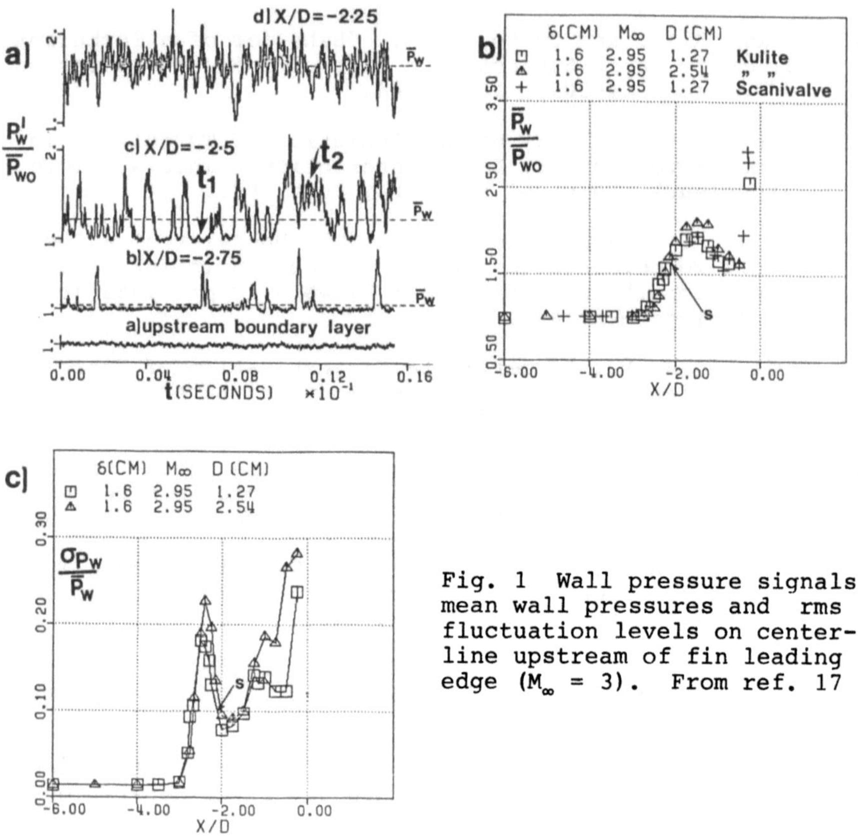

Fig. 1 Wall pressure signals, mean wall pressures and rms fluctuation levels on centerline upstream of fin leading edge ($M_\infty = 3$). From ref. 17

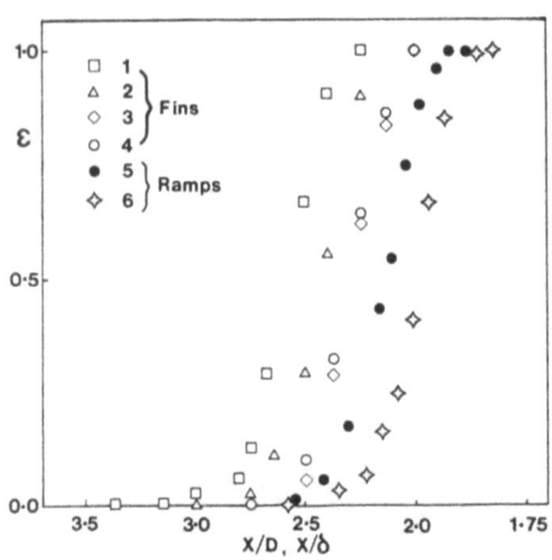

Fig. 2 Intermittency distributions for blunt fin and compression ramp flows. (X measured from fin leading edge and ramp corner line, $M_\infty = 3$). From ref. 17

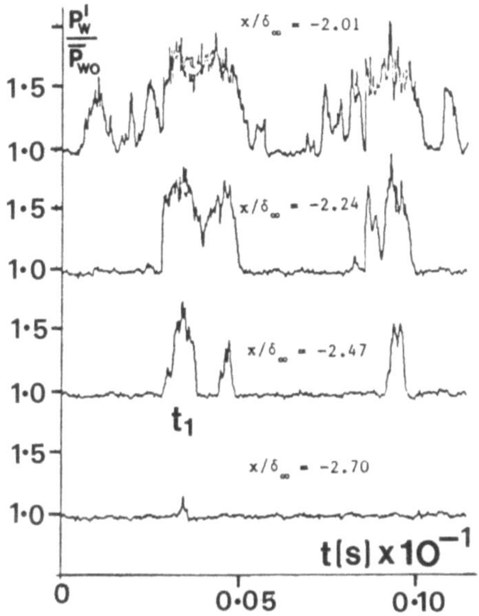

Fig. 3 Simultaneously sampled wall pressure signals with transducer aligned longitudinally in intermittent region. From ref. 3

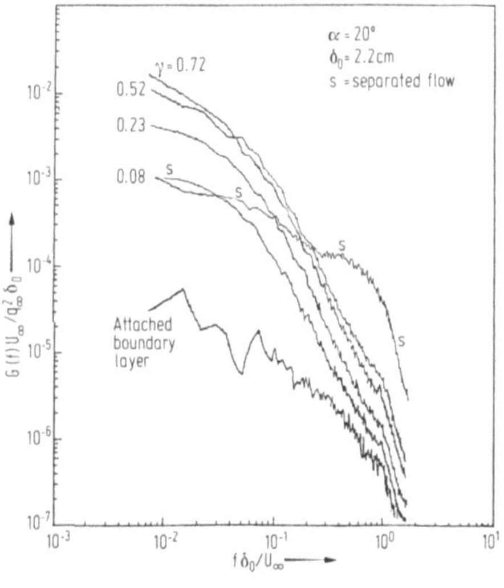

Fig. 4 Power spectra in the intermittent region (20° compression ramp at M_∞ = 3). From ref. 17

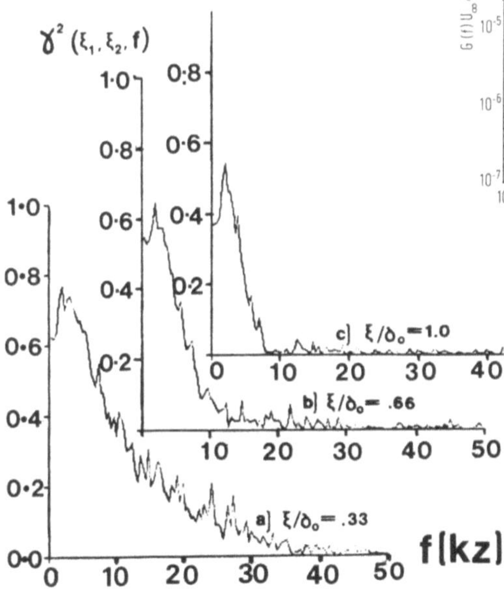

Fig. 5 Coherence function in 20° sharp fin flow (transducer aligned along upstream influence line, M_∞ = 3). From ref. 11

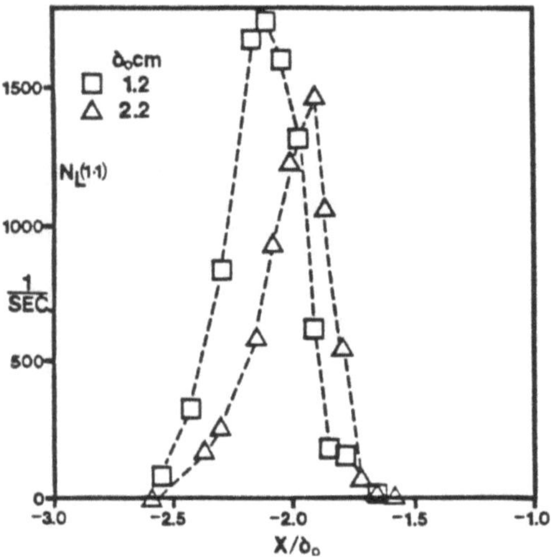

Fig. 6 Shock motion frequency in intermittent region (24° compression ramp, M_∞ = 3). From ref. 14

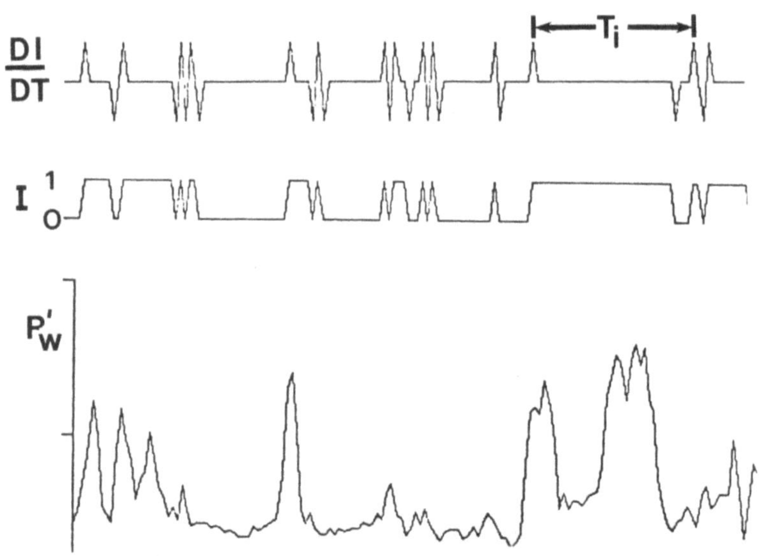

Fig. 7 Conditional sampling analysis of pressure signal and definition of one cycle by shock wave movement. From ref. 10

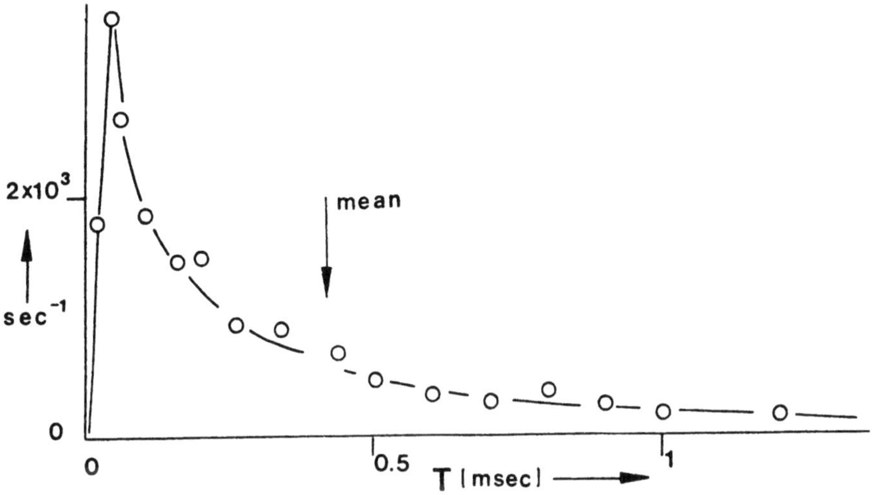

Fig. 8 Probability density distribution of the shock movement periods. From ref. 10

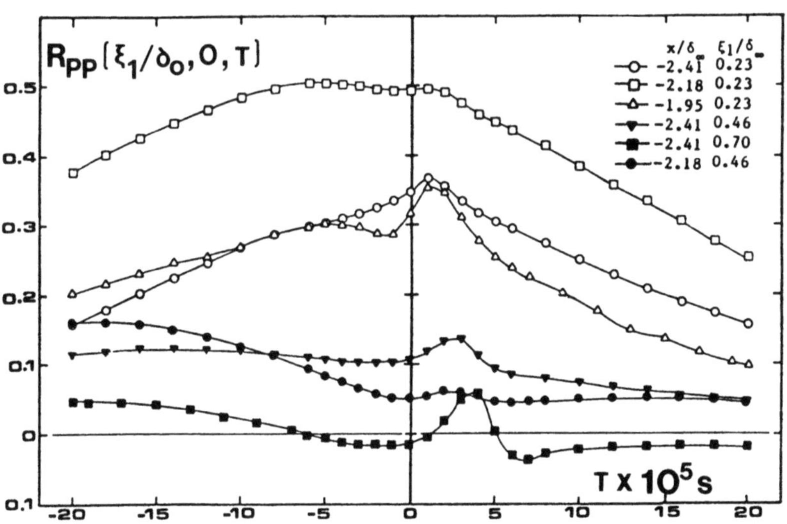

Fig. 9 Longitudinal space-time correlation in the intermittent region (24° compression ramp, $M_\infty = 3$). From ref. 3

Reduction of Low Frequency Buffet from a Bluff Body at Transonic Speeds

D. G. MABEY and P. LEE

Royal Aircraft Establishment,
Dynamics Laboratory,
Bedford,
UK

Summary

Large pressure fluctuations can be generated by shock induced separations on bluff bodies (such as external stores) at transonic speeds.

Recently an experimental aircraft with a bluff fairing experienced violent low-frequency excitation due to shock induced separation at transonic speeds. This low-frequency excitation was reproduced on models of the fairing in transonic wind tunnels.

The low frequency excitation on the model was then reduced by the installation of a few 'buffet breather' tubes. Besides stabilising the position of the shock induced separation, the mean pressure within the region of shock induced separation was altered.

Introduction

Recently, severe buffeting at transonic speeds was encountered on an experimental aircraft fitted with a bluff fairing close to the nose. The test pilot reported that the buffeting started suddenly as the Mach number exceeded M = 0.81, and quickly increased in severity; he was unable to judge whether the motion was primarily in a horizontal or a vertical plane but he estimated the frequency to be about 3 Hz. This corresponded with neither an aircraft stability mode nor a structural mode. Hence the buffeting could be explained only as the localised forced response to aerodynamic excitation (or buffet[1]). For M = 0.85 a frequency of 3 Hz corresponded with a frequency parameter based on the fairing width, w, and freestream speed, U, of about $fw/U = 0.006$, *ie* at least an order of magnitude lower than that experienced on a bluff fairing. (For a circular cylinder fw/u is about 0.2 at low speeds).

Turbulent Shear Layer/Shock Wave Interactions
IUTAM Symposium Palaiseau 1985
Editor: J. Délery

The low frequency parameter suggested a special form of aerodynamic excitation, provided by alternate shock oscillations between the port and starboard sides of the fairing. Such shock oscillations had been observed previously over a narrower Mach number range ($0.82 < M < 0.86$) on a 14% thick biconvex aerofoil[2], but at a higher frequency parameter ($fw/U = 0.021$). In addition, shock oscillationns have been observed on other aerofoils at a frequency parameter down to $fw/U = 0.008$ (Ref 3, Fig 11).

It was important both to confirm the hypothesis and then to devise a simple remedy to prevent the shock oscillation. [The external shape of the fairing could not be altered, because this was determined by the aircraft instrumentation pack.] With these objectives, wind tunnel measurements of unsteady pressures were made on simplified models of the fairing. These tests quickly confirmed the existence of low-frequency shock oscillations.

These shock oscillations were reduced by a new concept called 'buffet breathers', consisting of a series of tubes between the two sides of the fairing. Some details of the main experiment and the principal results are given. This concept, now patented[4], may not be restricted to transonic speeds and may have other applications which are described in the patent.

Nature of the Oscillatory Flow

The nature of the oscillatory flow was established by some quick tests made on a small solid steel model of the fairing mounted on the glass sidewall of the 2.5in × 2.5in Transonic Tunnel at the Cranfield Institute of Technology. These tests will be reported fully elsewhere[5]. The tests were made with fixed transition and at atmospheric total pressure and confirmed the existence of low-frequency excitation (a frequency of about 120 Hz on the 1/50 scale model corresponding with a frequency of about 2.4 Hz at full scale). Spark schlieren photographs showed that in the transonic speed range the flow separated from the shoulders of the fairing and formed a thick wake, which alternated in thickness from side to side of the fairing. The terminal shock system controlling these separations also alternated in strength, streamwise position and spanwise extent, from side to side of the fairing. Typical spark schlieren photographs show differing separation and shock configurations between the two sides of the fairing at M = 0.80 and 0.90.

The tests suggested also that 'buffet breathers' should be an effective means of attenuation. In an associated test, a narrow breather (a slot of 1% chord at 0.6 chord) on a 14% thick biconvex aerofoil (with chord = 32 mm) eliminated the very severe periodic shock oscillation, which involves large streamwise movements[2]. This test was of general interest because it suggested that the 'buffet breather' concept would be effective at frequency parameters as least as high as fw/U = 0.021.

The Buffet Breather

Fig 1 illustrates the principle of a 'buffet breather' installed in a bluff fairing. Suppose aerodynamic excitation is being provided at opposite points A and B by time-dependent pressure variations with a significant phase difference across the fairing, but with the same man pressure (*ie* when averaged over a long period). If points A and B are then connected by a short 'breather' tube (which need not be straight), the time-mean pressure may not be altered but the time-dependent pressures will be attenuated (by equalisation) due to the flow through the breather produced by the phase difference. The breather flow will be large when these time-dependent pressures are 180° out of phase (as they are probably on this bluff fairing at transonic speeds and as they are certainly on a circular shedding Karman vortices at subcritical Reynolds numbers).

With a large breather flow the local buffet excitation will be reduced. The breather should operate over a range of frequencies from quasi-static up to a somewhat ill-defined upper limit set by the acoustic response characteristics of the breather. For a given excitation frequency the limiting length and diameter of the breather can be determined by the method of Bergh and Tijdeman[6]. For a frequency of 3 Hz on the aircraft, a breather diameter of 50 mm and length (w) = 550 mm would provide only a few degrees of phase lag.

The precise location and spacing of the breather tubes is probably not important, although for maximum attenuation some careful research might be essential. With the large-scale shock-induced separations produced at transonic speeds, or with vortex shedding from circular cylinders, pressure fluctuations are correlated over quite large spanwise distances. Hence the number of breather tubes needed should not be excessive. For the present fairing, a spacing of about 0.25 w across the sides was somewhat arbitrarily selected. For both models four streamwise rows of holes were provided but the most effective row is that closest to the shock-induced separation at the shoulder. Installation of breathers on the aircraft was possible because there was little instrumentation inside the middle of the fairing.

The buffet breather is quite different in principle from a Helmholtz resonator. This has to be tuned to attenuate a single frequency (and would become excessively large at a frequency as low as 3 Hz). Helmholtz resonators are often useful to attenuate duct noise in ground installations. However they are inherently unsuited for buffet attenuation in flight because the spectrum of aerodynamic excitation provided by separated flows contains a wide range of frequencies and varies with speed. In addition, it has been suggested that on a fairing with an oscillatory flow, a pair of Helmholtz resonators might alter the local pressure fluctuations, without necessarily attenuating the integrated pressure fluctuations which excite buffeting[7].

The buffet breather is also different from the acoustic buzz suppressor patented by French and Minton in 1962. This vents two holes into a large plenum chamber, and thus is likely to be effective only at extremely low frequencies.
Bendock et al[8] state that although the device was effective in model tests, it was never used at full scale because of the large volume required for the plenum chamber.

Experimental Details

Two sets of unsteady pressure measurements were made on very rigid models. Model rigidity was an essential feature of the wind tunnel tests, to confirm the hypothesis that in flight the buffeting was a forced response. The initial series of tests were described in section 2.

More extensive tests were made at a total pressure of 0.64 bar on a 1/8.4 scale model of the fairing in the 2ft × 1½ft Transonic Tunnel at RAE Farnborough. Fig 2 shows the main aerodynamic features represented: the inclined wedge at the front of the fairing and the small radius of curvature at the shoulder. Downstream of the shoulder the sides of the fairing are parallel for a short distance, then the fairing tapers. The large radii at the bottom of the fairing on the aircraft are not represented on the model. The fairing was mounted on the closed roof of the tunnel, the boundary layer on the roof thus simulating the boundary layer on the curved-lower surface of the fuselage.

For simplicity and speed, only one pressure transducer was used for these tests. It was mounted in a small metal housing, which could be moved to different positions on the fairing in grooves machined in the sides of the model. The grooves were filled, and the transducer mounting secured with plasticene. The pressure transducer was a Kulite XCQL 093, 25 psi absolute type, used in conjunction with a special amplifier[9] which allowed the simultaneous measurement of steady and unsteady pressures. The steady signal was measured on a DC voltmeter and the unsteady signal was measured on a Bruel and Kjaer spectrum analyser, Type 2031. The bandwidth of the total rms pressure measurements, $\bar{p}$, extended from 0 to 20 kHz.

The Reynolds number based on the width of the model fairing was 0.6×10^6 at $M = 0.85$; in flight at sea level the corresponding Reynolds number on the aircraft was about 11×10^6. This large difference in Reynolds number posed the problem of achieving the correct thickness of the turbulent boundary layer at the shock in the model tests. It was considered that because the forward wedge of the fairing was immersed in the turbulent boundary layer on the tunnel roof, the sweep back angle of the wedge would ensure spanwise contamination of the boundary layer, and hence rapid natural transition. This assumption is consistent with the oil flow photographs, which show no evidence of laminar or transitional flow. If roughness had been applied to fix transition close to the leading-edge of the wedge, this would thus have produced an unrealistically thick turbulent boundary layer at the shoulder. In fact an excessively thick boundary layer was produced during the final test, when a piece of carborundum paper was glued round the leading-edge and found to make only little difference.

Results

Pressure Measurements

Fig 3 shows results of tests of a buffet breather on the 1/8.4 scale model. Three configurations are (Fig 3a):

No breathers,
2 breathers open, and
4 breathers open.

The severe curvature at the shoulder of the fairing produces local mean pressures much lower than the sonic condition (p/p_t = 0.528) at quite low speeds (Fig 3b). At a free stream Mach number M = 0.81, the local pressure stops falling, and suddenly increases as Mach number increases. This increase in pressure corresponds to the onset of severe local flow separation at the measurement point (due to an upstream movement of a separation and a shock) and to buffet onset in the flight tests. By about M = 0.86 the pressure starts to fall again (due to a downstream movement of separation and the shock) and at this point the effects of shock-induced separation are most severe. A rough measure of the severity of flow separation is the increase in p/p_t from M = 0.81 to 0.86. This increase is halved by opening only two breathers upstream of the measurement point and halved again by opening another two breathers. Hence some alleviation of buffet is suggested by the steady measurements.

The corresponding total broad band rms pressure fluctuations show a sudden increase from M = 0.81 to peak at M = 0.85, and then fall again as the supersonic region expands slowly round the shoulder (Fig 3c), consistent with the mean pressure.

The changes in the level of excitation, $\sqrt{nF(n)}$, at low frequency (Fig 3d) are more important for the flight buffeting problem and are much larger than the changes in total rms pressure fluctuations. For f = 21 Hz (which corresponds to 2.5 Hz on the aircraft, close to the pilot's estimate) the buffet intensity is $\sqrt{nF(n)} = 0.05$ with no breathers open. This is approaching the level found at the reattachment point of a bubble and recognised[1] as an upper limit of excitation (in the absence of resonance). With two breathers open upstream of the bubble the level falls to about $\sqrt{nF(n)} = 0.03$. With the addition of two other breathers open the level falls to about $\sqrt{nF(n)} = 0.015$. Hence there is a large progressive change in the low-frequency excitation as the breathers are opened. (There is probably a much smaller progressive change in the high-frequency excitation). These results are not particular to the single transducer position considered (close to the fuselage to accentuate shock/boundary layer interaction effects). When the pressure transducer is displaced along the shoulder away from the fuselage the mean pressure is increased (lower supervelocities) and the severity of the rms pressure fluctuations is decreased. Hence the choice of a transducer position close to the fuselage for the main series of tests was justified.

Adding two breathers downstream of the transducer, (and therefore away from the separation point) produces only a small improvement in the mean pressures, but no further reduction in the rms pressure fluctuations or the low frequency excitation.

Increasing the boundary layer thickness by wrapping carborundum paper round the leading edge makes a small improvement in the mean pressures and reduces the rms pressure fluctuations slightly. (This is as observed in previous tests[2].)

Hopefully, corresponding with an improvement in buffet excitation shown by Fig 3d, the aircraft buffeting at very low frequencies will be reduced from 'heavy' to 'light'. This prediction will be checked during the test flights.

Flow Visualisation

The flow about this fairing at transonic speeds has several interesting features.

Surface oil flow tests in the CIT 2½in × 2½in Tunnel[5] show the separation on the side of the fairing which provides the low-frequency excitation and the large area of the fairing/fuselage boundary layer interaction. This interaction is identical with the fuselage wing-root vortex system discussed by Stanbrook[10].

The steadiness of the separated flow on the side of the fairing is crucial to the buffet problem. The separations are well defined and pose some interesting topographical problems. The flow on the front face of the wedge is identical with the breathers closed and with four breathers open: this is consistent with the supersonic expansion of the flow round the shoulder. A straight line roughly in the middle of the shoulder marks the forward limb of the lambda shock and is caused in the usual way by the local thickening of the boundary layer upstream of the main shock. (The straightness of this thin line indicates that natural transition did occur close to the leading-edge of the wedge, because laminar or transitional boundary layer/shock interactions are always ragged.) Shock-induced separation occurs downstream of this thin line. The distance between the forward shock and the main shock increases progressively as the distance from the tunnel roof increases and the severity of shock/boundary layer interaction decreases. The relatively small differences between the surface oil flow observations with and without buffet breathers is consistent with the fact that time-averaged observations often give a poor indication of radical differences in the characteristics of time-dependent flows.

Conclusions

The cause of severe buffeting at transonic speeds on an experimental aircraft, fitted with a bluff fairing close to the nose, was identified in wind tunnel tests on models of the fairing as being due to alternate low-frequency shock oscillations between the port and starboard sides of the fairing. These shock oscillations caused time-dependent pressure variations which had a significant phase difference between the two sides of the fairing. It was found that connecting both sides of the fairing (where the mean pressures were the same) by a number of tubes stabilised the position of the shock-induced separation and reduced the level of low-frequency pressure fluctuation. This device has been termed a 'buffet breather'.

The principle of the buffet breather may not be restricted to transonic speeds, and may be applicable to a wide range of aerodynamic problems. These are specified in the patent application[4].

Symbols

f	frequency (Hz)
Δf	analyser bandwidth (Hz)
M	Mach number
$\sqrt{nF(n)}$	dimensionless level of aerodynamic excitation at a particular frequency = $\Delta p/q\sqrt{\varepsilon}$
$\overline{p}$	total broadband rms pressure fluctuation
Δp	rms pressure fluctuation in a frequency band Δf at frequency f
q	free stream kinetic pressure
U	free stream velocity
w	width of bluff body or fairing
ε	analyser bandwidth ratio ($\Delta f/f$)

References

1. Mabey, D.G.; Beyond the buffet boundary. Aero Journal, 77, 748, p 201-215 1973.

2. Mabey, D.G.; Welsh, B.L.; Cripps, B.E.; Periodic flows on a rigid 14% thick biconvex wing at transonic speeds. RAE Technical Report 81059 1981.

3. Lambourne, N.C.; Some instabilities arising from the interactions between shock waves and boundary layers. AGARD Report 182 1958.

4. Mabey, D.G.; Improvements in or relating to fluid buffet excitation alleviation. Patent Applications 84-19158 1984.

5. Mohan, S.R.; CIT unpublished tests 1983.

6. Bergh, H.; Tijdeman, H.; Theoretical and experimental results for the dynamic response of pressure measuring systems. NLR TR 238 1965.

7. Webb, D.; Private communication.

8. Bendock, J.G.; Hines, H.E.; Piercy, T.G.; Ram jet/ram inlet lecture. (Figs 49 and 50). AGARD LS 136 1984.

9. Welsh, B.L.; Pyne, C.R.; A method to improve the temperature stability of semi-conductor strain gauge pressure; transducers. RAE Technical Report 77155 1977.

10. Stanbrook, A.; Experimental observation of vortices in wing body junctions. RAE Technical Report 2589, ARC 19630 1957.

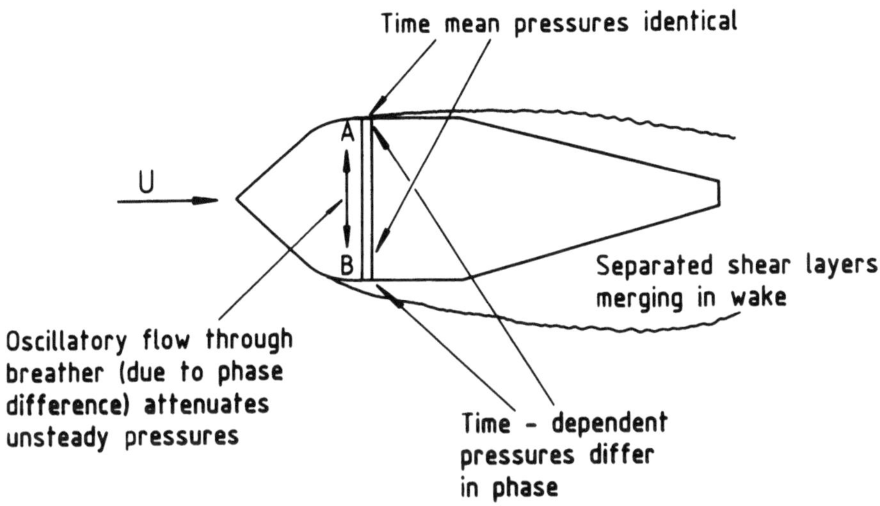

Fig.1. Principle of buffet breather

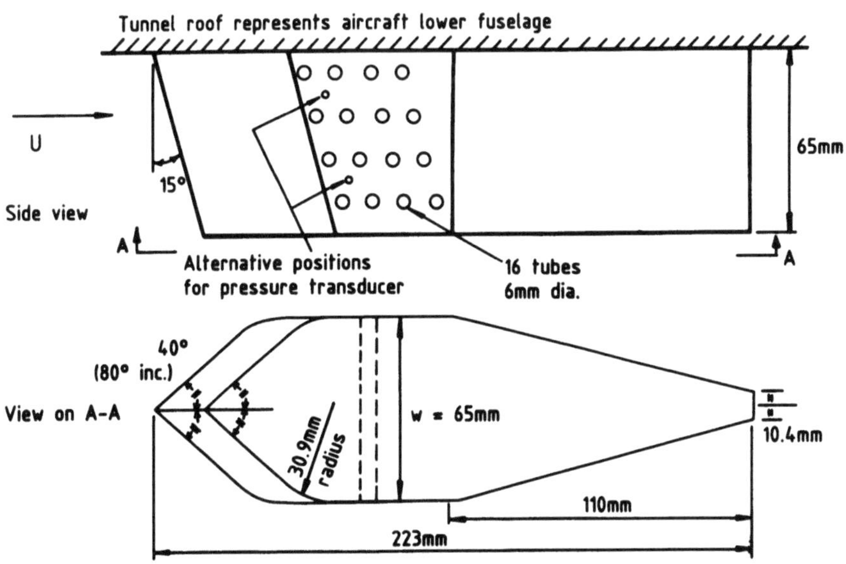

Fig.2. Simplified 1/8.4 model of aircraft fairing

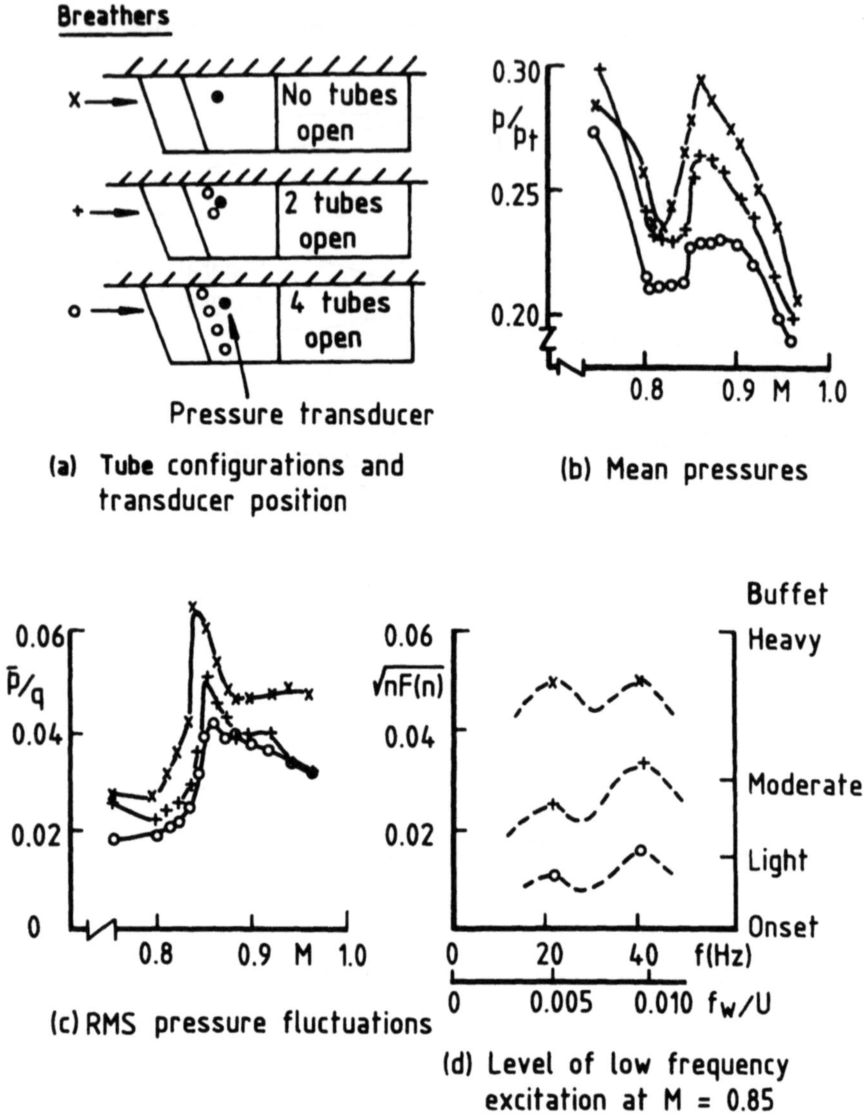

Fig.3. Variation of pressures at root of fairing with forward breathers

Turbulence Alteration due to Shock Motion

M. Y. Hussaini*
Institute for Computer Applications in Science and Engineering
NASA Langley Research Center, Hampton, Virginia 23665

F. Collier
Virginia Polytechnic Institute & State University
Blacksburg, Virginia 24061

D. M. Bushnell
NASA Langley Research Center, Hampton, Virginia 23665

Introduction

Research within the past five years (e.g., References 1 to 5) identified a multitude of mechanisms whereby shock interactions can affect turbulence. These mechanisms include: (1) direct vorticity amplification, (2) production of vorticity from incident sound and entropy modes, (3) bulk compression and (4) shock-induced streamline curvature as "extra rates of strain," (5) instantaneous shock focusing, and (6) direct effects of shock motion/dynamics/instability. The present paper addresses three facets of this last issue, the effects of shock motion upon turbulence/"unsteady vorticity."

The first of these subjects involves the direct production of fluctuating velocity and Reynolds stress due to shock oscillations. Possible sources of shock wave motions/ oscillations include: (a) shock-wave instability, (b) local (and propagating) motions driven by incident turbulence fields, and (c) overall flow field instabilities, including bifurcating shock-wave systems [6]. For the ideal gas case planar shock waves are stable [7] and therefore the sources of shock motion for

---

*Research was supported by the National Aeronautics and Space Administration under NASA Contract No. NAS1-17170 while this author was in residence at ICASE, NASA Langley Research Center, Hampton, VA 23665.

Turbulent Shear Layer/Shock Wave Interactions
IUTAM Symposium Palaiseau 1985
Editor: J. Délery

conventional aerodynamic application are limited to turbulence-induced oscillations and overall flow-field fluctuations. For the usual shock-boundary layer interaction case the shock oscillations are turbulence driven, [8], [9] including reflection from a turbulence-distorted sonic line, and propagate, in a decaying fashion, along the wave system (e.g., [10]). The present work provides first-order estimates for the production of downstream velocity perturbations engendered by such oscillating shock waves. The problem is modeled as an oscillating wedge in a supersonic stream. The contribution of this first portion of the present paper is therefore merely a reinterpretation of previous solutions [11], [12] in terms of the oscillating shock-vorticity production mechanism for the shock-turbulence interaction problem.

In contrast to this first case of shock motion, which acts to increase velocity fluctuation amplitude, the second shock motion issue examined herein, eddy shocklets [13] constitutes the current best candidate explanation for the observed reductions [14] in perturbation levels [15] and entrainment rate [16] for high-speed free shear layers. These flows exhibit perturbation Mach numbers in excess of 1 and therefore can support transient shock structures associated with migrating eddies. In the present paper this physics is modeled by a two-dimensional computation of an eddy which, initially, is convecting subsonically in a supersonic local flow with a Mach number difference (eddy-to-flow) greater than unity. An eddy shocklet forms during the subsequent eddy acceleration with consequent eddy deformation. This eddy-shocklet-induced transient eddy deformation provides at least a partial explanation for the factor of three decrease in free shear layer entrainment rate between low speed and Mach 5 [17].

For the eddy shocklet case a slow-moving concentration of vorticity produces a transient shock wave. The final problem of shock motion considered herein is really the antithesis of this case--the production of organized transient vorticity due to shock motion through a region of entropy spottiness [3],

[18], [19]. This situation is especially prevalent in combustion problems where detonation occurs at multiple sites within the flow and can be a potent source of increased mixing/combustion rate (as can the shock-wave instabilities which can occur in these non-ideal gas cases, (e.g., [20])).

In summary, the present paper provides physical and computational models and quantitative estimates for three cases of shock motion-turbulence interaction; production of velocity fluctuations due to shock oscillations, eddy alteration due to transient eddy shocklets, and eddy production due to transient shock-entropy spottiness interactions.

Discussion of Computational Models and Results

The classical problem of an oscillating wedge in a supersonic flow appears to have been first formulated by Carrier [11]. His motivation was to study the stability of the shock wave to small perturbations. In particular, he intended to resolve the nonuniqueness of the shocked solutions of Euler equations. Since then there have been numerous studies of this problem with the view of obtaining aerodynamic forces and stability derivatives. Our motivation differs from theirs in that we use this formulation to model the effect of shock oscillation on the generation of vorticity and "Reynolds stress." The relevant equations are those of conservation of mass, momentum and energy, known as the Euler equations for an ideal gas. For the present illustrative calculations they are linearized about the basic nonlinear steady flow past a wedge of 5° semi-vertex angle in a Mach 5 stream at zero incidence. The wedge is pivoted at a point a distance 0.1 units downstream of its apex, and oscillates with small angle of attack equal to $\exp(ict)$. Our derivation is similar to that of Carrier [11] and Van Dyke [12] and hence the details are omitted. Figures 1 and 2 display shock locus, instantaneous vorticity and pressure fields, and Reynolds stress distribution behind the shock for the cases of low and high frequencies $c = 1$ and 10, respectively. The lengths have been scaled with respect to

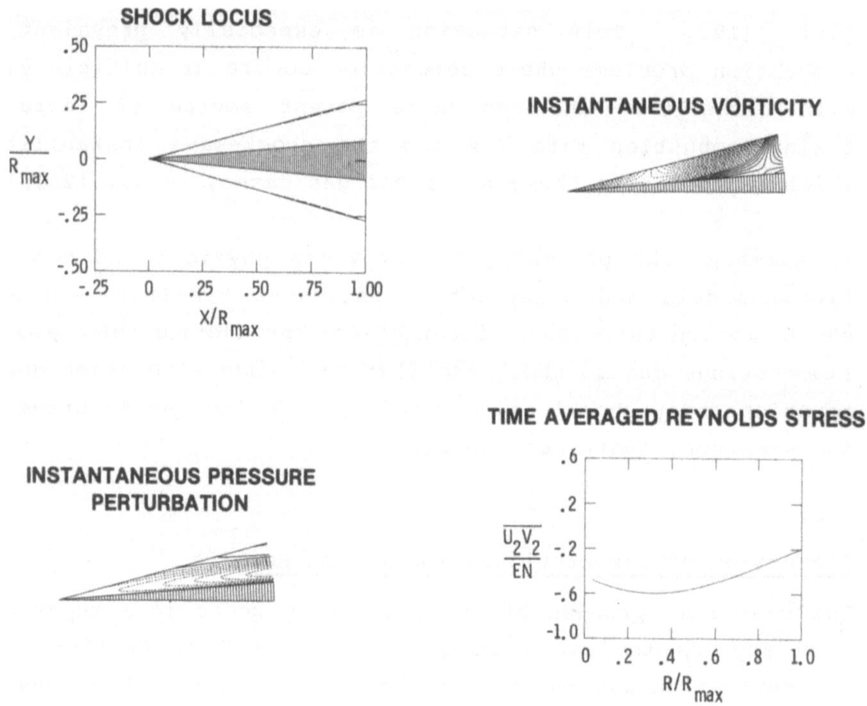

Figure 1. Oscillating shock wave, c = 1.

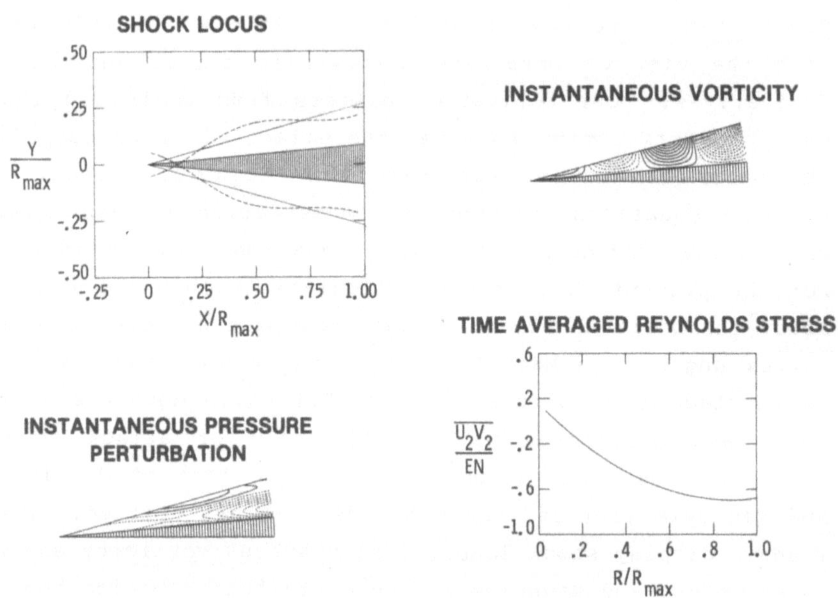

Figure 2. Oscillating shock wave, c = 10.

the wavelength $R_{max} = 2\pi(M^2 -1)/(cM)$, M being the free stream Mach number, and the Reynolds stress is scaled with respect to the perturbation energy. The results shown correspond to time $t = 2\pi/c$. The effect of frequency is quite apparent, the high frequency causes large ripples in the shock shape and resultant large Reynolds stresses. The spatial variation in Reynolds stress apparent in the figure is presumably due to the mismatch in the convection speed between the forced motion due to and near the wedge and the resultant propagation of that forced motion along the shock. In other words, the convection speed of the shock wave disturbance can be greater or less than that corresponding to the frequency of the "driven" shock motion. Since the ideal gas shock systems are inherently stable in the absence of globally unstable flow fields, the shock perturbations introduced in the present calculation do not amplify but may require extensive downstream distance to decay, i.e., shock motion induced fluctuating flows can propagate over extensive regions of the local viscous and inviscid flow.

For the simulation of the eddy shocklet problem, the two-dimensional unsteady compressible Navier-Stokes equations are solved by MacCormack's method. The Cartesian velocity components of the vortex are

$$u = U_c - \varepsilon y/(\sigma^2 + r^2)/q_{max} \quad ,$$

$$v = \varepsilon x/(\sigma^2 + r^2)/q_{max} \quad ,$$

where

$$r = \sqrt{\{(x-x_c)^2 + (y-y_c)^2\}} \quad ,$$

$$q^2_{max} = \max(u^2 + v^2);$$

U_c is the convection velocity of the vortex, and (x_c,y_c) is the initial position of the vortex center; the velocities are made nondimensional by the mainstream velocity. Results from two different parametric simulations are presented. The

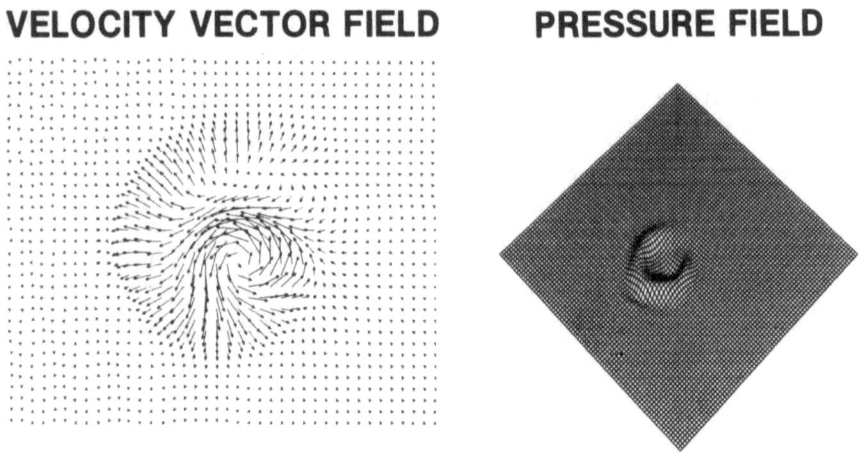

Figure 3. Eddy shocklet.

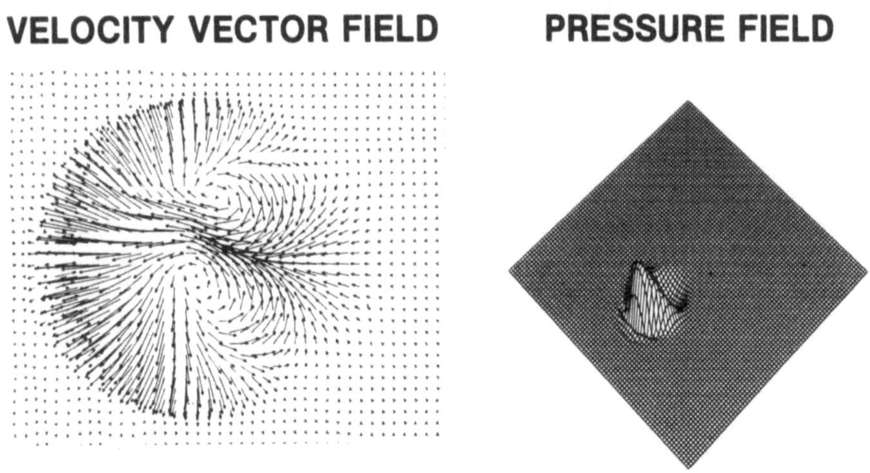

Figure 4. Eddy shocklet.

essential parameters of the problem are the Mach number of the incident flow the initial convection speed and strength of the vortex. Figure 3 shows the results corresponding to free stream Mach number of 2.5, initial vortex Mach number of 2.25 and vortex strength ε equal to .2. This is the case of high convection speed of the vortex, and it simply diffuses as it accelerates. The initial sense of the eddy circulation is preserved. Figure 4 displays results of the transient response of an eddy of strength ε equal to .2 and convection speed equal to 1.25 exposed to an incident flow of Mach number 2.5. The low initial convection speed along with the high initial Mach number produces an eddy shocklet. Note that in this case an eddy bifurcation occurs--a vortex of opposite circulation is generated--and the length scale of the original vortex is reduced. The important implication is that eddy shocklets (and/or large difference beteween the local flow speed and eddy convection speed) can reduce turbulent mixing through the combined effects of (a) reduced turblence scale, and (b) production of counter fluctuating vorticity. This possibly explains the observed decrease in mixing rates in free shear flows as Mach number increases from 1 to 5. The essential mechanism involves the instantaneous "inviscid" pressure field induced about the front (or back) of the eddy (depending on whether the eddy is propagating at less than or greater than the local flow speed). This pressure field, which is obviously enhanced by increased stream to eddy velocity difference and increased Mach number, will always counter the initial eddy circulation over a portion of the eddy contour and, for long enough times and weak enough eddies (or large enough velocity difference and Mach number) the formation of counter vorticity and consequent eddy splitting will occur. This production of counter vorticity is one of the most effective means of altering the structure of turbulent shear flows, and such a mechanism is not currently included in any of the turbulent modelling approaches. To the best knowledge of the present authors this is the first observation of such a mechanism.

To simulate the motion of a shock wave through a region of temperature inhomogeneity, an infinite two-dimensional shock wave is generated at $x = 0$ so that it moves into the region $x > 0$. In the region between $x = L < 0$ and the shock wave $x = X_s(y,t)$, the two-dimensional unsteady Euler equations are solved by MacCormack's method [3]. The temperature distribution is assumed to be given by

$$T = 1 + \varepsilon \exp[(x-x_0)^2 + y^2]/(2\,\sigma^2),$$

where $(x_0,0)$ is the center of temperature spot and σ is its width. The pressure is assumed constant and the density is obtained from the equation of state. A Mach 2 shock wave is

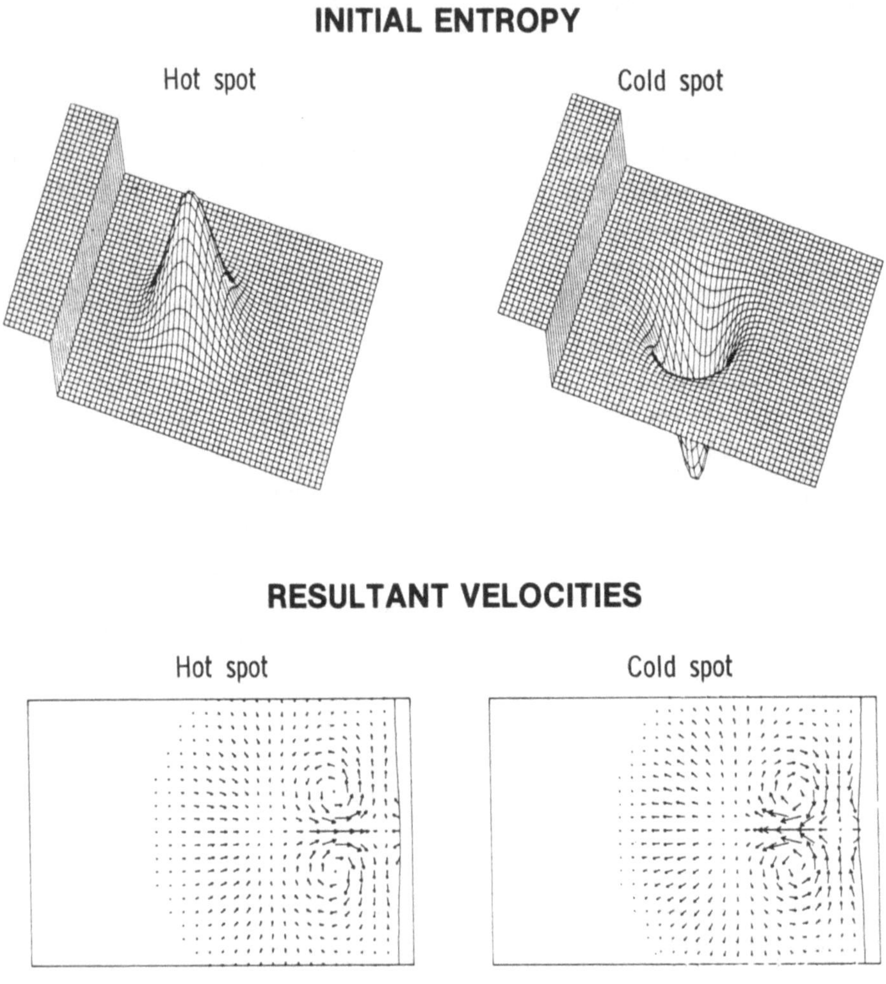

Figure 5. Shock wave/temperature spot interaction.

generated and it advances to the right. In Figure 5, the top row displays the surface plot of the entropy spot with a maximum temperature perturbation $\varepsilon = 0.25$; the entropy discontinuity denotes the shock front; the bottom row shows the vector plots of the disturbance flow field after interaction; the solid line bounding the flow field on the right is the position of the shock at time $t = 0.5$ units; the sense of rotation of the vortex pair in the hot spot case agrees with experimental observation [18]. The interaction of the Mach 2 shock with equivalent hot and cold spots produced different amounts of vorticity, the vorticity in the cold spot case being a third higher in amplitude than in the hot spot case.

Conclusions

1. Shock motions are extremely efficient in converting mean flow energy into fluctuation energy.
2. High frequency induced shock motions produce intense vorticity fields from the incident mean flow, as indicated by production of high intensity Reynolds stresses.
3. Subsonic eddy motion in a supesonic stream such as occurs due to eddy migration in a supersonic shear layer creates transient eddy shocklets. These eddy shocklets cause, in the two-dimensional case, (1) eddy spliting and consequent reduction in eddy length scale, and (2) production of dynamic opposing vorticity which would tend to cancel a portion of the dominant fluctuating vorticity in the shear layer. This eddy segmentation and partial reversal occurs even in cases without eddy shocklet formation. What is required, in the present calculations, is a relatively weak initial eddy strength and a large eddy to local stream velocity difference at high stream speeds (i.e., essentially a supersonic/hyersonic phenomena).
4. We conclude from the study of shock motion through a hot spot that such interactions can be a potent source of turbulence production or enhancement, especially in reacting flows.

References

[1] Delery, J.M.: Experimental investigation of turbulence properties in transonic shock/boundary layer interactions. AIAA J., Vol. 21 (1983), pp. 180-185.

[2] Anyiwo, J.C. and D.M. Bushnell: Turbulence amplification in shock wave boundary layer interactions. AIAA J. Vol. 20 (1982), pp. 893-899.

[3] Zang, T.A., M.Y. Hussaini, and D.M. Bushnell: Numerical computations of turbulence amplification in shock wave interactions. AIAA Paper No. 82-D293, AIAA 20th Aerospace Sciences Meeting, January 11-14, 1982, Orlando, Florida. Also AIAA J., Vol. 22 (1984), pp. 13-21.

[4] Hayakawa, K., Smits, A.J., and Bogdonoff, S.M.: Hot-wire investigation of an unseparated shock wave-turbulent boundary layer interaction. AIAA Paper No. 82-0985, 1982.

[5] Parthasarathy, S.P., Harstad, K., and Massier, P.F.: Jet oscillations caused by vorticity interactions with shock waves. Presented at the AIAA 7th Aeroacoustics Conference, October 5-7, 1981, Palo Alto, California. AIAA Paper No. 81-1974.

[6] Jungowski, W.M.: Some self-induced supersonic flow oscillations. Progress Aerospace Sciences, Vol. 18, No. 1978, pp. 151-15.

[7] Majda, Andrew: The stability of multi-dimensional shock fronts. Memoirs of the American Mathematical Society, Number 275. American Mathematical Society, January 1983, Vol. 41.

[8] Plotkin, Kenneth J.: Shock wave oscillation driven by turbulent boundary-layer fluctuations. AIAA J., Vol. 13, August 1975, pp. 1036-1040.

[9] Grande, E. and Oates, G.C.: Unsteady flow generated by shock-turbuelnt boundary layer interactions. Presented at the AIAA 11th Aerospace Sciences Meeting, Washington, D.C., January 10-12, 1973. AIAA Paper No. 73-168.

[10] Whitham, G.B.: A new approach to problems of shock dynamics. Part 1 - Two-dimensional problems. J. Fluid Mech., Vol. 2, 1957, pp. 145-171.

[11] Carrier, G.F.: The oscillating wedge in supersonic stream. J. Aero. Sci., Vol. 16, 1949, pp. 150-152.

[12] Van Dyke, Milton D.: On supersonic flow past an oscillating wedge. Quart. Appl. Math., Vol. XI, No. 3, 1953, pp. 360-363.

[13] Oh, Y. H.: Analysis of two-dimensional free turbulent mixing. Presented at the AIAA 7th Fluid and Plasma Dynamics Conference, Palo Alto, California, June 17-19, 1974. AIAA Paper No. 74-594.

[14] Birch, Stanley F. and Eggers, James M.: A critical review of the experimental data for developed free turbulent shear layers. Free Turbulent Shear Flows, Vol. 1, Conference Proceedings, NASA SP-321, 1972.

[15] Wagner, Richard D.: Mean flow and turbulence measurements in a mach 5 shear layer. Part II - Hot-wire measurements of the mean flow and turbulence intensities. Fluid Mechanics of Mixing ASME, 1973.

[16] Morrisette, E. Leon and Birch, Stanley F.: Mean flow and turbulence measurements in a mach 5 shear layer. Part I - The development and spreading of the mean flow. Fluid Mechanics of Mixing ASME, 1973.

[17] The 1980-81 AFOSR-HTTM-Stanford Conference on Complex Turbulent Flows: Comparison of Computation and Experiment, Vol. II, Taxonomies, Reporters' Summaries, Evaluation, and Conclusions. Proceedings of the 1981 Conference, Stanford University, Stanford, California, September 14-18, 1981, pp. 814-836.

[18] Dosanjh, D.S. and Hamernik, R.P.: Interaction of an advancing shock front with a concentrated heat (entropy) source. Phys. Fluids, Supp. I, 1969, pp. I136-I139.

[19] Picone, J. M., Oran, E. S., Boris, J. P., and Young Jr., T. R., Theory of vorticity generation by shock wave and flame interactions in Dynamics of Shock Waves, Explosions, and Detonations, J. R. Bowen, N. Manson, A. K. Oppenheim, and R. I. Soloukhin (eds.), Vol. 94, Progress in Astronautics and Areonautics.

[20] Swan, G.W. and Fowles, G.R.: Shock wave stability. Phys. Fluids, Vol. 18, No. 1, January 1975, pp. 28-35.

Properties of Wall Pressure Fluctuations in a Separated Flow over a Compression Ramp

J.P. DUSSAUGE *, K.C. MUCK, J. ANDREOPOULOS

Gas Dynamics Laboratory, Princeton University, Princeton, N.J.
* Permanent address : I.M.S.T., U.A. C.N.R.S. N° 130,
Univ. Aix-Marseille II, Marseille, France.

Summary

An experimental study of the unsteadiness of a shock/boundary layer interaction is presented. The interaction is produced by a 24° compression ramp at Mach number 2.84. Multi-channel measurements of wall pressure fluctuations are performed. It is found that the shock has significant large scale "rippling"; the shock induced fluctuations produce a maximum in the space time correlation for negative time delay, indicating a strong modification of transport properties in the interaction. At last, the analysis of frequency measurements suggests that shock oscillation could be triggered by the bursting mechanism in the incoming boundary layer.

Introduction

In shock/boundary-layer interactions, it can be important to know the structure of the shock within the boundary-layer. The unsteadiness and the three-dimensionality of a shock wave can affect the properties of the turbulence amplified or generated in the interaction (see for example Anyiwo, Bushnell [1]); moreover, the effect of a "flapping" shock, whose motion is externally driven, can be very different from a "rippling" shock whose unsteadiness would be produced by the incident field itself. In the past, the unsteadiness of shock waves in various separated supersonic flows was noticed and some measurements of wall pressure fluctuations were performed with high frequency transducers [2], [3], [4], [5]. However, there is a lack of detailed measurements concerning the instantaneous shock structure, as it can be deduced from wall pressure.

The present study gives some experimental information about that point, in a shock/boundary-layer interaction produced by a 24° compression ramp at Mach number 2.84,

Turbulent Shear Layer/Shock Wave Interactions
IUTAM Symposium Palaiseau 1985
Editor: J. Délery

where the shock system is unsteady [4],[5]. Mean flow measurements are given in ref. [6].

Experimental arrangement

The experiment was made in the Princeton University 203x203mm high Reynolds number blowdown wind tunnel. The model was a 24° compression ramp. In order to minimize 3-d perturbations, there was a gap on either side of the model, and side fences were used [7]. The parameters defining the turbulent boundary-layer a head of the interaction are given in Table 1.

M_0	2.84 ± 0.03
U_0	$5.75 \pm 15\ ms^{-1}$
δ_0	$24 \pm 2\ mm$
δ_0^*	$6.4 \pm 0.2\ mm$
θ_0	$1.3 \pm 0.1\ mm$
$\bar{\tau}_{w0}$	$2.4 \times 10^4 \pm 3\%\ Nm^{-2}$
c_f	0.00100 ± 0.00015

0 denotes upstream undisturbed condition

Wall pressure measurements were performed with four Kulite differential transducers (Model XCQ-062-25D; diameter 0.7 mm) mounted flush with the tunnel wall. The bandwith of the amplifiers was 100 KHz. The signal was filtered, and sampled digitally at 100 KHz per channel. Many pressure fluctuation measurements in supersonic boundary-layers suffer from the poor spatial resolution of the transducer. An estimate of the damping of high frequencies due to the finite size of the sensor was made from low-speed studies, for no equivalent theory is available for supersonic flow. Corcos's work [8], if applied to our case, indicates a bandwidth limitation of 100 kHz in the upstream boundary layer. That implies that spectra and correlation measurements should be unaffected at low frequencies, up to 2 U_0 / δ_0 (U_0 upstream external velocity ; δ_0 boundary layer thickness). Since we are mainly interested in the unsteadiness of the shock system, this restriction is not severe : previous measurements [4] [5] showed that these fluctuations occured at low frequencies, typically 0.1 U_0 / δ_0 .

Results and discussion

Fig. 1 presents a sketch of the flow, deduced from mean measurements [6]. The r.m.s. wall pressure σ_{pw} as given by

Dolling et al.[4] is shown in Fig. 2

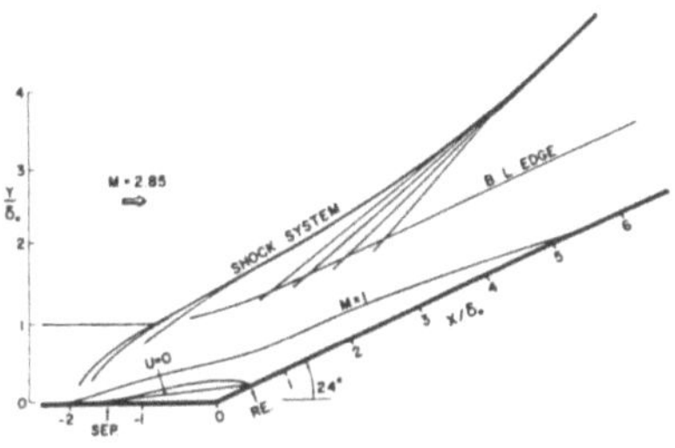

Fig.1 : Sketch of the flow, from Settles et al [6].

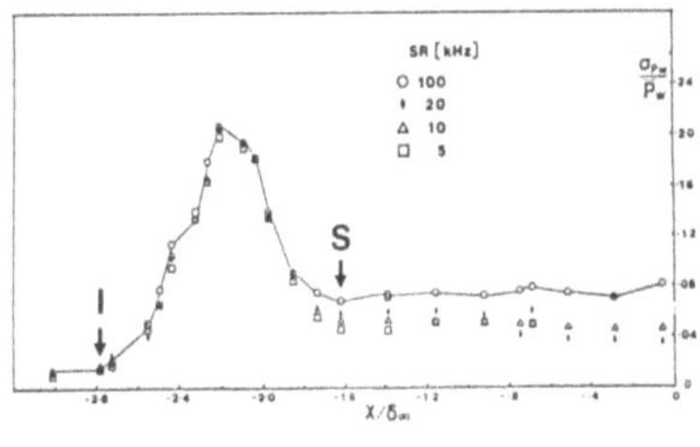

Fig. 2 : r.m.s. presssure fluctuations σ_{pw} from Dolling et al [5].

I indicates the limit of the zone where an intermittent signal is found; S is the mean separation point. The presents results were obtained in three zones : the upstream boundary-layer, the shock zone, and the separated region.

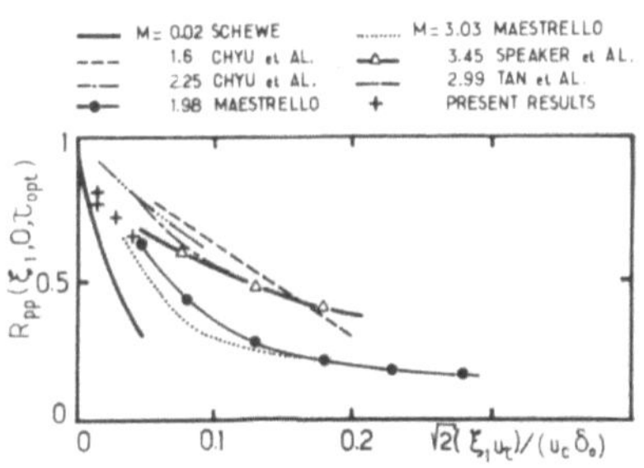

Fig.3 : Optimum space time correlation in the initial boundary layer. Data : References [9] to [13].

Space time correlations were measured in the upstream boundary layer. Their optimum values are compared to the results of other authors [9][10][11][12] in Fig. 3. The optimum correlation coefficient R $(\xi_1, 0, \tau_{opt})$ is plotted versus $u_\tau \xi_1 / U_c \delta_0$ which represents the ratio of two time scales : ξ_1/U_c is the convection time of a turbulent structure; δ_0/U_τ is a time scale of energy containing eddies of size δ_0 (here the boundary layer thickness) and of velocity scale the friction velocity U_τ. The present results are in good agreement with the other data.

Fig.1 suggests the existence of a single shock ahead of separation, producing a 10° turning of the flow; the other 14° of the total turning would be effected gradually downstream by a compression fan. Typical recordings of pressure fluctuations measured near the root of the shock are given on Fig.4. The signal is intermittent essentially between two mean levels p_1 and p_2 .

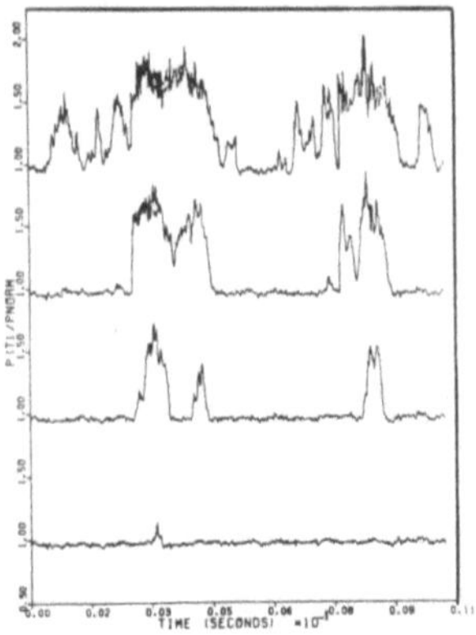

Fig. 4 : Time-history plots of the simultaneously recorded 4 channel (streamwise) fluctuating wall pressure in the intermittent region. Upstream transducer (bottom) located at x/δ_0 = -2.7 ; separation between tranducers ξ_1/δ_0 = 0.23

It seems very likely that the sharp front can be associated with a single shock passing over the transducer ; p_2 is practically constant in the intermittent region and equal to the mean pressure at the separation point. The ratio p_2/p_1 can be determined : $p_2/p_1 \simeq 1.7$. This value is consistent with the features given by Fig. 1 : shock tables indicate an initial Mach number of about 1.5, for a 10° turning of the flow through the shock and a pressure ratio of 1.7. In our case, the point where the Mach number has such a value is located at distances $y \simeq 0.5\,\delta^*$ (δ^* displacement thickness), i.e. in a zone close to the wall, where the velocity fluctuations predominantly produce the wall pressure fluctuations.

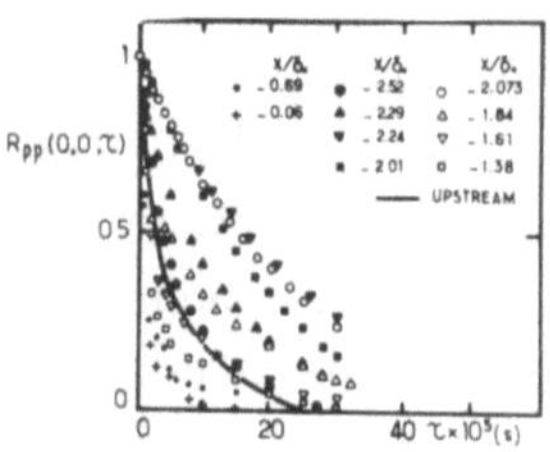

Fig. 5 : Pressure autocorrelation in the interaction.

Fig. 5 shows the pressure autocorrelation in the interaction; when the signal is intermittent the autocorrelation increases and keeps significant values up to $\tau = 30.10^{-5}$ s.; this shows an important contribution of low frequencies. The largest correlation is obtained for $-2.24 \leq x/\delta_0 \leq -2$; in this zone, is maximum and the intermittency of the signal is about 0.5, so that it may be inferred that the low frequencies are produced by the shock going forth and back over the transducer. Further downstream, where the signal is no more intermittent and has higher and higher frequencies, the autocorrelation falls below the upstream curve.

Longitudinal space time correlations were measured in various points of the flow. In the intermittent part the results depend on the pressure fluctuations upstream and downstream of the shock and on the unsteadiness of the shock. Besides, the upstream pressure fluctuations can be associated with convective phenomena in the direction of the flow ; in the separated zone, the pressure field depends on structure moving in the main direction of the flow (external layer) and in the opposite direction (near the wall in the separated bubble); the shock motion can be represented by low frequency oscillations. We may then expect to observe space-time correlations of a complicated shape, with extrema for positive and negative time delay. However, by low-passing the signal, it is possible to enhance the contribution of the shock oscillations.

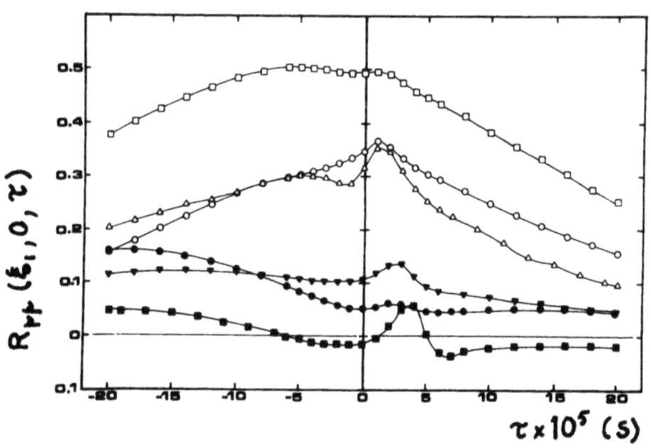

Fig. 6 : Longitudinal space time correlations

Symbol	O	□	Δ	▼	■	●
x/δ_0	-2.41	-2.18	-1.95	-2.41	-2.41	-2.18
ξ_1/δ_0	0.23	0.23	0.23	0.46	0.70	0.46

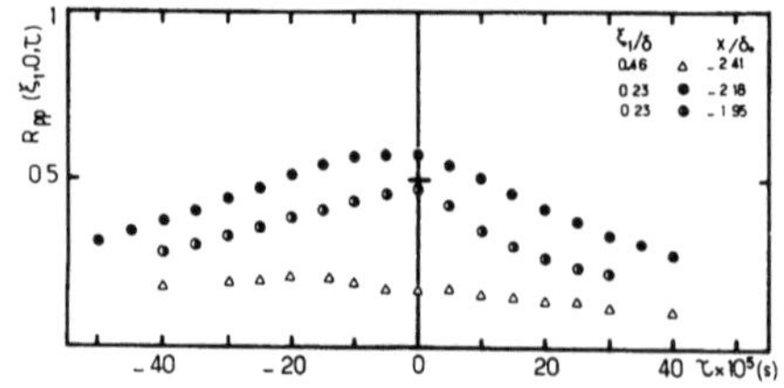

Fig. 7 : Low-pass filtered correlations (10 KHz).

A first result is given in Fig. 6, that shows the correlations in the intermittent zone : the upstream transducer is located at the very beginning of the intermittent region ; for a small separation (ξ_1/δ_0 =0.23) the level of the correlation is much lower than in the upstream boundary layer, suggesting that the upstream and downstream fluctuations and those induced by the shock itself are weakly correlated. For large separations (ξ_1/δ_0 =0.46 ; 0.70) there is a significant correlation for large negative τ ; filtered data (Fig.7) (Low-pass filter 10 kHz) show a similar trend. When the intermittency is strong (x/δ_0 = -2.18 ; ξ_1/δ_0 = 0.23), the broad band correlations are rather flat near the origin ; however, it seems that there are two maxima : one for negative τ , which can be observed on filtered data (Fig. 7) and is probably due to the shock motion ; a second one for positive τ ($\tau \simeq 10 \mu s$) corresponding to the eddy convection in the main flow direction.

As the shock oscillates, the maximum for negative τ is not associated with transport phenomena, but probably means that the shock motions in the reverse direction (corresponding to the sharp leading front in the plot of Fig.4) are strongly correlated. In the filtered measurements (Fig.7), it is not easy to decide where the maximum of the correlation is located. For example, at x/δ_0 =-1.95, ξ_1/δ_0 =0.23, the maximum can be evaluated for τ = 0. This would correspond to in-phase fluctuations in the zone between the maximum of σ_{pw} and the mean separation point.

A similar interpretation can be deduced from Coe's measurements [3] performed in a shock/boundary-layer interaction ahead of a 45° ramp at Mach 2 : for low frequencies the coherence function of pressure fluctuations is high and their phase is practically null.

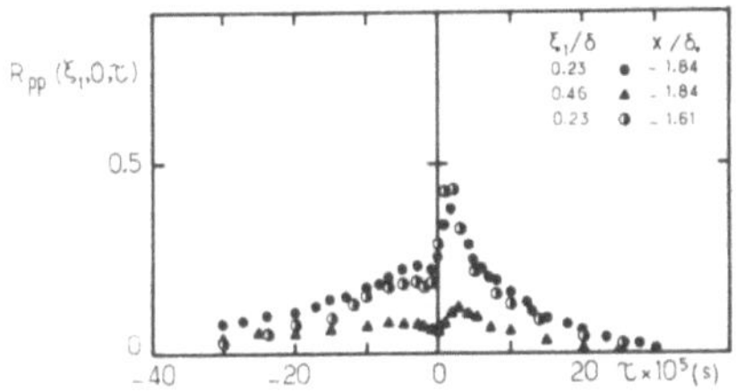

Fig. 8 : Space-time correlations near the mean separation point.

Measurements were performed at the vicinity of the separation point (Fig. 8). Their maxima are lower than in the incoming flow. Two maxima are found, for negative and positive τ ; the latter is the more important. This can probably be explained by the development of high turbulence intensities above the separated bubble associated with eddies moving in the main flow direction ; this strongly perturbed zone seems to contribute predominantly to the wall pressure. In the separated region ($x/\delta_0 \simeq -0.8$) the maximum at negative τ practically disappears [14].

Spanwise measurements were performed and showed a lower correlation level that in the upstream boundary layer.

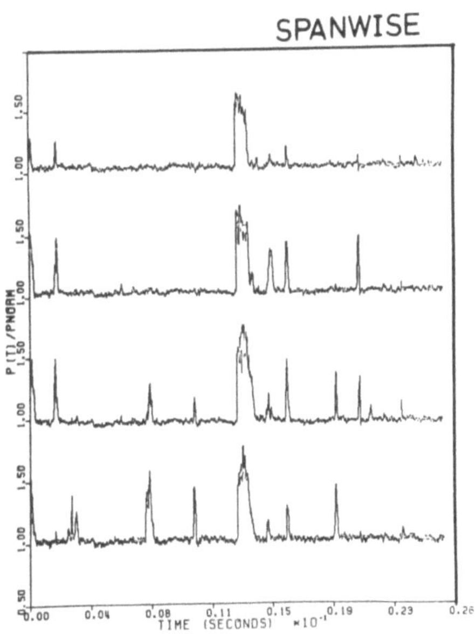

Fig. 10 : Time-history plots of the wall pressure recorded from four transducers placed spanwise. Intermittent region, x/ = -2.41.

The time history plots of the signal of four transducers placed spanwise are shown in Fig. 10. They confirm that the shock is not two-dimensional, but has ripples with space scales as small as (if not smaller than) $0.23\delta_0$ and as large as (if not larger than) $0.69\,\delta_0$.

Some spanwise non uniformities were observed in the autocorrelation coefficients determined in various spanwise locations (Fig. 11), for broad-band an low passed signal.

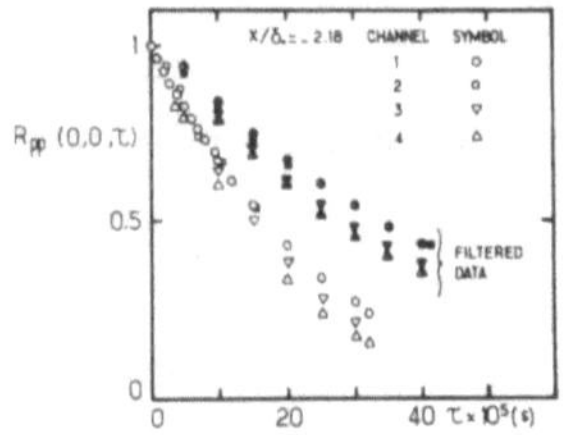

Fig. 11 : Spanwise distribution of autocorrelation coefficient. Separation between transducers ξ_2/δ_0 =0.23.

The intermittency factor given in Fig. 12 was computed and shows some strong spanwise non unformities ; this agrees with surface oil visualizations given by Settles et al. [15] which show that the separation line has a wave-like shape.

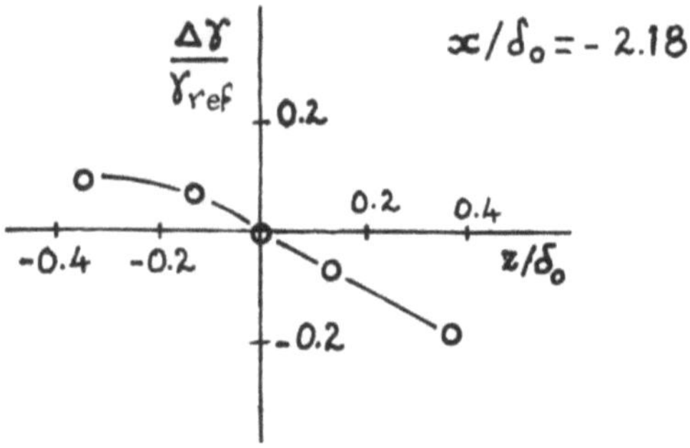

Fig. 12 : Spanwise distribution of the intermittency factor. γ_{ref} is the intermittency factor on the centerline ; $\Delta\gamma$ is the departure from this value.

These observations and the fact that the streamline is concave suggests that vortices of Taylor Görtler type are formed in the interaction.

Further analysis of the instantaneous pressure signals indicated that a typical shock wave oscillations frequency is scaled with the outer layer parameters of the incoming boundary layer. This non-dimensionalized frequency seems to be independent of the shock wave strength (wedge angle) and is equal to the bursting frequency of the upstream boundary layer ; the velocity of the shock wave was determined and was found at the order of the turbulence intensities. These results suggest that the shock wave unsteadiness may be triggered by the bursting mechanism of the initial boundary layer.

Conclusions

From the analysis of the wall pressure fluctuations, an overall picture of the shock system oscillations in a shock/boundary-layer interaction has been proposed : two main regions are found, the intermittent and the separated zone. In the intermittent region, the fluctuating pressure is made up of two distinct levels, i.e. the upstream level and that just dowsnstream of a single shock-wave, so that the intermittency is due to the shock system oscillations. Typical values of the frequency of these oscillation was found to correspond to the bursting frequency in the incoming boundary layer ; the shock wave unsteadiness may be triggered by the bursting mechanism in the upstream boundary layer. The spanwise distribution of various statistical properties suggest than Taylor Görtler vortices can be formed in the part of the flow of concave curvature. In the intermittent and separated regions, space time correlations show a peak on negative time delays ; this peak is more evident in the filtered data. This indicates an important contribution of fluctuations in the upstream direction, resulting probably in high values of shearing and stretching of the vortical structures.

References

[1] Anyiwo J.C., Bushnell D.M. : A.I.A.A. Jl.,Vol.20 N°7,1982.

[2] Kistler A.L. : J.A.S.A., Vol.36 N° 3, 1964.

[3] Coe C.F. : N.A.S.A. SP-207, 1969.

[4] Dolling D.S., Murphy M. : A.I.A.A. Paper 82-0986,1982.

[5] Dolling D.S., Or C.T.: A.I.A.A. Paper 83-1715, 1983.

[6] Settles G.S., Vas I.E., Bogdonoff S.M. : A.I.A.A. Jl., Vol.14, N° 12, 1976.

[7] Settles G.S. : Ph.D. Thesis, MAE Dpt., Princeton Univ. N.J., 1975.

[8] Corcos G.M. : J.A.S.A., Vol. 35 N° 2, 1963.

[9] Schewe G. : J. Fluid Mech., Vol. 134, 1983.

[10] Chyu W.J., Hanly R.D. : AIAA Paper 68-77, 1968.

[11] Maestrello L. : Boeing Scientific Research, Laboratories document, D1-82-0719, 1968.

[12] Speaker W.V., Ailman C.M. : NASA CR-486, 1966.

[13] Tan D., Tran T., Bogdonoff S.M. : A.I.A.A. Paper, Jan. 1984.

[14] Muck K.C., Dussauge J.P., Bogdonoff S.M. : A.I.A.A. Paper 85-0179, 1985.

[15] Settles G.S., Fitzpatrick T.J., Bogdonoff S.M. : A.I.A.A. Paper 78-1167, 1978.

Acknowledgments :

This work was supported by AFOSR Grant N°85-0126 and AFOSR Contract N° F 49620-84-C-0086 monitored by Dr. J. Wilson and Dr. J. Mc Michael. The stay of J.P. Dussauge at Princeton University was supported by a bourse de voyage de l'O.T.A.N. which is gratefully acknowledged. The authors are indebted to Prof. S.M. Bogdonoff and Prof. A.J. Smits for their encouragements and stimulating discussions.

A Shock-Wave/Free Turbulence Interaction

J.F.DEBIEVE, J.P. LACHARME

Institut de mécanique statistique de la turbulence, L.A. No. 130 of the C.N.R.S. 12, Avenue du Général Leclerc, Marseille, France

Summary

In an unsteady flow, the conditions of dynamic compatibility on both sides of a discontinuity cause the shock wave velocity to appear as an additional unknown. The shock wave turbulence can therefore be examined from two points of view: first, as a fixed discontinuity, the turbulence amplification can be summarized using a few notions of rapid distortion theory; second, the shock wave disturbances and the consequences on the apparent fluctuation level can be considered as a problem of intermittence. These two aspects are combined and interrelated, but the relative importance of intermittency effects depends mainly on the ratio $c'/\Delta\bar{c}$ which represents the incident turbulence level and the mean step through the shock. This study examines the experimental case of free turbulence subjected to a shock wave. The evolution of the field of turbulence is characterized on the basis of scale and level measurements, and the shock wave disturbances are described using appropriate signal processing techniques. The results are compared with those obtained in two different types of phenomenological analysis: amplification by rapid distortion and secondly an intermittence diagram associated with rippling of the shock wave.

I. Experiment Procedures

A flow was generated to obtain $M_\infty = 2.3$ with a stagnation pressure of 0.3 atm. An oblique shock wave was produced by a compression corner ($\sigma = 6°$) without separation. An air injector used for turbulence generation was placed upstream the sonic neck.

Turbulent Shear Layer/Shock Wave Interactions
IUTAM Symposium Palaiseau 1985
Editor: J. Délery

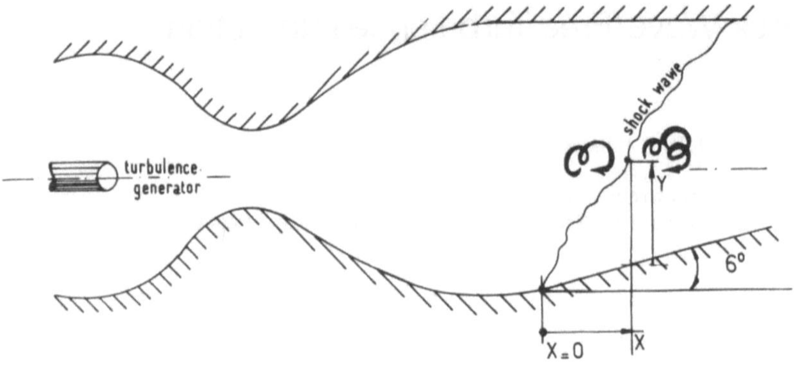

Fig. 1. Schematic diagram

In the planar portion of the nozzle upstream the shock wave, the mean field and turbulent field are practically homogeneous.

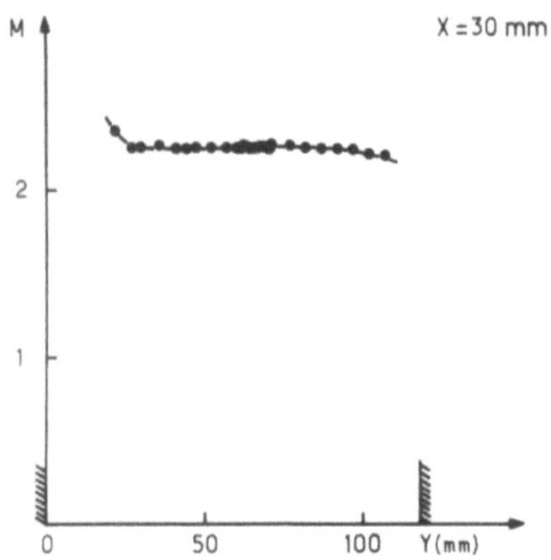

Fig. 2. Transversal Mach number distribution.

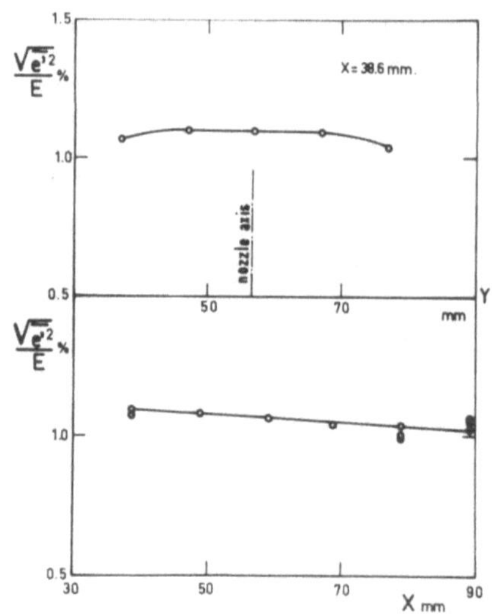

Figs. 3. and 4. Transversal and axial distribution of the fluctuating voltage.

Measurements of voltage fluctuations taken using a yawed hot-wire anemometer show a state of isotropy in the plane perpendicular to the flow.

II. Experiment Results Involving the Field of Turbulence

The evolution of the field of turbulence was described, first of all, by taking axial measurements of the voltage fluctuations. In the region of interaction, a large increase is observed in the r.m.s value which then drops downstream the shock. The span of this evolution, which reached approximately 1 cm, would appear to be ten times less without the turbulence generator.

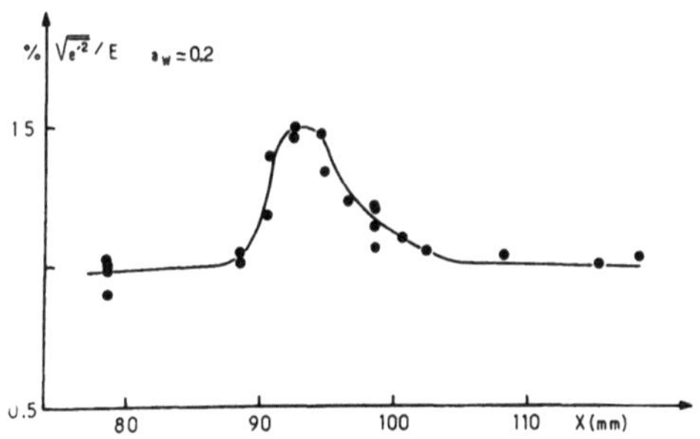

Fig. 5. Axial evolution of $\sqrt{e'^2}/E$

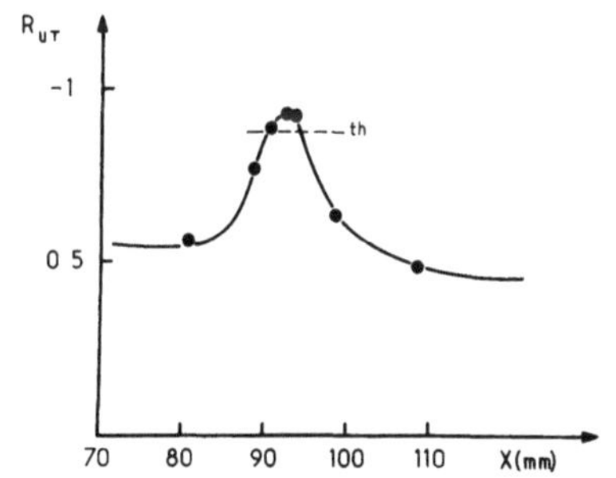

Fig. 6. Velocity temperature correlation coefficient.

For greater exactitude, measurements were then taken of turbulent intensities for the stagnation temperature, momentum, velocity, temperature, Mach number and the correlation coefficient. This data was obtained using a fluctuation diagram method.

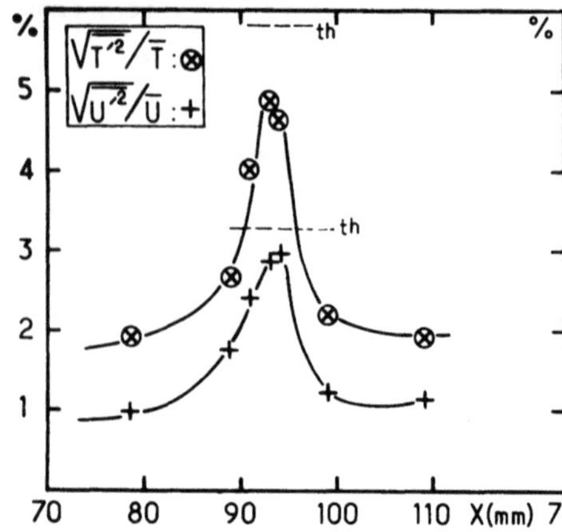

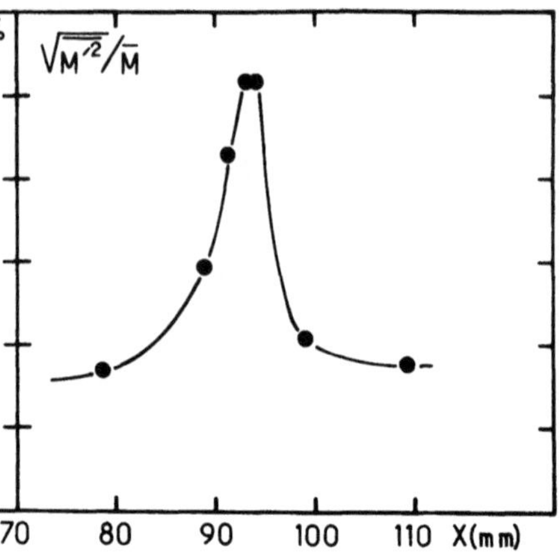

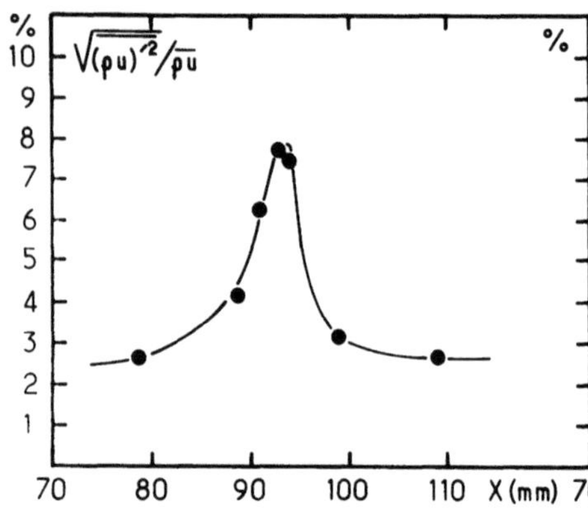

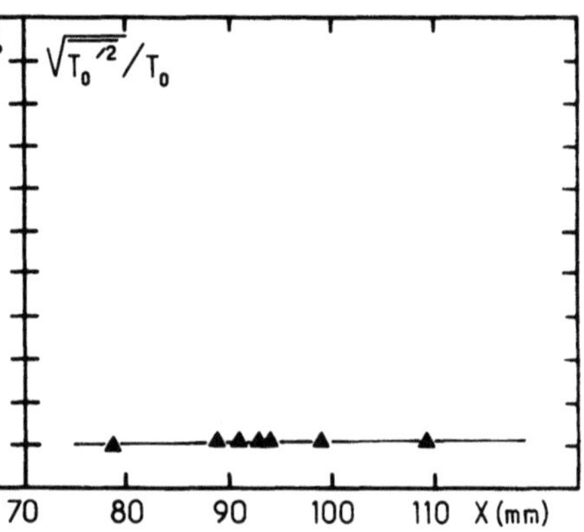

Fig. 7. Temperature and velocity turbulence intensity.

Fig. 9. Mach number turbulence intensity.

Fig. 8. Momentum turbulence intensity

Fig. 10. Stagnation temperature turbulence intensity.

The fluctuation levels of velocity and temperature continue to show a strong Reynolds analogy relationship ($T_o \simeq C^{st}$). Only the stagnation temperature fluctuations show certain particularities in behavior through the shock.

As concerns the spectral characteristics, there are no drastic changes observed in the upstream spectrum and the downstream spectrum of the fluctuating signal. In the maximum turbulence zone, however, a low-frequency power injection appears with a slope that is noticeably higher.

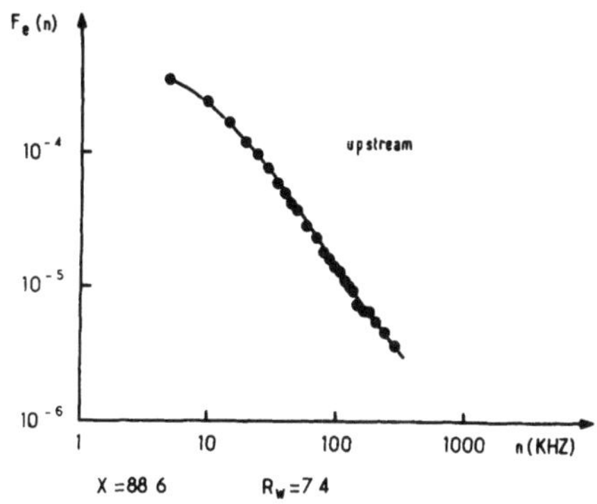

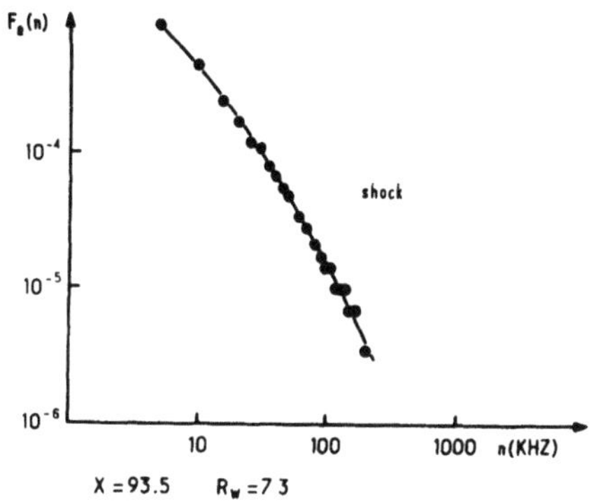

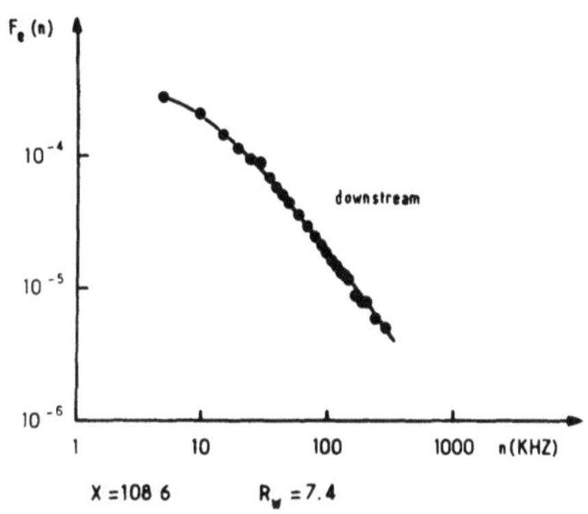

Fig. 11. Upstream spectrum of the fluctuating signal.

Fig. 13. Shock spectrum of the fluctuating signal.

Fig. 12. Downstream spectrum of the fluctuating signal.

Nonetheless, by defining the spectral regions involved, the same qualitative observation may be made for turbulent intensities of different quantity, i.e., there are no modifications in the stagnation temperature spectra, but changes did occur in the velocity and temperature spectra of the region of interaction, with an increase in slope and a low-frequency power supply $n < 20$ kHz that also leads to an increase in the integral scales of Taylor. (It should be noted, however, that in the region of interaction, the scales were undoubtedly underestimated).

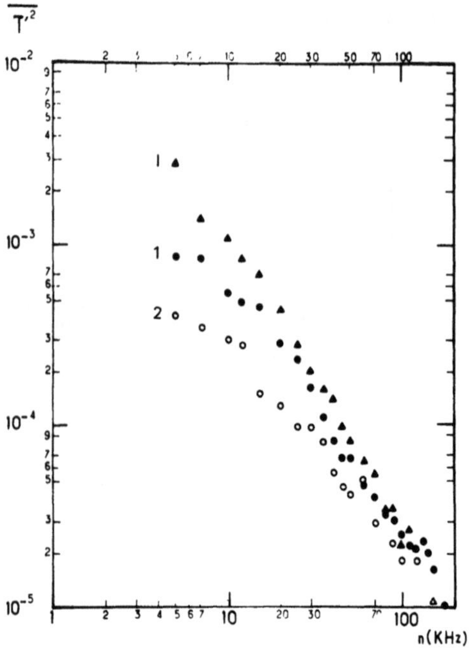

Fig. 14. Temperature spectrum.

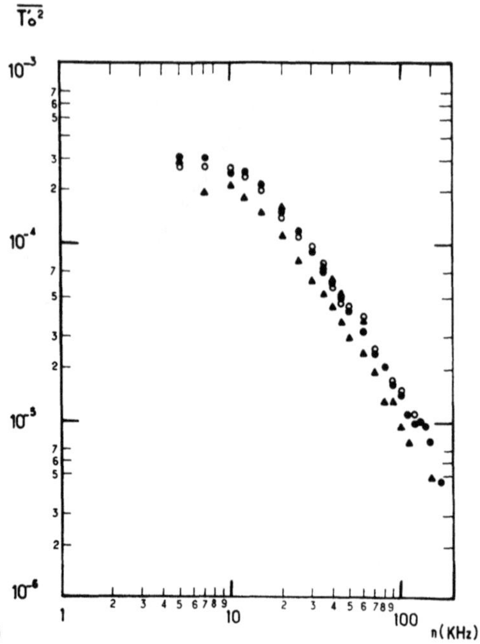

Fig. 15. Stagnation temperature spectrum.

It may be added that on the basis of the Taylor hypothesis, the spatial scale of velocity associated with the incident field of turbulence is approximately $\Lambda \simeq .5$ cm.

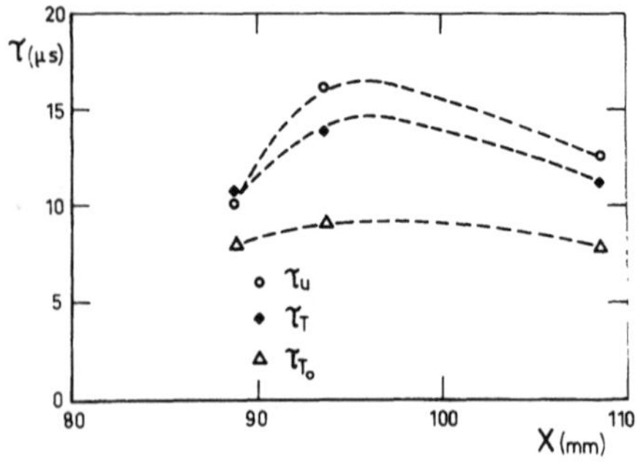

Fig. 16. Taylor's integral scales

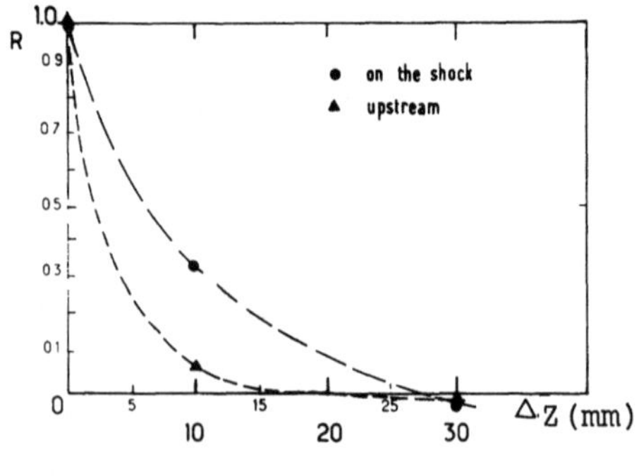

Fig. 17. Correlation coefficient at two points on the surface of the shock wave.

III. Description of Shock Wave Disturbances

The second aspect of the problem involves the influence of the incident field of turbulence on the shock wave. An attempt was made to describe these disturbances experimentally by determining the spatial scales associated with the size of the shock wave disturbances and the excursion of the shock wave, and also by finding the time scale associated with the shock wave oscillation.

At a maximum point of turbulence, the fluctuating signal displayed a great degree of intermittence corresponding with shock wave disturbances. This appeared on the probability density function as two distinct bumps representing the upstream and downstream states of turbulence.

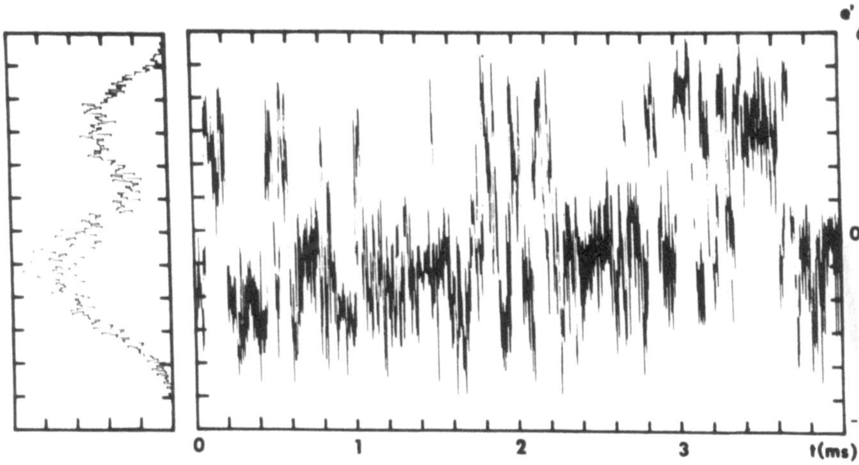

Fig. 18. Probability density of the intermittent signal.

Appropriate processing of these signals, making it possible to separate the influence of the shock wave from that of the turbulence, reveals two types of information. First, the intermittence coefficient γ , or the probability of the presence of the upstream state, can be reasonably adjusted to an "erf" function.

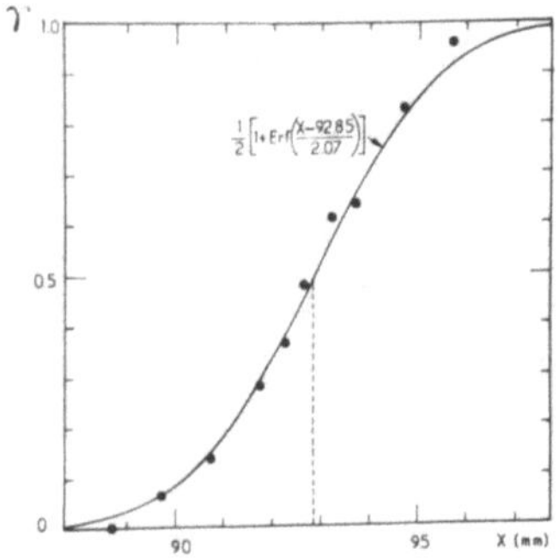

Fig. 19. Intermittency coefficient

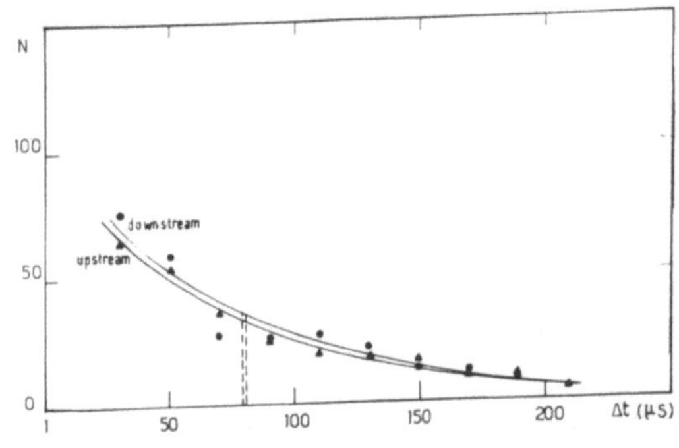

Fig. 20. Statistical frequency of the duration of a state

From this it is possible to represent the probability density of the presence of a shock wave as a normal distribution with its center at $\overline{X}$ = 92.8 mm and a standard deviation of 2.07 mm. The result is a scale of excursion of the shock wave calculated using ± two standard deviations, $\Delta X \simeq$ 8 mm. This Gaussian distribution restricts the choice of an oscillation model for the shock wave.

Second, the statistical frequency Δt representing the duration of a state between the crossing of two consecutive shock waves was approximated in the region of maximum turbulence by an exponential distribution (Fig. 20). The mean duration of both states is essentially equal to 80 μs, for a mean oscillation frequency of the shock wave of 9kHz.

As concerns the spread of the deformations, information is revealed by correlations made at two points on the "mean" surface of the shock wave: first, the spatial correlation coefficient, taken perpendicularly with respect to the plane X, Y (Fig. 17); second, the spatiotemporal correlation coefficient taken from the line with the greatest slope on the shock wave surface, with an optimum time delay.

The value of these coefficients drops quite abruptly whenever the difference between the two points is greater than two centimeters, excluding the possibility of either a strong oscillation of the entire shock wave or the propagation of a high-amplitude sine wave at the surface of the shock wave. The spectrum of the signal in the region of intermittency was a continous broadband spectrum and did not exhibit any particular resonance frequency. The size of the shock wave disturbances observed showed that both perpendicular disturbances (obtained from $\gamma(x)$) and tangential disturbances reached the approximate scales of the incident field of turbulence.

IV. Analysis Results

Increases in the turbulence level during turbulence-shock wave interaction have already been estimated by certain authors [1, 2, 3, 4] using rapid distortion theories. In previous works [3] we proposed this type of analysis, where the result was expressed by the equation

$$T_2 = \bar{U}_{1N}/\bar{U}_{2N}\ K T_1 K^*$$

where

$$K = \{I - (\bar{U}_{2N} - \bar{U}_{1N})/\bar{U}_{2N}\ NN^*\}$$

This draws a relationship between the upstream Reynolds tensor T_1 and the downstream Reynolds tensor T_2 with respect to the intensity of the shock wave $(U_{2N} - U_{1N})$ and its geometry (normal N). In the case of the boundary layer on the compression corner subjected to the same shock, where $U'/\Delta U \simeq .6$ for $Y/\delta = .4$, a reasonable agreement is found between experiments and the velocity variance calculated as above. Assuming the hypothesis that the initial Reynolds tensor is isotropic, this type of estimation may be applied to the case of free turbulence examined in this study. The step between the upstream and downstream velocity turbulence intensity only reaches 25%, i.e., a value that is less than the measured peak result (300%). The step calculated in this manner is not incompatible with the experimental observation, but it clearly appears that the shock wave intermittence phenomena are preponderant in this case ($U'/\Delta U \simeq .15$) and conceal the amplification due to rapid distortion mechanisms.

The strong variations of the field of turbulence occur in the region of the shock wave excursion (ΔX being slightly less than 1 cm). An intermittence diagram therefore provides a means of separating the contribution of the upstream (1) and downstream (2) fields of turbulence from that of the shock wave.

For example, the resulting covariances $(\overline{U'T'})_r$ could be expressed as:

$$(\overline{U'T'})_r = (1-\gamma)\ {}^{1}\overline{U'T'} + \gamma\ {}^{2}\overline{U'T'} + \gamma(1-\gamma)\ \Delta\bar{U}\Delta\bar{T} \qquad (1)$$

Or, in the case of variance involving the parameter c'/ c:

$$(\overline{c'^2}/\overline{\Delta c}^2)r = (1-\gamma)\ {}^{1}\overline{c'^2}/\overline{\Delta c}^2 + \gamma\ {}^{2}\overline{c'^2}/\overline{\Delta c}^2 + \gamma(1-\gamma)$$

As for the spectra, it has been shown that the duration Δt of a state is represented by an exponential distribution. This would suggest using a Poisson distribution to calculate the probability of the occurrence of shock fronts. The Poisson distribution leads to a spectrum slope of -2, which is consistent with the increase in slope observed on the spectra measured in the region of intermittence.

The results obtained on the basis of equation (1) have been plotted by a dotted line in Figures 6, 7 and 9. They were calculated at maximum turbulence ($\gamma = 0.5$), and the value of the conditional levels of turbulence during interaction were estimated on the basis of the upstream and downstream values. The fact that $\overline{T'_o{}^2}$ remains constant appears normal, since there is no additional mean stagnation temperature gradient that occurs with the shock wave that would be susceptible of producing amplification or intermittence phenomena ($\sqrt{\overline{T'_o{}^2}}/\Delta\bar{T}_o \gg 1$).

For the most part, the evolution observed may be represented by this non-dynamic intermittence diagram where the additional reaction mechanism of pumping by externally driven shock oscillation [4, 5] does not seem to play a role of major importance.

Conclusion

These measurement and analysis results provide a description of the evolution of a dynamic and thermal field of turbulence as it relates to characteristics of shock wave disturbances. We have shown the importance of the value of the parameter $c'/\Delta\bar{c}$ in describing the turbulence level variation c in the interaction; two situations may be qualitatively defined: first, an intermittence problem ($c'/\Delta c \ll 1$), showing a peak in the evolution of the turbulence level; second, an eventual amplification of the turbulence level which forms a step and is significant of rapid distortion analysis. In this case, the amplification through the terms of production extracts energy from the shock wave. When the condition of distortion $U'\Delta L/U\Lambda \ll 1$ is fulfilled, the terms of production in the Reynolds stress equation are then preponderant if $(U'/\Delta\overline{U})(\Delta L/\Lambda) \ll 1$. This inequality shows that on several occasions it is possible for the amplification through production to be combined with the effect of intermittence. The examination of these two typical situations has contributed meaningful information to phenomenonological analysis of the interactions between turbulence and a shock wave.

References:

1. Ribner, H.S: Convection of a Pattern of Vorticity Through a Shock Wave. N.A.C.A. Rept. 1164, 1954.

2. Mc Kenzie, J.F.; Westphal, K.O.: Interaction of Linear Waves with Oblique Shock Wave. Ph. of Fl. Vol. 11, N° 11, 1968.

3. Debiève, J.F.; Gouin, H.; Gaviglio, J.: Evolution of the Reynolds Stress Tensor in a Shock Wave-Turbulence Interaction. First Asian Congress of Fluid Mechanics, Institute of Science, Bangalore, India, 1980.

4. Anyiwo, J.C.; Bushnell, D.M.: Turbulence Amplification in Shock Wave-Boundary Layer Interaction. A.I.A.A. Vol. 20, No. 7, 1982.

5. Morkovin, M.V.: Note on the Assessment of Flow Disturbances of a Blunt Body Traveling at Supersonic Speeds Owing to Flow Disturbances in Free Stream. App. Mech. Vol. 27, 1960.

Acknowledgments:

This work was supported by an ONERA/IMST research convention.

Other Aspects of Interaction Flows

An Overview of some Investigations of Pressure and Thermal Distributions Induced by Trailing Edge Controls on Hypersonic Aircraft

LOUIS G. KAUFMAN II
Grumman Corporation
Bethpage, NY 11714 USA

CHARLES B. JOHNSON
NASA Langley Research Center
Hampton, VA 23665 USA

Abstract

Detailed surface heat transfer and pressure distributions have been obtained in three-dimensional shock-wave boundary-layer interactions flow regions. The data described were obtained on fundamental shapes: planar wings with trailing edge flaps or spoilers and planar or cylindrical center bodies, representative of the aft portion of hypersonic aircraft. An overview of the work is presented; details of the projects are available in many reports in the open literature. Analytic, empiric methods are advanced for predicting the extent of separation and the increased heat transfer and pressure loads in three-dimensional separated flow regions.

Symbols

b	span
h	step height, or heat transfer rate coefficient
HL	hinge line
ℓ_{sep}	length of separation
M	Mach number
p	pressure
Re	Reynolds number based on local flow conditions
x,y,z	rectangular coordinates
γ	ratio of specific heats for air
ε	flow deflection angle, or elevon angle
θ	shock wave angle

Introduction

Jointly with U.S. Air Force, Army, Navy and NASA personnel, we investigated various types of shock-wave boundary-layer interactions and their effects on high-speed vehicles in order to develop empiric methods for predicting these effects (Refs. 1-3). Our investigations of high-speed flow separation effects include: flows over cavities, flows past transverse jets, incident shock waves, self-induced shock impingement on blunt fins, and separation induced by trailing edge controls (e.g. Refs. 4-8).

Turbulent Shear Layer/Shock Wave Interactions
IUTAM Symposium Palaiseau 1985
Editor: J. Délery

This paper addresses the type of interaction flow induced by trailing edge controls on hypersonic aircraft. Detailed heat transfer, pressure and photographic data, obtained on fundamental flat plate wing models with various trailing edge controls, were used to develop empiric methods for predicting separated flow effects on high speed aircraft. The data were obtained in hypersonic wind tunnels at: Arnold Engineering Development Center (AEDC), NASA Langley Research Center, and Wright-Patterson Air Force Base (Refs. 9-15).

Finite span steps

Heat transfer data were obtained for finite span spoilers (forward facing steps) on sharp leading edge flat plates in tunnel B at AEDC (Ref. 9). Step spans were varied from 3 to 10 times as large as the step height, and the step heights were 3 to 4 times larger than the undisturbed turbulent boundary-layer thickness. Streamwise distributions of heat transfer rate coefficient ratios at selected spanwise stations are shown in Fig. 1. The average heat transfer to the surface in the separated flow region is approximately 2.5 times larger than for the undisturbed flows over the flat plate surfaces. Local regions of peak heating occur slightly upstream of the step face at locations approximately 0.5 step height inboard from the side edges of the steps. The peak heating amplification factor diminishes with increasing step span, going from 9.0 for the three-inch step to 7.0 for the ten-inch step. However, the location of peak heating to the plate surface, with respect to the face and side edges of the steps, does not vary with step span.

A method was devised for calculating the curved extent of separation ahead of finite span steps based on the observation that the difference between the shock wave angle and the effective flow-deflection angle at the onset of separation is a minimum when turbulent boundary-layer separation occurs ahead of a forward facing step (Ref. 9). The shape of the separation line on the plate surface follows from the assumption that hypersonic, turbulent boundary-layer separation is driven by flow conditions in the outer, inviscid flow and that the inner separated flow region can be approximated by a linear source distribution with its strength dependent upon the extent of separation along the centerline.

The extent of separation along the centerline ahead of a forward facing step depends on the step height and the effective angle of the streamline dividing the separated flow from the outer, inviscid flow. The oblique shock wave angle, θ, inviscid flow deflection angle, ε, and local Mach number, M_1, are related by oblique shock relations (Ref. 16). Setting the partial derivative of tan $(\theta-\varepsilon)$ with respect to θ equal to zero yields (Ref. 9):

$$\frac{\gamma-1}{2} M_1^2 = \frac{1-2 \sin^2\theta}{\sin^2\theta} \tag{1}$$

Assuming a straight dividing streamline:

$$\ell_{sep} = \frac{h}{\tan.\varepsilon} \tag{2}$$

The region of separated flow ahead of finite span steps on a plate surface is calculated assuming that a line source distribution exists along the step, perpendicular to the flow. The line source distribution is a function of the step span, height, and free stream conditions. Superimposing this line source on a uniform free-stream function, the curved separation line shape can be calculated (cf Ref. 9).

Steps on Swept Wings

The variation of separation length with running length Reynolds number is shown in Fig. 2, (Refs. 10 and 11). Laminar separation increases in extent with increasing Reynolds number, this trend is reversed when transition occurs in the free shear layer prior to reattachment. At still larger Reynolds numbers, transition occurs upstream of separation and the extent of separation becomes insensitive to further increases in Reynolds number.

Because of its dependence on running length Reynolds number, the extent of laminar separation ahead of a step on a swept wing to laminar reattachment at the step shoulder would resemble the curved line indicated in Fig. 3. If the boundary layer were totally turbulent prior to separation, then the extent of separation is independent of Reynolds number and would occur along a fairly straight line, such as the dashed line in Fig. 3. For a swept wing with a sharp leading edge at zero angle of attack, the flow over the wing is streamwise and transition occurs in a region parallel to the leading edge. Turbulent separation can occur inboard while outboard the separation can be characteristically laminar. In between, the separated flow will be characteristically transitional, and may be unsteady. Fairing a simple "S" curve through the transition region to join the laminar and turbulent separation lines, one obtains the estimated separation line indicated by the x's in Fig. 3. Although the existence of a cross flow and early boundary layer transition in a free shear layer make questionable the use of a 2-D strip-type analysis, these are secondary effects and should not destroy the value of this method for qualitatively estimating the effects of transition on separation ahead of controls on swept wings.

For laminar separation the plateau pressure p_p can be estimated using (Refs. 10 and 11):

$$\frac{p_p}{p_1} = 1 + 1.22\, M^2 [(M^2 - 1)\, Re_{s\infty}^{-1/4} \qquad (3)$$

where p_1 is the undisturbed flow pressure acting on the flat plate surface and Re_s is the Reynolds number based on local undisturbed flow conditions and distance from the plate leading edge to the location of separation. If the boundary layer becomes turbulent prior to separation, then the pressure rise in the separated flow region can be estimated using:

$$\frac{p_p}{p_1} = \begin{matrix} 1 + 2.24\, M^2 / [8 + (M-1)^2] & \text{for } M < 3.4 \\ 0.091\, M^2 - 0.05 + 6.37/M & \text{for } M > 3.4 \end{matrix} \qquad (4)$$

Separation lines calculated using oblique shock relations (Ref. 16), eqs. (2-4), and the method described above agree well with experimental data (Fig. 4). The greatest uncertainty in the method is estimating the location of transition. The results of many investigations of transition were used in estimating transition locations for these experiments (Refs. 10 and 11).

Wing-control- center body

Sketches of two types of 3-D interaction flows that have been investigated are shown in Fig. 5 (Refs. 12 and 13). For elevon deflection angles less than 25° there is little flow separation upstream of the hinge line as long as the boundary layer on the wing surface is turbulent. The shock wave emanates from the hinge line. For larger elevon deflections the boundary layer separates from the wing surface upstream of the hinge line; a shock wave emanates from this dividing streamline location. In either case, the region of increased loads on an adjacent aft fuselage surface is much larger than would be anticipated using inviscid theory (Fig. 5).

On the wing surface there are no significant effects for elevon deflection angles less than 25° (Refs. 12-15), but the pressure on the wing surface is increased in the separated flow region upstream of the 30° elevon and spoilers (steps). In all cases the load on the control surface is reduced (Fig. 6). Heat transfer to the wing surface in the turbulent separated flow region ahead of the 30° elevon is twice as large as for attached flow, and is considerably further aggravated in the separated flow regions ahead of spoilers (Ref. 12).

The addition of a center body inhibits inboard venting of any separated flow upstream of the trailing edge control (Refs. 12 and 15). There is no flow separation for elevon deflection angles less than 25° and center bodies have little effect on the flow over the wing and elevon surfaces. For larger elevon deflection angles, sufficient to cause flow separation, center bodies (representative of a portion of an aft fuselage) dramatically change the flow over the wing and elevon surfaces. Planar end plate results are predictable but those resulting from a curved shock, emanating from a cylindrical center body, interacting with the separated flow region are not predictable (Ref. 12).

Wing sweep affects both magnitude and distribution of heat transfer rate coefficients on the wing and elevon surfaces. Depending on sweep, elevon deflection angle and center body, maximum heating may occur along the edges of the elevon, in the middle of the elevon, or elsewhere on the elevon surface. The extent of separation is fairly constant ahead of the spoilers on the unswept and 50° swept wings, for which the boundary layers were turbulent prior to separation. However, the extent of separation is much more extensive on the outboard portion of the 70° swept wing where the boundary layer is transitional or laminar prior to separation (Ref. 10 and 11). In all cases, the surface oil flow and phase change coating patterns are quite similar for the same configuration (Ref. 12).

The region on the adjacent aft fuselage section affected by the trailing edge control is much larger than would be anticipated using inviscid flow analyses (Refs. 12-15), (3.7 times larger for a 20° elevon), (Fig. 7). Empiric methods have been developed that may be used to estimate the pressure distribution induced on the end plate surface by the deflected elevon (Refs. 13 and 17). Although the area on the end plate affected by the elevon is much larger than would be anticipated using inviscid analyses, the total side force induced on the end plate is somewhat less than that calculated using inviscid flow analyses (Ref. 15). If the induced pressure distributions are known, then heating magnification factors can be calculated using empiric methods (Refs. 13 and 17).

The blurred oil accumulation line in Fig. 8 is typical of unsteady flow. In this case the small amount of fluctuation in the turbulent boundary layer separated flow region ahead of the 30° elevon is accentuated on the end plate surface.

Thermocouple and phase change coating heat transfer data agree quite well and are repeatable. Surface oil flow and phase change coating patterns are quite similar for the same model configuration, and adequately delineate the surface regions affected by the trailing edge control.

Summary

An empiric method has been advanced for predicting the extent of turbulent boundary layer separation ahead of finite span steps on planar wings. The average heat transfer to the wing surface in the turbulent separated flow region is 2.5 times as large as for undisturbed flow over the flat plate wing. Local regions of peak heating occur slightly upstream of the step face at locations approximately 1/2 step height inboard from the edges of the steps, regardless of step span.

Another empiric method was postulated for predicting the extent and shape of the separated flow region ahead of spoilers on swept wings, with emphasis on the effects of boundary layer transition. The method agrees with experiment for the subject test conditions, but depends on estimating adequately the location of boundary layer transition. The strip-type analysis correctly predicts extensive regions of laminar separation on the outboard portions of the swept wing and less extensive regions of separation on the inboard portions of the wing. Although more extensive, the increased load on the wing surface in the laminar separated flow region is less than that in the turbulent separated flow region.

Trailing edge controls affect adjacent aft fuselage surfaces over regions that are much more extensive than would be anticipated using inviscid flow analyses but the total side force induced on the aft fuselage is less than would be anticipated. Regions of induced forces and increased heating rates are predictable for planar surfaces using empiric analyses but not for general, curved, bodies. Increased heating rates may be calculated as long as the pressure rise is known. Results presented in the references enlarge the data base required to refine empiric analytic methods for estimating increased surface pressure and thermal loads in complex three dimensional interaction flow regions.

References

1 Kaufman, L.G. II, Meckler, L. and Hartofilis, S. A. , "An Investigation of Flow Separation and Aerodynamic Controls at Hypersonic Speeds," J. of Aircraft, AIAA, Vol 3, No. 6, Nov 1966.

2 Korkegi, R.H., "Survey of Viscous Interactions Associated with High Mach Number Flight," AIAA J., Vol. 9, No. 5, May 1971.

3 Kaufman, L.G. II and Gleitsmann, F.G.W., "Descriptive Survey of High Speed Flow Separation," AFWAL TM 83-174-FIMG, Wright-Patterson AFB, March 1983 Grumman R&D Center Report RM-782, December 1983.

4 Clark, R.L., Kaufman, L.G. II and Maciulaitis, A., "Aeroacoustics Measurements for Mach 0.6 to 3.0 Flows Past Rectangular Cavities," AIAA Paper 80-0036, Jan 1980.

5 Kaufman, L.G. II, Maciulaitis, A. and Clark, R.L., "Mach 0.6 to 3.0 Flows Over Rectangular Cavities," AFWAL TR 82-3112, Wright-Patterson AFB, May 1983.

6 Kaufman, L.G. II, "Hypersonic Flows Past Transverse Jets," J. Spacecraft and Rockets, Vol. 4, No. 9, Sept 1967.

7 Johnson, C.B. and Kaufman, L.G. II, "Incident Shock Interactions with Boundary Layers," J. Spacecraft and Rockets, AIAA, Vol 12, No. 6, June 1975.

8 Kaufman, L.G. II, Korkegi, R.H. and Morton, L.C., "Shock Impingement Caused by Boundary-Layer Separation Ahead of Blunt Fins," AIAA Jr, Vol 11, No. 10, Oct 1973.

9 Kaufman, L.G. II and Kirchner, R.D., "High Speed Flow Separation Ahead of Finite Span Steps," AFFDL TR 78-30, Wright-Patterson AFB, April 1978.

10 Kaufman, L.G. II and Freeman, L.M., "Separation Ahead of Steps on Swept Wings," J. of Aircraft, AIAA, Vol 13, No. 12, Dec 1976.

11 Kaufman, L.G. II and Freeman, L.M., "Separation Ahead of Controls on Swept Wings," ARL TR 75-0134, Wright-Patterson AFB, June 1975.

12 Kaufman, L.G. II and Johnson, C.B., "Aerodynamic Heating Rate Distributions Induced by Trailing Edge Controls on Hypersonic Aircraft Configurations at Mach 8," Grumman Aerospace Corp, R&D Center Report RM-799, NASA TM 8753, Aug 1984.

13 Kaufman, L.G. II and Johnson, C.B., "Methods for Estimating Pressure and Thermal Loads Induced by Elevon Deflections on Hypersonic Vehicle Surfaces with Turbulent Boundary Layers," NASA TM 83130, Langley, Sept 1981.

14 Johnson, C.B. and Kaufman, L.G. II,"High Speed Interference Heating Loads and Pressure Distributions Resulting from Elevon Deflections," J. Aircraft, AIAA, Vol 17, No. 3, March 1980.

15 Kaufman, L.G. II and Johnson, C.B., "Pressure and Thermal Distributions on Wings and Adjacent Surfaces Induced by Elevon Deflections at Mach 6," NASA TP 1356, Langley Research Center, March 1979.

16 Ames Research Staff, "Equations, Tables, and Charts for Compressible NACA Report 1135, 1953.

17 Hayes, James R., "Prediction Techniques for the Characteristics of Fin Generated Three Dimensional Shock Wave Turbulent Boundary Layer Interactions," AFFDL TR 77-10, Wright-Patterson AFB, May 1977, available from DTIC as AD A042 024.

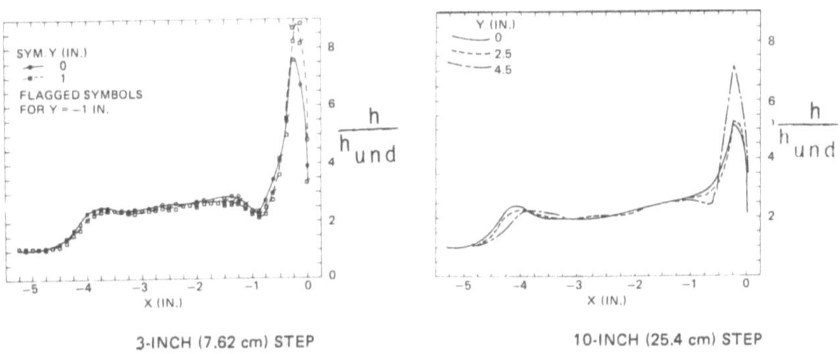

Fig. 1 Streamwise Surface Heat Transfer Rate Coefficient Ratios for Mach 5 Flows Ahead of Steps

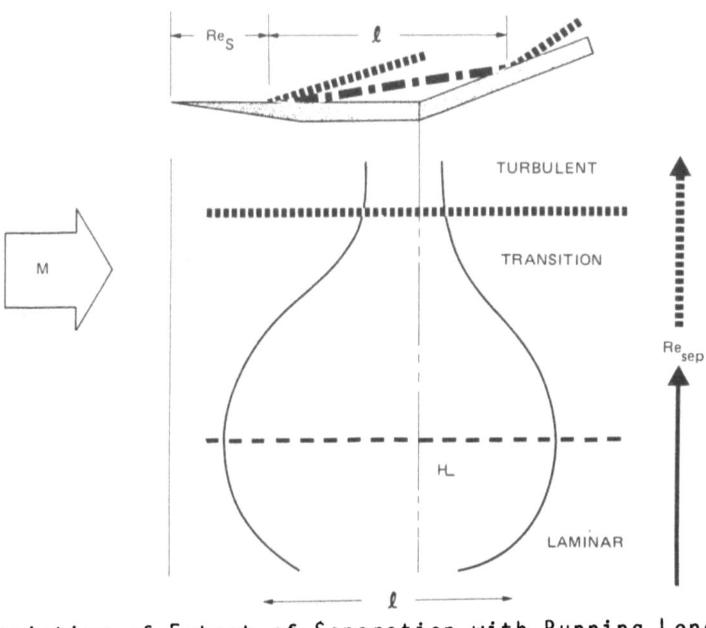

Fig. 2 Variation of Extent of Separation with Running Length Reynolds Numbers

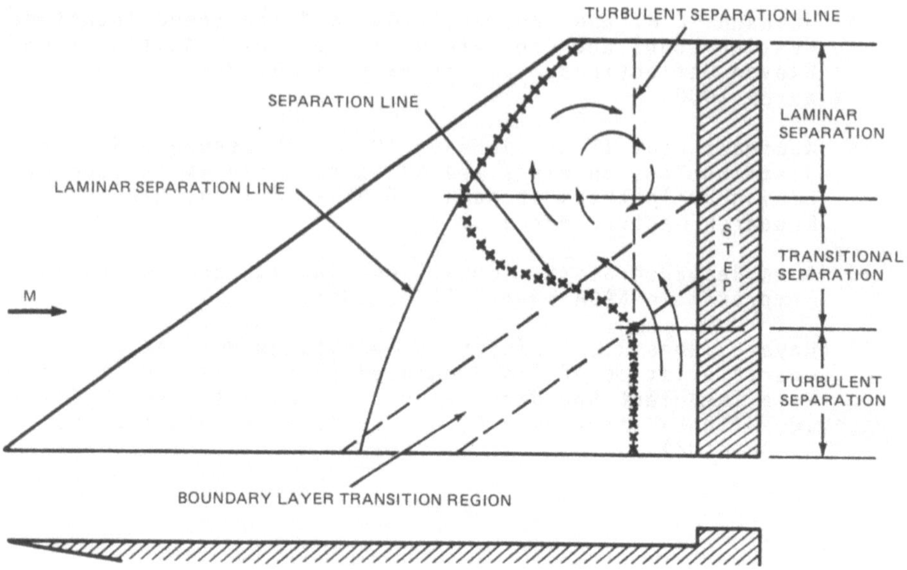

Fig. 3 Separation Line Shapes on a Swept Wing Ahead of a Forward Facing Step

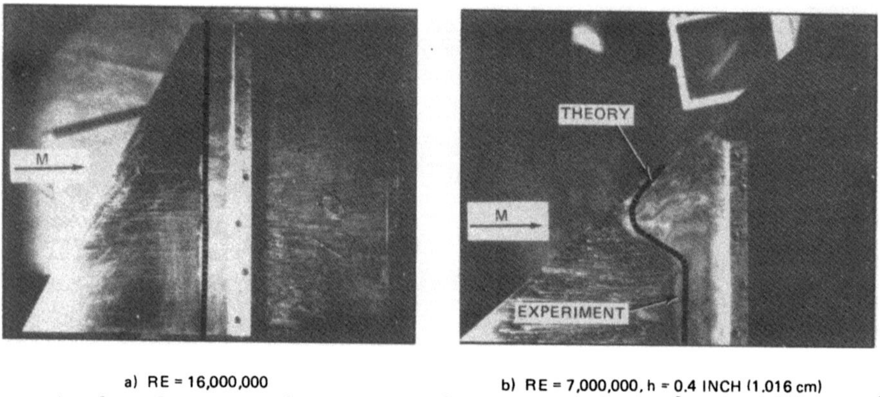

Fig. 4 Oil Flow and Calculated Separation Line Shapes Ahead of Step on Swept Wing

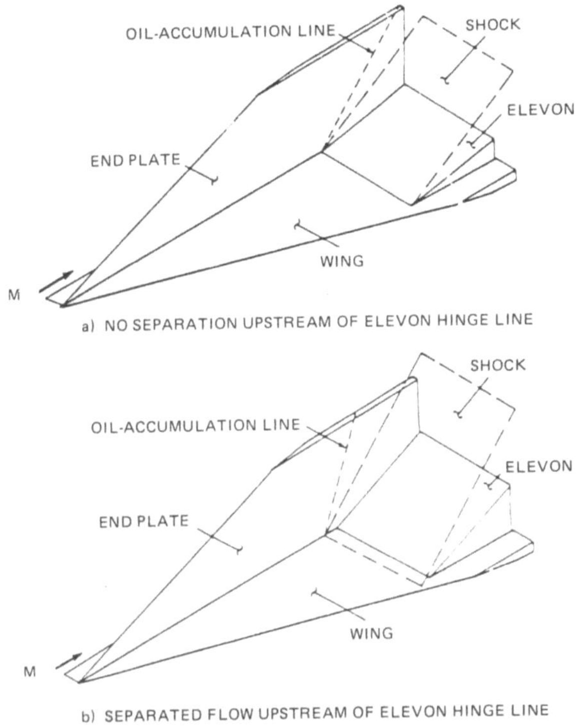

Fig. 5 Two Types of 3-D Interaction Flow Models for Turbulent Boundary Layers Upstream of Elevons

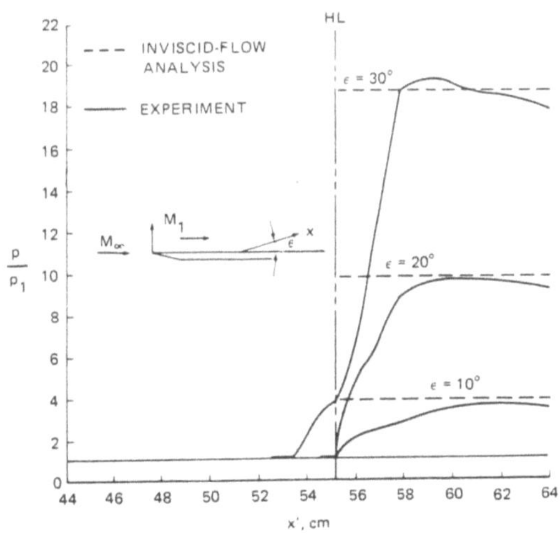

Fig. 6 Streamwise Pressure-Ratio Distributions on Unswept Wing and Elevon Surfaces for $M_1 = 5.9$

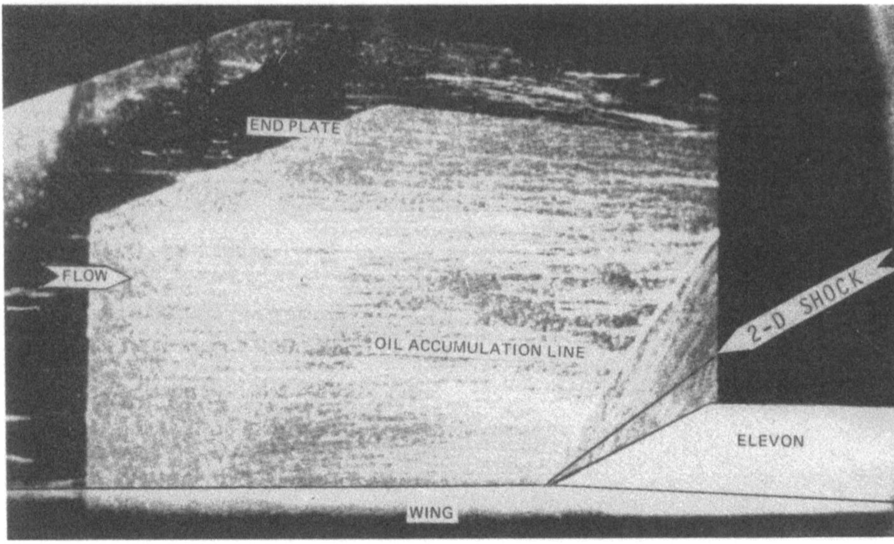

Fig. 7 Oil Flow, 20° Elevon, Wing with End Plate

Fig. 8 Oil Flow, 30° Elevon, Wing with End Plate

Fluid Turbulence: Deterministic or Statistical

Sin-I Cheng

Department of Mechanical and Aerospace Engineering
Princeton University
Princeton, New Jersey 08544
U.S.A.

Summary

The deterministic view of turbulence suggests that the classical theory of fluid turbulence may be treating the wrong entity. We explore the physical implications of such an abstract mathematical result, and provide a constructive computational demonstration of the deterministic and the wave nature of fluid turbulence. The associated pressure disturbance for restoring solenoidal velocity is the primary agent, and its reflection from solid surface(s) the dominant mechanism of turbulence production. Statistical properties and their modeling must address to the statistics of the uncertainties of initial boundary data of the ensemble.

INTRODUCTION

Studies of non-linear dynamics [1,2,3] suggest that repeating solution bifurcation of an unstable system lead deterministically to chaos, called "turbulence" or "Strange Attractor", with unusually complex but stable topological structure. A spectrum of superposed periodic oscillations is expressly "excluded". The classical statistical theory of turbulence in terms of such spectral function [4] is thus in question. Both the deterministic and the statistical theory presume that fluid turbulence is described by the Navier-Stokes equations system at sufficiently large Reynolds numbers. With an exclusive interest in the statistical correlations the classical statistical theory begins with the Ergodic Hypothesis, develops the dynamic equations of correlations in physical and then in spectral spaces that need closure postulate to effect a solution. The deterministic view denies not only the spectral representation even as approximation in a finite interval, but also the mathematical details of the classical approach.

Turbulent Shear Layer/Shock Wave Interactions
IUTAM Symposium Palaiseau 1985
Editor: J. Délery

The ergodic hypothesis equates the long time to the space averages:

$$\langle g \rangle = \lim_{T\to\infty} \frac{1}{T} \int_{t}^{t+T} g(x,t')\, dt'.$$

$$= \lim_{L\to\infty} \frac{1}{2L} \int_{X-L}^{X+L} g(x',t')\, dx'. \qquad (1)$$

for any $g(x,t)$ at position x and time t. When the spectral representation is introduced, L has to be "cut-off" to preserve bounded space integrals. There is no such explicit relation between the two averages except for the class A strange attractors [2,3], with

$$\frac{\partial}{\partial t} \langle g \rangle = 0 \qquad (2)$$

and

$$\langle g \rangle = \lim_{L\to\infty} \frac{1}{2L} \int_{X-L}^{X+L} f(x',t')\, g(x',t')dx' \qquad (3)$$

where $f(x,t)$ is the probabilistic measure of the statistics over the uncertain data. Shear flow turbulence near a solid wall might indeed be class A, since a typical topological structure of class A strange attractors is the horseshoe manifold, reminiscent of the chaotic activities observed in turbulent shear flows. If fluid turbulence is such a class A strange attractor, the dynamic equations of spatial correlations studied in the statistical theory should contain the probabilistic measure. Then the determination of these spatial correlations will inevitably involve the assignment (or closure) of this probabilistic measure in some form. Turbulence modeling should therefore address to uncertainties in the initial-boundary data of the ensemble under consideration. A correct solution of the Navier-Stokes Equations should, however, represent an individual realization of turbulent flows whether they are strange attractors of class A or not.

Navier Stokes Equations and Vorticity Transport

The Navier Stokes system of equations:

$$\frac{\partial}{\partial t} \underset{\sim}{q} + \underset{\sim}{q} \cdot \nabla \underset{\sim}{q} = - \frac{1}{\rho} \nabla p + \mu \nabla^2 \underset{\sim}{q} \quad (4)$$

$$\nabla \cdot \underset{\sim}{q} = 0 \quad (5)$$

defines the evolution of velocity $\underset{\sim}{q}$ and pressure p of an incompressible fluid where the fluid density ρ is constant to be taken as unity. Incompressibility leads to the degenerated form of the continuity equation (5), so that the evolution of p is such as to keep the velocity solenoidal everywhere and at all times during its evolution through (4). Processes of iterative correction of pressure in the solution of (4) and (5) will, however, diverge without repeated smoothing and filtering [5]. It is therefore commonly desired to avoid the determination of pressure in computing the flow field developments.

The curl of (4) gives the vorticity transport equation:

$$\frac{\partial}{\partial t} \underset{\sim}{\omega} = \nabla \times (\underset{\sim}{q} \times \underset{\sim}{\omega}) - \frac{1}{Re} \nabla \times \nabla \times \underset{\sim}{\omega} \quad (6)$$

the solution of vorticity from (6) would be independent of pressure if its initial-boundary data could be so formulated. Many such approximate solutions have been developed. With the advanced vorticity, the induced velocity can be obtained from Biot-Savart Law or solved from:

$$\nabla^2 \underset{\sim}{q} = - \nabla \times \nabla \times \underset{\sim}{q} = - \nabla \times \underset{\sim}{\omega} \quad (7)$$

Thus, velocity and vorticity are iteratively corrected with (6) and (7). Despite repeated smoothing and filtering, successes are limited, especially if one attempts to obtain pressure and to check if $\nabla \cdot \underset{\sim}{q}$ and $\nabla \cdot \underset{\sim}{\omega}$ are zero as they should be.

The transport of vorticity through convection and stretching $[\nabla \times (\underset{\sim}{q} \times \underset{\sim}{\omega})]$, and through diffusion $1/Re\ \nabla^2 \underset{\sim}{\omega}$ would not create any residual divergence from solenoidal initial data. Any residual divergences will, however, grow exponentially with time in an "unstable laminar flow". Smoothing and filtering will limit their growth, but not reduce them to zero. With (5) violated, (6) does not follow from (4). The divergence of the curl of a vector fails to vanish. Such computational results are merely numerical chaos [5].

Spurious oscillations, are common in the computational solution of nonlinear problems. They distort severely the small scale structure of the flow field that controls dissipation and the energy cascading process according to the statistical theory. Therefore, it is considered necessary to introduce "Subscale Modeling" to properly represent the turbulent activities at scales smaller than the computational mesh. In this manner, turbulence modeling becomes the essential part of computational simulation of fluid turbulence, both for the statistical correlations and for the instantaneous, large scale motions. The computed solutions must provide convergent discrete approximations to "all" scales.

The strange attractor theory claims that the mathematical "Route to Turbulence" is successive solution bifurcations into more complex but stable topological structures. The large scale structures due to early bifurcations should be little affected by those small scale structures due to later stage bifurcations. Successive bifurcations take place within the earlier bifurcations and at closely packed small spatial and temporal scales. The global features of turbulent shear flows can then be simulated compuationally based on the Navier-Stokes equations without turbulence modeling so long as the large scale features are adequately resolved by the computational grids. For such computational simulation of global features due to the first few bifurcations of unstable laminar flows, it is adequate even if the computed solution at fairly coarse mesh sizes can provide merely asymptotic discrete approximations [6]. A constructive computational demonstration to test the thesis of deterministic vs. statistical theory of turbulence is thus feasible.

Computational Formulations

We choose to solve computationally the Navier-Stokes equations in the form of vorticity transport and continuity to bring out explicitly the role of continuity relation. To maintain solenoidal velocity and vorticity, we resolve $\tilde{q}$ and $\tilde{\omega}$ into rotational and potential components

$$\tilde{q} = \tilde{q}_r + \nabla \phi \tag{8}$$

$$\tilde{\omega} = \nabla \times \tilde{q} = \zeta + \nabla\psi \tag{9}$$

With

$$\nabla^2 \phi = - \nabla \cdot \tilde{q}_r \tag{10}$$

$$\nabla^2 \psi = - \nabla \cdot \zeta \tag{11}$$

where $\tilde{q}$ is solved from (7), and $\tilde{\omega}$ is advanced from (6).

We discretize all differential equations with simple forward time and centered space differencing schemes without introducing any computational artifices. The details of the iterative solution of the nonlinear discrete system are illustrated in the flow diagram.

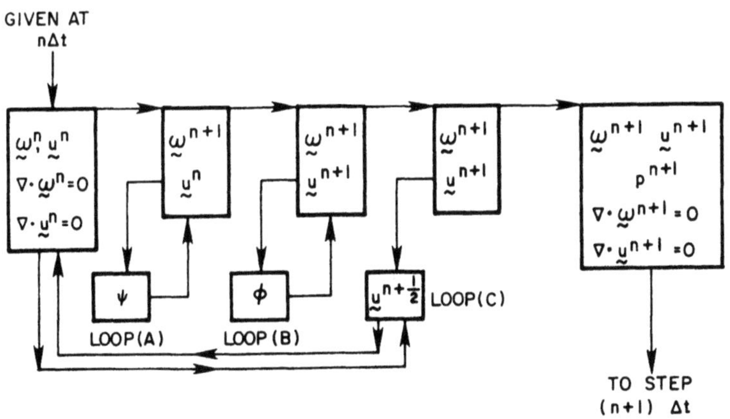

The inflow into the computational region is fixed at the undisturbed values. The lateral and the outflow boundaries are treated approximately with lateral and downstream extrapolations. The solid wall is treated with the nonslip condition on the mean boundary. Vorticity boundary condition on the solid surface at rest should be

$$\nabla \times \tilde{\omega} = - \mathrm{Re}\, \nabla p \tag{13}$$

To avoid iteration on pressure p explicitly, we used the conventional consistency condition on solid boundary

$$\nabla \times \underset{\sim}{\omega} = - \nabla^2 \underset{\sim}{q} \tag{13a}$$

with $\underset{\sim}{q}$ evaluated from the latest available solenoidal velocity.

When solenoidal velocity and vorticity are achieved at the end of the quasi-linear iteration cycle(s) (loop c), the advanced state of the pressure field is calculated from

$$\nabla^2 p = \frac{1}{4} (e_{ij} e_{ij} - \omega_{ij} \omega_{ij}) \tag{14}$$

which is the divergence of the momentum equations (4) where e_{ij} is the rate of strain tensor and ω_{ij} the vorticity tensor.

Neumann boundary conditions in the form of $\partial/\partial n = 0$ along the boundary normal n are used for the solution of all the Poisson equations, (7), (10), (11) and (14) so as not to alter the boundary formulations satisfied by the previous iterants. An exception is made in the solution of $\underset{\sim}{q}_r$ from (7) on the solid boundary where $\underset{\sim}{q} = 0$ is maintained on wall as $\nabla\phi$ evolves in the iterative processes.

The computation for the development of a given initial disturbance is terminated when and if the approximate lateral and downstream boundary begins to distort the development of the flow field in the interior. For those computed cases with developing local turbulent region, this occurs in about 100 time steps, with each time step corresponding roughly to the characteristic transit time of the flow field.

We computed the development of several types of impulses of velocity and/or vorticity applied to a cluster of 6 mesh points in the center portion of the computational field next to the stationary wall with a uniform shear flow at Re = 3000 between two parallel plates. The disturbed velocity and/or vorticity field is not solenoidal and hence non-physical. They are primarily adopted here for convenient reference. This is necessary because we do not know how to specify a simple solenoidal velocity/vorticity disturbance. Our computation begins, therefore, with the equilibration of the associated pressure disturbances into a complicated disturbance. It consists generally of a solenoidal velocity/vorticity perturbation and a smooth pressure perturbation over the undisturbed uniform shear flow.

Computational Results

We computed many cases on two sets of coarse meshes, (15/30 x 15x15) on IBM 3081 computer at Princeton University. Identical qualitative results are obtained for each case. The developments of different initial disturbances eventually fall into two categories:

(i) A dominant horse-shoe vortex evolves, lifting away from the stationary solid wall, spreading vorticity over the entire field of computation, and eventually decaying at later times in some organized manner. The flow field remains distinctly laminar although vortical and somewhat complicated. The undisturbed uniform shear flow was nearly reestablished in some cases.

(ii) The highly disturbed vortical region remains highly complex, grows and hugs onto the wall while drifting downstream. It evolves into a complex structure like turbulence locally with an r.m.s. vorticity level substantially higher than that in the neighbouring apparently laminar region. The spreading boundary of the local turbulent region becomes computationally "sharp". Some such local turbulent regions soon become similar to those turbulent spots observed in experiments. We list a few cases below: [7]

CASE NO.	ARTIFICIAL IMPULSIVE DISTURBANCE	MAX MAGNITUDE OF PHYSICAL DISTURBANCES AFTER EQULIBRATION	LOCAL TURBULENT REGIONS	REMARKS
1	$\delta u = 1$	10^{-2}	yes	spot
2	$\delta v = 1$	10^{-2}	no	horseshoe
3	$\delta v = -1$	10^{-2}	yes	spot
4	$\delta w = 1$ or -1	10^{-2}	no	---
5	$\delta u = 0.1$	10^{-3}	no	---
6	$\delta \omega_x = 1$	10^{-3}	no	---
7	$\delta \omega_x = 10$	10^{-2}	yes	mixed

We recomputed case 1 at a finer mesh 45x25x25 on Cray 1 computer at NCAR. The computed results and those with the coarse 30x15x15 mesh on IBM 3081 at common spatial and temporal steps, agree quantitatively to 10^{-4} with at least one or two significant figures in the turbulent region. They are reproducible and deterministic. The computations in CRAY 1 are conducted at 64 bits word vs. IBM's 32 bits. CRAY is a pipeline machine while IBM is serial. We have also altered our algorithm and changed over to NCAR's fast Poisson solver. Our results are therefore "insensitive" to consistent perturbations and provide meaningful approximations to the deterministic solution of the evolution problem of the Navier Stokes system. The qualitative similarity of computed results with those of experimental observations affirms the "Deterministic" view of Laminar-Turbulent Transitional flows.

Local turbulent (or laminar) regions, bounded by propagating front of laminar-turbulent transition are widely observed in turbulent shear flows, often expressed in terms of "Intermittency". Such phenomena are said to be transitional, not present in the "fully" turbulent flow according to the classical statistical theory. The deterministic view suggests the contrary that such large scale phenomena will persist, perhaps less frequent and less easily observed, upon further bifurcations at increasing flow Reynolds numbers on the Route to "Ultimate" turbulence. Now the Voyger pictures of Jupiter atmosphere revealed the large scale structures at $Re > 10^{12}$. Thus, all turbulent flows on Earth would have to be considered transtional within the frame of the classical statistical theory of turbulence. It appears more natural to consider the large scale structures of early bifurcations as the permanent features of fluid turbulence in space-time as is implied in the deterministic view.

Turbulence Mechanisms - Associated Pressure Disturbances

The potential functions ϕ and ψ were introduced as a mathematical device to permit iterations of the dynamic states to be carried out in physical or solenoidal subspace exclusively. It has been shown mathematically that for iterative convergence, the mass sources or the residual divergences should be accompanied by associated pressure disturbances introduced into the momentum equations [5]. For periodic (or other trivial) boundary value problems, this is $1/2\ q\nabla\cdot q$. For nontrivial boundary value problems, especially those containing drag body or surface(s), convergence can be achieved only in some weak sense with the associated pressure disturbances in the form of Dirac functions. They appear to bring about "instantaneous" local changes in flow velocity, but are actually equivalent to the presence of a "velocity potential" [8]. Likewise, the residual divergences in vorticity can be eliminated by the potential function ψ, generated by a suitable distribution of equal and opposite pairs of impulses with appropriate local moment of forces.

In our computed results, the equilibrated associated pressure disturbances have, in eliminating the residual diverengences, almost annihilated the artificial disturbances, leaving a physical disturbance only 10^{-2} as large. The importance of eliminating the residual divergence is clear. We note that the vorticity transport and interaction are largely kinematic, independent of the presence of any potential force fields. There should be no surprise that the solutions of (6) and (7) will generally contain residual diveregences of velocity and vorticity, arising from the incorrect content of the potential force field necessary for maintaining solenoidal velocity. The physical flow field will have to contain associated pressure disturbances along with vorticity transport and interaction in the course of its evolution. There is, however, no absolute need of any reference to vorticity. A sudden change in velocity due to external disturbances is equivalent to the application of an impulsive pressure force locally to the undisturbed flow field from the momentum considerations alone, [8]. The impulsive forces promptly relax into an additive velocity potential and an associated smooth pressure field according to equations (4), (5), (8), (10) and (14) in the incompressible limit.

For a real fluid with very small compressibility, the propagation speed of pressure waves is large but finite. The relaxation of impulsive pressures to the equilibrated potential state involves extensive wave propagation and interaction, but will be accomplished in a time period much shorter than the discrete time interval of our computational simulation. While the intermediate iterants need not represent physical states, the converged iterants at successive discrete time steps can be a valid global approximation of the propagating potential waves. A typical pressure distribution in a transverse section across a calculated turbulent spot is shown in Fig. 1. Such waves cause rapid rise and fall of local velocities (and vorticities) along their evolution paths. If the root mean square (r.m.s.) of the fluctuating vorticity is taken as a measure of local turbulent activity, the r.m.s. of the associated pressure disturbances is a legitimate measure, and hence, an essential mechanism of fluid turbulence.

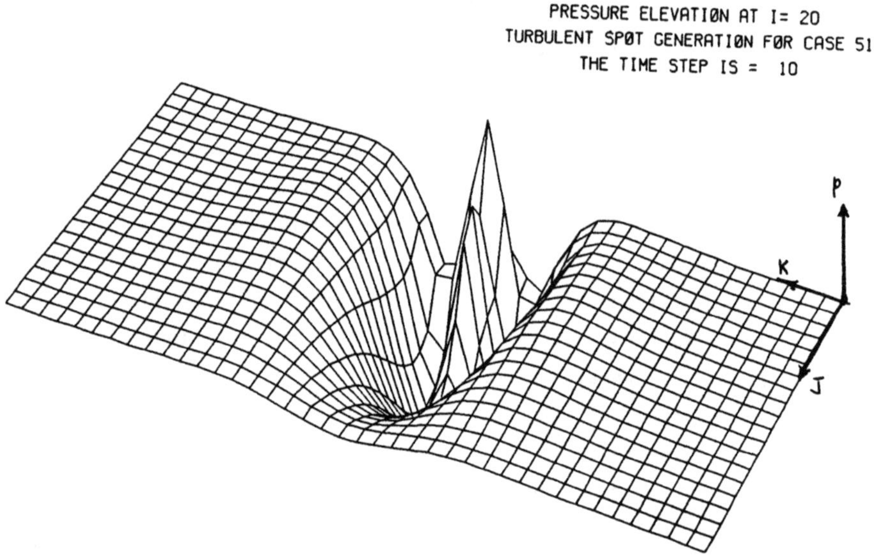

Figure 1. Typical Pressure Disturbance in the Transverse Section Across a Turbulent Spot in a Uniform Shear Flow at Re = 3000

An oblique incidence and reflection of a pressure wave at a solid surface generates a flow velocity parallel to the nonslip wall surface, almost twice as large as it generates mid-stream. The fluid velocity must vanish on the solid surface. Thus, each reflection of an incident wave generates on the surface a vorticity significantly larger than what the wave generates mid-stream. The successive incidence and reflection of numerous such waves therefore generates large vorticity fluctuations on the solid surface and becomes the most effective means of "turbulence production". Within our limited temporal resolution, this equilibrated vorticity production in physical space is represented by the $\nabla\psi$ component parallel and next to the solid surface. This large turbulence production next to the solid surface is carried away by convective motion and gradient type diffusion. The production mechanism $\langle u'_i\, u'_j \rangle \partial Ui/\partial Xj$, representing the energy exchange with the mean flow through turbulent stresses is secondary. A developing local turbulent region is therefore expected to hug onto the solid wall. A pressure wave weakens as it propagates further away from a point source. The same disturbance nearer to the solid surface is hence more effective in generating local turbulent regions than one further away. Convection of a disturbance away from the solid surface is a weakening influence, unfavorable for generating and sustaining a local turbulent region. All these are observed in our computed results as the contrasting behaviors of the developments of unit impulses of $v > 0$ and $v < 0$ and of the up-drift and downdrift arms of a vortical impulse (case 7) and others. Finally, the wave interaction mechanism of turbulence production is essentially inviscid. Thus transitional and turbulent boundary layers over a solid surface is more likely at higher Reynolds numbers with less dissipation.

For a compressible fluid, the continuity relation stands as an evolution equation for density. When combined with an equation of state $p = p(\rho,e)$ and an evolution equation of energy e, the formulation of the wave propagation processes is explicit. With laminar solution unstable, the uncertainties in the initial boundary data will generate pressure waves that need not remain potential but grow into propagating N-waves. Repeated passages of these waves bring about repeated solution bifurcations, on Route to Turbulence. An N-wave contains a shock i.e. a non-potential, propagating pressure discontinuity. With potential waves alone, the propagating boundary of a local turbulent region is continuous in pressure but discontinuous in vorticity. With both types of bifurcations and with time scales of wave propagation comparable to those of flow field evolution, turbulence in compressible fluids may appear much more complex physically. The deterministic and wave nature of fluid turbulence is, nevertheless, mathematically more transparent in compressible fluids than in the incompressible limit. The flow fields of shock wave-boundary layer interaction appears particularly suitable for the study of such transitional processes. Thus, we conclude:

i) Fluid turbulence is a complex nonlinear wave phenomenon of successive solution bifurcations in unstable laminar flows.

ii) Fluid turbulence is deterministic, but sensitive to the uncertainties in the initial-boundary data. Its temporal evolution can be computationally simulated.

iii) Statistical properties in a turbulent field and their modeling is ensemble specific rather than universal.

iv) The incidence and reflection of pressure waves from a solid surface is an important turbulence production mechanism.

Acknowledgement

The author wishes to thank the National Science Foundation of U.S.A. for supporting this work through NSF Grants MAE80-10876 and MAE83-12094 and thank the National Center of Atmospheric Research at Boulder, Colorado, U.S.A. in providing us with their computational facility.

References

1. Ruelle, D. and Takens, F., "On the Nature of Turbulence", Commun. Math. Phy. 20, 167 (1971).

2. Smale, S., "Differential Dynamical Systems", Bull. Ann. Math. Soc. 73 747, (1967).

3. Ruelle, D., "Measures Describing a Turbulent Flow", Articles in Nonlinear Dynamics, Ann. New York Acad. Sci. Vol. 387 (1980).

4. Hinze, J.O., "Turbulence" second Ed. McGraw Hill, Inc., 1975.

5. Cheng, S.I., "Physial and Numerical Turbulence" 3rd Symposium on "Numerical and Physical Aspects of Aerodynamcic Flows", Calif. State Univ. Long Beach. Calif., Jan. 1985.

6. Cheng, S.I. & Sato, C., "Computational Fluid Dynamics, Convergent or Asymptotic," Invited Lecture, 11th International Math. and Comp. Simu., Oslo, Norway, August, 1985.

7. Cheng, S.I., "A New Look on Turbulent Shear Flow", 4th Inter. Symposium on Turbulent Shear Flow, Karlsruhe, Germany September 1983.

8. Lamb, H., Hydrodynamics, 6, Dover Publications, 1935.

MIX
Papier aus verantwortungsvollen Quellen
Paper from responsible sources
FSC® C105338

If you have any concerns about our products,
you can contact us on
ProductSafety@springernature.com

In case Publisher is established outside the EU,
the EU authorized representative is:
Springer Nature Customer Service Center GmbH
Europaplatz 3, 69115 Heidelberg, Germany

Printed by Libri Plureos GmbH
in Hamburg, Germany